Ballou 3/12/93

ENCYCLOPEDIC DICTIONARY OF
Chemical Technology

by
Dorit and Herman Noether

Dorit and Herman Noether
20 Greenbriar Drive
Summit, NJ 07901

Library of Congress Cataloging-in-Publication Data

Science
Ref.
TP
9
N64
1993

CIP Pending

Printed in the United States of America
ISBN 1–56081–329–3 VCH Publishers
ISBN 3–527–26696–8 VCH Verlagsgesellschaft

Printing History:
10 9 8 7 6 5 4 3 2 1

Published jointly by

VCH Publishers, Inc.
220 East 23rd Street
New York, New York 10010

VCH Verlagsgesellschaft mbH
P.O. Box 10 11 61
D–6490 Weinheim
Federal Republic of Germany

VCH Publishers (UK) Ltd.
8 Wellington Court
Cambridge CB1 1HZ
United Kingdom

Preface

This encyclopedic dictionary of Chemical Technology is designed for nontechnical and for specialized chemical practitioners. Since chemical technology pervades modern society, words used in the technical literature are often needed for communication with the public. Such terms must be defined clearly in a widely understood language. We consider this dictionary a "poor man's encyclopedia" whose purpose is to present short descriptions of technical terms, chemical processes, and products.

The book is to serve as an interface between technologists and the lay public; between experts in different fields; between science and technology. It is a bridge between general and specialized dictionaries as well as between dictionaries and encyclopedias.

This volume is primarily a tool for communicators: educators at all levels, librarians, reporters, writers, and personnel affiliated with technical staff in plants and research laboratories. In addition it will serve workers in nontechnological fields such as economists, political scientists, and lawyers.

Chemical technology is central to basic human needs: shelter, food, and clothing. It provides the materials used in energy production ranging from coal purification and petroleum refining to nuclear power. Chemical technology furnishes fertilizers and pesticides for agriculture; detergents and pharmaceuticals for health; catalysts for economical reactions; paper, ink and computer chips for communication. It supplies engineering plastics and fire-proofing for buildings; light-weight materials for aerospace; explosives for construction; dyes for synthetic fabrics; oxidizers for waste treatment; preservatives for food. Our dictionary covers these topics and numerous others that are involved in human activities and their interactions.

We included illustrations and brief paragraphs to help in the understanding of chemical technology. Individual illustrations focus on processes and the machinery applicable to many technologies; they were chosen to demonstrate the operation of a chemical or physical process rather than to stress any given application. The process illustrations show that industrial operations are designed to consider yield, economic use of energy, starting materials, and absence of waste. These considerations are generally not covered in science dictionaries or books.

Because many excellent dictionaries on chemistry are available we have limited descriptions of individual chemicals to their industrial production and use. Generally we have not included physical properties such as boiling and melting points or indices of refraction. These can be found in handbooks.

While chemical technology is intimately related to the economics of nations and to international trade we have not included costs or treaties. Nor have we included educational concerns or human factors. We have generally restricted names to generics and have avoided the use of trade names. In processes we have usually retained the British system because at this time plant personnel still use Fahrenheit and BTUs. We have not included details of metallurgy, mining of ores.

One unique feature is the inclusion of tables of acronyms used for chemicals, processes, and institutions. With the proliferation of abbreviations, the technical literature is often difficult to follow without access to such glossaries.

Summit, NJ

July, 1992

D.L.N.

H.D.N.

Contents

Abbreviations for Compounds/Groups Used in Formulas

A	adenine (in nucleic acids)
A	alanine (in peptides and proteins)
Ac	acetyl–, MeC0
acac	acetylacetonate (ligand)
Ala	alanine
Ar	aryl (group)
Arg	arginine
as–	asymmetrical
Asn	asparginine
Asp	aspartic acid
Asx	asparginine or aspartic acid
B	asparginine or aspartic acid (in peptides and proteins)
binap (BINAP)	2,2'-*bis*(diarylphosphino)–1,1'–binaphthyl (bidentate chiral ligand)
bipy	2,2'–bipyridyl
boc	*t*–butyloxycarbonyl–
n–bu	*n*–butyl; $Me(CH_2)_3-$
t–bu, *tert*-bu	*t*–butyl; *tert*-butyl; Me_3C-
Bz	benzyl; $C_6H_5CH_2-$
C	cytosine (in nucleic acids)
C	cysteine (in peptides and proteins)
chiraphos (CHIRAPHOS)	$Ph_2PC(H)(Me)C(Me)Ph_2$ (chiral ligand)
cod (COD)	1,5–cyclooctadiene (ligand)
cot (COT)	1,3,5,7–cyclooctatetraene (ligand)
cp	cyclopentadienylide; $C_5H_5^-$
cuen	cupriethylenediamine (ligand)
Cys	cysteine
Cys_2	cystine
D	aspartic acid (in peptides and proteins)
diars	phenylenebis(dimethylarsine); $o-C_6H_4(AsMe_2)_2$; (ligand)
dien	diethylenetriamine (ligand)
diglyme	diethylene glycol dimethyl ether (ligand)

diop (DIOP)	phosphine (chiral bidentate ligand)

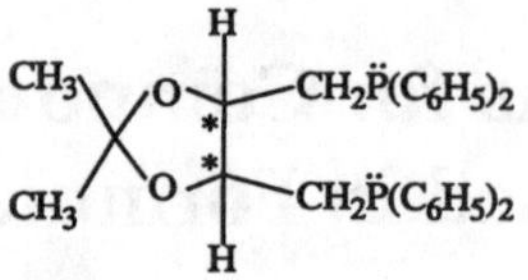

DIOP

dipamp (DIPAMP)	[o–OMePhP(Ph)CH$_2$–]$_2$ (chiral bidentate ligand)
diphos (DIPHOS)	[C$_6$H$_5$]$_2$ PCH$_2$CH$_2$P[C$_6$H$_5$]$_2$, 1,2–*bis*(diphenylphosphino)ethane (bidentate ligand)
dipy	2,2'–dipyridyl (ligand)
dppe	see diphos
E	glutamic acid (in peptides and proteins)
en	ethylenediamine (ligand)
et	ethyl, C$_2$H$_5$–
F	phenylalanine (in peptides and proteins)
Fp	FeCp(CO)$_2$ where Cp is cyclopentadienylide anion, C$_5$H$_5^-$
G	guanine (in nucleic acids)
G	glycine (in peptides and proteins)
Gln	glutamine
Glu	glutamic acid
Gly	glycine
glyme	ethylene glycol methyl ether
H	histidine (in peptides and proteins)
hfac	hexafluoroacetylacetonate (ligand)
His	histidine
i–	prefix for *iso*–
I	isoleucine (in peptides and proteins)
Ile	isoleucine
K	lysine (in peptides and proteins)
L	ligand
L	leucine (in peptides and proteins)
Leu	leucine
Lys	lysine
m–	prefix for *meta*–

2

M	methionine (in peptides and proteins)
Me	methyl, CH_3-
Mes	mesityl, 1,3,5–trimethylphenyl
Met	methionine
$n-$	prefix for *normal*
N	asparginine (in peptides and proteins)
norphos	4,5 di–$P(C_6H_5)_2$ norbornene (chiral ligand)
$o-$	prefix for *ortho*
obbz	oxamidobis(benzoate) (ligand)
$p-$	prefix for *para*
PAMP (CAMP)	$CH_3(C_6H_5)PC_6H_{11}$ (monodentate chiral ligand)
Pc	phthalocyanine (ligand)
Ph	phenyl, $C_6H_5^-$
Phe	phenylalanine
phen	1,10–phenanthroline (ligand)
pn	propylenediamine (ligand)
Pr	propyl, $C_3H_7^-$
Pro	proline
prophos (PROPHOS)	$[C_6H_5]_2PCHC(CH_3)CH_2P[C_6H_5]_2$ (bidentate chiral ligand)
py	pyridine (ligand)
Q	glutamine (in peptides and proteins)
R	arginine (in peptides and proteins)
R	alkyl group
rib	ribose
S	serine (in peptides and proteins)
salen	(o–$HOC_6H_4CH=NCH_2)_2$; *bis*(salicylaldehyde) ethylenediamine (ligand)
Ser	serine
$t-$, *tert*–	prefix for *tertiary*
T	thymine (in nucleic acids)
T	threonine (in peptides and proteins)
Thr	threonine
tn	1,3–diaminopropyl, trimethylene diamine (ligand)
tren	$(H_2NCH_2CH_2)_3N$, triaminotriethylamine (ligand)
trien	triethylenetetramine $(H_2NCH_2CH_2NHCH_2)_2$ (ligand)
Trp	tryptophan
Tyr	tyrosine

U	uridine (in nucleic acids)
V	valine (in peptides and proteins)
Val	valine
W	tryptophan (in peptides and proteins)
X	a halogen
Y	tyrosine (in peptides and proteins)
Z	glutamine or glutamic acid (in peptides or proteins)

Acronyms for Chemicals (*) and Methods

A *	adenine
AA	Atomic Absorption (spectroscopy)
AA *	acrylic acid; anthranilic acid
aa *	α-amino acids
A acid *	1,7-dihydroxy-naphthalene-3,6-disulfonic acid
AAO *	acetaldehyde oxime
AAS *	acrylic ester rubber
AAS	atomic absorption spectroscopy
AAT *	alanine aminotransferase
ABFA *	azo-*bis*-formamide
ABR *	acrylate-butadiene rubber
ABS *	acrylonitrile-butadiene-styrene resin
ABS *	alkylbenzene sulfonate
AC *	allyl chloride
AC	asbestos cement
AC	alternating current
ACE *	alcohol, chloroform, ether mixture
ACE *	angiotensin-converting-enzyme
ACh *	acetylcholine
AChE *	acetylcholine esterase
ACM *	acrylate rubber
ACN *	acrylonitrile
ACN	acetone cyanohydrin process
ACP *	calcium hydrogenphosphate (used in foods)
ACR	advanced cracking reactor
ACTH *	adrenocorticotropic hormone
ADA *	acetonedicarboxylic acid
ADC *	allyl diglycol carbonate
ADC, A/D	analog-to-digital converter
ADI	Acceptable Daily Intakes
ADN *	adiponitrile
ADP *	adenine diphosphate
ADWC	automatic dishwasher compound
AE *	alcohol ethoxylate
AEDP *	2-amino-2-ethyl-1,3-propanediol
AEM *	2-aminoethyl mercaptan methacrylate
AEM	Analytical Electron Microscopy
AES *	alcohol ether sulfates
AES *	acrylonitrile-ethylene-propylene-styrene quarterpolymer
AES	Atomic Emission Spectroscopy

5

AES	Auger Electron Spectroscopy
AET *	aminoethylthioisourea dihydrobromide
AETT *	acylethyl tetramethyl tetralin
AFC	alkali fuel cell
AFP *	α-fetoprotein
AFS	Atomic Fluorescence Spectroscopy
AGES *	alcohol glyceryl ether sulfonates
AGP *	α-$_1$-acid glycoprotein
AGS *	adipic, glutaric, succinic acids
AGS *	alkyl glyceryl sulfonate
AI	active ingredient
AIA	acid insoluble ash
AIB *	ammonium isobutyrate
AIBN *	2,2-azobisisobutyronitrile
AID *	N-acetylimidazole
AIDS	acquired immune deficiency syndrome
AIP *	aluminum isopropoxide
AM *	acrylamide
AMMA *	acrylonitrile-methyl methacrylate copolymer
AMP *	2-amino-2-methyl-1-propanol
AMP *	adenosine monophosphate (adenylic acid)
A5MP *	adenosine-5-monophosphate (muscle adenylic acid)
AMPD *	2-amino-2-methyl-1,3-propanediol
AN *	acrylonitrile
AN *	ammonium nitrate
ANA *	α-naphthyl acetic acid
ANFOL *	ammonium nitrate-fuel oil explosives
ANOVA	overall analysis of variance
ANPO *	α-naphthylphenyl oxazole
ANTU *	α-naphthylthiourea
AO *	α-olefins; antimony oxide
AO *	active oxygen; antioxidant
AOM	active oxygen method
AOS *	α-olefin sulfonate
AP *	aminopyridine
AP *	ammonium perchlorate
APA *	antipernicious anemia factor (vitamin B$_{12}$)
APA *	available phosphoric acid
6-APA *	6-aminopenicillanic acid
APAP *	acetyl-para-amino-phenol (acetaminophen)
APC *	ammonium perchlorate
APE *	alkylphenol ethoxylates
APG *	alkyl polyglycoside
APL	angleplied laminates
APMA *	N-(3-aminopropyl)methacrylamide
APO *	triethylenephosphoramide

6

APP *	ammonium polyphosphate fertilizers
APP *	atactic polypropylene
APS	appearance potential spectroscopy
AR *	acrylic rubber (British Standards Institution)
ARC	antireflection coating
ARPES	angle-resolved photoelectron spectroscopy
ART	asphalt rseid treating
AS *	alcohol sulfate
ASA*	acrylate-styrene-acrylonitrile terpolymer
ASA *	alkenylsuccinate anhydrides
ASCII	American Standard Code for Information Interchange
A-stage	early cure stage for thermoset resins
ATA *	aminotriazole
ATE *	aluminum triethyl; triethyl aluminum
ATH *	aluminum trihydrate
ATM *	aluminum trimethyl; trimethyl aluminum
ATP *	adenosine triphosphate
ATPase *	adenosine triphosphatase
ATR	attenuated total reflectance
AU *	polyester type of polyurethane rubber
AVA	added-value-agriculture
AVLIS	atomic vapor laser isotope separation
AVT	all volatile treatment
AWT	advanced wastewater treatment
AZT *	3'-azido-3'-deoxythymidine (azidothymidine)
BAA *	battery acid
BACT	best available control technology (environmental)
BAL *	2,3-dimercaptopropanol (British anti-Lewisite)
BAP *	benzyl-*para*-aminophenol; benzyl adenine
BAP *	*N*-benzyl-2-aminopyridine
BaP *	benzo(a)pyrene
BB	blood-brain barrier
BBC *	bromobenzylcyanide
bbl	barrel
BBO *	2,5-dibiphenylyloxazole
BBP *	butyl benzyl phthalate
BCD	binary coded decimal
BCWL *	basic carbonate white lead
BDO *	1,4-butanediol
BEK *	butyl ethyl ketene
BET	Brauner, Emmett and Teller equation
BFE *	bromotrifluoroethylene
BFPO *	*bis* (dimethylamino)fluoro phosphine oxide
BFS	beam-foil spectroscopy
BHA *	butylated hydroxyanisole

7

BHC *	1,2,3,4,5,6, hexachlorocyclohexane
BHC *	benzenehexachloride
BHT *	butylated hydroxytoluene; 2,6-di-*tert*-butyl-*para*-cresol
BIIR *	brominated isobutene-isoprene (butyl) rubber
BIPP *	bismuth, iodoform, paraffin mixture (wound dressing)
BLEVE	boiling liquid expanding vapor explosion
BMC	bulk molding compound
BNOA *	2-naphthoxyacetic acid
BOC *	*tert*-butyloxy carbonyl
BOD	biological oxygen demand
BOD5	amount of dissolved oxygen consumed in 5 days
BOM	built on mask
BON *	β-oxynaphthoic acid; 3-hydroxy-2-naphthoic acid
BOPOB *	*bis*[2-(5-*para*-biphenylyloxazolyl]) benzene
BPA *	bisphenol A
bpcd	barrels per calender day
BPL *	bone phosphate of lime
BPL *	β-propiolactone
BPMC *	2-*sec*-butyl phenyl-*N*-methyl carbamate
BPO *	benzoyl peroxide
bpsd	barrels per stream day
BR *	butyl rubber
BR *	*cis*-1,4-butadiene rubber; *cis*-1,4-polybutadiene
BR *	butyl ricinoleate
BS *	butadiene-styrene copolymer (*see* SB)
BS	British Standard
B-stage	intermediate stage in curing thermoset resins
BT *	bacillus thuringiensis' pesticide
BTB *	bromothymol blue
BTDA *	3,3',4,4'-benzophenone tetracarboxylic dianhydride
BTMSA *	*bis*(trimethylsilyl)acetylene
BTNENA *	bi-trinitroethylnitramine
BTNEU *	bi-trinitroethylurea
BTX *	benzene,toluene, xylene
BVE *	butyl vinyl ether; vinyl *n*-butyl ether
BWR	boiling water reactor
BZ *	nonlethal gas, causes temporary disorientation
BzH *	benzaldehyde
C *	cytosine
CA *	cellulose acetate
CA *	cortisone acetate
CA	controlled atmosphere
CAB *	cellulose acetate butyrate
CAD	computer-assisted design

CAE	computer-assisted engineering
CAF *	chloramphenicol
CAM	computer-assisted manufacture
cAMP *	cyclic adenosine-3',5'-monophosphate
CAN *	cellulose acetate nitrate
CAP *	cellulose acetate propionate
CAP *	chloracetophenone
CAP *	chloramphenicol
pCBA *	*para*-carboxybenzaldehyde
CBA	chemical blowing agent (plastics industry)
CBM *	chlorobromomethane; bromochloromethane
CBN *	cubic boron nitride
CBN *	chlorobutyronitrile
CBP *	chlorinated biphenyl
CCD	charge-coupled devices
CD *	cyclodextrin
CD	compact disk
CDA *	completely denatured alcohol
CDAA *	α chloro-*N*,*N*-diallyl acetamide
CDDT *	cyclododecatriene
CDEA *	2-chloro-*N*,*N*-diethylacetamide
CDEC *	2-chloroallyl diethyldithiocarbamate
cDNA *	complementary DNA
CDEC *	2-chloroallyldiethyldithioethylcarbamate
cDNA *	complementary DNA
CDP *	cytidine diphosphate
CDP *	cresyl diphenyl phosphate
CDT *	cyclododeca-1,5,9-triene
CDTA *	*trans*-1,2-diaminocyclohexanetetraacetic acid
CE	capillary electrophoresis
CE	Compton edge
CEA *	carboxyethyl acrylate
CED	cohesive energy density
CES *	cyanoethyl sucrose
CF *	cresol-formaldehyde resin
CF	citrovorum factor
CFC *	chlorofluorocarbons
CFE *	chlorotrifluoroethylene
CFE *	polychlorotrifluoroethylene resins
CFRP *	carbon-fiber-reinforced plastics
CFRTP	carbon-fiber-reinforced thermoplastics
CFSTR	continuous flow stirred tanks with recycle
CFT	crystal field theory
CGS	centimeter-gram-second units
ChAT *	choline acetyltransferase
CHC *	epichlorine-ethylene oxide rubber

9

CHDI *	cycloheaxanediisocyanate
CHDM *	1,4-cyclohexanedimethanol
CHP *	N-cyclohexyl-2-pyrollidone
CHR *	epichlorine rubber (see CO)
CI	chemical ionization
CI	Colour Index (British)
C.I.	Colour Index
CID	collision-induced dissociation
CID	charge injection devices
CIPC *	chloro-IPC; isopropyl N-(3-chlorophenyl)
CLF	compressed liquid fuel
CLS	characteristic loss spectroscopy
CM *	carboxymethylated cotton
CMA	cylindrical mirror analyzer
CMC	critical micelle concentration
CMC *	CM-cellulose; carboxymethylcellulose
CME	crucible-melt extraction
CMOS *	carboxymethyl oxysuccinate
CMOS	complementary metal oxide semiconductor
CMP *	cytidine monophosphate; cytidylic acid
CMPP *	2-(4-chloro-2-methylphenoxyl)
CMU *	chlorophenyldimethylurea
CN *	cellulose nitrate
CNG	compressed natural gas fuel
CNR *	carboxynitroso rubber (tetrafluoroethylene-trifluoronitrosomethane-unsaturated terpolymer
CNS	central nervous system
CO *	poly(chloromethyl)oxirane; epichlorohydrin rubber (see CHR)
COC	Cleveland Open Cup, flash point determination
COD	chemical oxygen demand
COPE	Cope process (Claus oxygen-based process expansion)
COSY	(homonuclear) correlated spectroscopy
COV *	sulfuric acid, 95–96% by weight
CP *	cellulose propionate
CP *	cyclophosphoramide
CP	crude protein
CP	chemically pure
CP *	cyclopentadiene
CPC *	Cu-phthalocyanine
CPE *	chlorinated polyethylene
CPFV	canned, frozen, and preserved fruits and vegetables
CPI	chemical process industry
CPM	critical path method
CP/MA.nmr	cross-polarized magic angle spinning nmr
CPMC *	o-chlorophenyl-methyl carbamate

CPR *	cyclonene-pyrethrene-rodentone
CPU	central processing unit
CPVC *	chlorinated polyvinyl chloride
CR *	chlorinated rubber; chloroprene rubber (British Standards Institute)
cRNA *	complementary RNA
CRT	cathode ray tube
CS *	casein
CSF	cerebrospinal fluid
CSM *	chlorosulfonated polyethylene
C stage	final stage in curing thermoset resins
CSTR	constant stirred reactor
CT	computerized tomography
CTA *	cellulose triacetate
CTD	charge transfer device
CTE	coefficient of thermal expansion
CTEM	conventional transmission electron microscopy
CTFE *	poly(chlorotrifluoro)ethylene (*see* PCTFE)
CTP *	cytidine triphosphate
CTPR *	carbon-reinforced plastics
CTS *	anhydrous aluminum sodium sulfate
CVD	chemical vapor deposition
CYAD *	cyano acetic acid
CZ	Czochralski process
DAA *	diacetone alcohol
DAA *	diacetone acrylamide
DAC *	diallylchlorendate
DAC, D/A	digital-to-analog converter
DADHT *	diacetyl dioxohexahydrotriazine
DAF *	diallyl fumarate
DAHQ *	2,5-*tert*-amylhydroquinone
DAIP *	diallyl isophthalate
DAM *	diallyl maleate
dAMP *	deoxyadenosine monophosphate
DANS *	4-*N*,*N*-dimethylamino-4'-nitrostilbene
DAP *	diallyl *o*-phthalate
DAP *	diammonium phosphate
DAS *	4,4'-diamino-2,2'-stilbenedisulfonic acid
DBCP *	1,2-dichloro-3-bromopropane
DBDPO *	decabromodiphenyl oxide
DBEP *	dibutoxyethyl phthalate
DBM *	dibutylmaleate
DBMC *	4,6-di-*tert*-butyl-*meta*-cresol
DBP *	dibutylphthalate
DBPC *	di-*tert*–butyl-*p*-cresol
DBS *	dibutyl sebacate

11

DBSA *	4,4'oxy-*bis*-benzenesulfonylhydrazide
DCA *	deoxycorticosterone acetate
DCB *	1,4-dichlorobutane
DCC *	*N,N'*-dicyclohexylcarbodiimide
DCHP *	dicyclohexyl phthalate
DCMO *	carboxin; 5,6-dihydro-2-methyl-1,4-oxathiin-3-carboxanilide, DMEO
DCNA *	2,6-dihydro-4-nitroaniline
DCO *	dehydrated castor oil
DCP *	dicapryl phthalate
DCP *	dicyclopentadiene
DCPA *	dimethyl-2,3,4,5-tetrachloroterephthalate
DCPC *	dichlorophenyl methylcarbinol; di(*para*-chlorophenyl)ethanol
DCU *	*N,N'*-dicyclohexylurea
DDAc *	poly(diallyldimethylammonium chloride)
DDB *	dodecylbenzene
DDBSA *	sodium dodecylbenzenesulfonate; dodecylbenzenesulfonic acid
DDD *	dichlorodiphenyldichloroethane
DDDA *	dodecanedioic acid
DDDM *	dichlorophene; 2,2'-dihydroxy-5,5'-dichlorodiphenylmethane
DDE *	dichlorodiphenyldichloroethylene
DDH *	dichlorodimethylhydantoin
DDM *	4,4'-diaminodiphenylmethane
DDM *	*n*-dodecyl mercaptan
DDNP *	diazodinitrophenol
DDP *	dodecyl phthalate
DDPS *	dichlorodiphenylsulfone monomer
DDQ *	2,3-dichloro-5,6-dicyanobenzoquinone
DDS *	sulfonyldianiline; diaminodiphenylsulfone
DDT *	dichlorodiphenyltrichloroethane
DDTA	derivative differential thermal analysis
DDVP *	dimethyl dichlorovinyl phosphate; dichlorovos
DE	diatomaceous earth
D.E.	dextrose equivalent
DEA *	diethanolamine
DEAC *	diethyl aluminum chloride
DEAE *	diethylaminodiethyl- (derivatives)
DEC *	β-diethylaminoethylchloride hydrochloride
DEET *	diethyltoluamide
DEG *	diethanolglycine
DEG *	diethylene glycol
DEGN *	diethylene glycol nitrate
DEMO *	*see* DCMO

DEP *	diethyl phthalate
DEPC *	diethylpyrocarbonate
DEPC *	γ-diethylaminopropyl chloride HCl
DES *	diethylstilbestrol
DET *	diethyltoluamide
DETA *	diethylene triamine
DF	degrees of freedom
DFDT *	difluorodiphenyltrichloroethane
DFP *	diisopropyl fluorophosphate
DFT	discrete Fourier transform
DHA *	dihydroacetone
DHP *	dihexyl phthalate
DHPs *	dihydroxybenzenes
DHS *	dihydrostreptomycin
DIBA *	diisobutyl adipate
DIBAC *	diisobutylaluminum chloride
DIBAL.H *	diisobutylaluminum hydride
DIBP *	diisobutyl phthalate
DIC *	β-diisopropylaminoethyl chloride HCl
DIDA *	diisodecyl adipate
DIDG *	diisodecyl glutarate
DIDP *	diisodecyl phthalate
DINA *	dioxyethylnitramine dinitrate
DINP *	diisononyl phthalate
DIOA *	diisooctyl adipate
DIOP *	diisooctyl phthalate
DIOS *	diisooctyl sebacate
DIOZ *	diisooctyl azelate
DIP	dual-in-line package
DIPA *	diisopropanolamine
DIPE *	diisopropyl ether
DM *	diphenylamine chloroarsine
DMA *	dimethylamine; dimethylacetamide; disodium methyl arsenate
DMAC *	dimethylacetamide
DMAEC *	dimethylaminoethylchloride
DMAMP *	2-dimethylamino-2-methyl-1-propanol
DMB *	dimethoxybenzene
DMC *	dichlorodiphenyl methyl carbinol; penicillamine; di(*para*-chlorophenyl)ethanol
DMC *	β-dimethylaminoethyl chloride HCl
DMD *	dimethylmetadioxane
DMDHEU *	dimethyloldihydroethylene urea
DMDT *	dimethoxydiphenyltrichloroethane; methoxychlor
DME *	dimethyl ether
DMEP *	di(2-methoxyethyl) phthalate

DMF *	*N,N*-dimethylformamide
DMFA *	*N,N*-dimethylformamide
DMG *	dimethyl glutarate
DMH *	dimethyl hydantoin
DMHF *	dimethyl hydantoin formaldehyde; methylol dimethylhydantoin
DMP *	dimethyl glutarate
DMPA *	*O*-(2,4-dichlorophenyl *O*-methyl isopropyl-phosphoramidothioate; dimethylol propionic acid
DMPC *	1-dimethylamino-3-propyl chloride
DMPPO *	poly(2,6-dimethyl-1,4-phenylene oxide)
DMSO *	dimethylsulfoxide
DMT *	dimethyl terephthalate
DMU *	dimethylurea
DMU *	dimethylolurea
DNA *	dinonyl adipate
DNA *	deoxyribonucleic acid
DNase *	deoxyribonuclease
DNBP *	dinitro-*ortho*-*sec*-butylphenol
DNC *	dinitrocresol
DNFB *	2,4-dinitrofluorobenzene
DNHZ *	di-*n*-hexyl azelate
DNOBP *	2-*sec*-butyl-4,6-dinitrophenol
DNOC *	4,6-dinitro-*ortho*-cresol
DNOCHP *	dinitro-*ortho*-cyclohexyl-phenol
DNODA *	di(*n*-octyl,*n*-decyl)-adipate
DNODP *	di(*n*-octyl,*n*-decyl)-phthalate
DNP *	dinonyl phthalate
DNPC *	*N,N'*-di-β-naphthyl-*para*-phenylenediamine
DNPT *	dinitrosopentamethylenetetramine
DNT *	dinitrotoluene
D.O.	dissolved oxygen
DOA *	dioctyl adipate
D.O.C.	dichromate oxygen consumed
DOF *	di(2-ethylhexyl)fumarate
DOIP *	dioctylisophthalate
DOM *	di(2-ethylhexyl)maleate
DOM *	hallucinogenic drug
DOP *	dioctyl phthalate
L-dopa *	dihydroxyphenylalanine
DOS *	dioctyl sebacate
DOTG *	di-*ortho*-tolylguanidine
DOTP *	dioctyl terephthalate
DOTT *	di-*ortho*-tolylthiourea
DOZ *	dioctyl azelate
D.P.	degree of polymerization

DPA *	diphenyl amine
DPA *	diphenolic acid
DPCF *	diphenyl cresyl phosphate
DPG *	diphenylguanidine
DPH *	1,6-diphenylhexatriene
DPIP *	diphenyl isophthalate
DPN *	diphosphopyridine dinucleotide; nicotinamide adenine dinucleotide
DPO *	2,5-diphenyloxazole
DPO *	*bis*(2-ethylhexyl)phthalate
DPOF *	diphenyloctyl phosphate
DPPD *	*N,N'*-dipenyl-*para*-phenylenediamine
DQN *	diazoquinone novolac
DRAM	dynamic random access memory
DRIFT	diffuse reflectance infrared Fourier transfer spectroscopy
DSA	dimensionally stable anode
DSC	differential scanning calorimetry
DSP *	disodium phosphate
DTA	differential thermal analysis
DTBP *	di-*tert*-butyl peroxide
DTDP *	ditridecyl phthalate
DUP *	diundecyl phthalate
DVB *	divinylbenzene
DVM	digital voltmeter
DXE *	dixylyl ethane
EAA *	ethylene-acrylic acid copolymer; poly(ethylene-co-acrylic acid)
EADC *	ethylaluminum dichloride
EAK *	ethyl amyl ketone
EAM *	ethylene-vinyl acetate copolymer
EASK *	ethylaluminum sesquichloride
EBDM *	ethylene glycol dinitrate
EBT	Elmer Bumpy torus
EC *	ethyl cellulose
EC	exclusion chromatography
E. C.	Enzyme Commission Code
ECB *	ethylene copolymer blends with bitumen
ECD	electron capture detector
ECH *	epichlorohydrin
ECTEOLA-Cellulose	reaction product of epichlorohydrin triethanolamine and cellulose
ECTFE *	ethylene-chlorotrifluoroethylene copolymer
ED	electrodialysis; electron diffraction; energy dispersive
EDA *	ethylene diamine
EDB *	ethylene dibromide

EDC *	ethylene dichloride
EDM	electric discharge machines
EDNA *	ethylenedinitramine, haleite
EDS	Exxon Donor Solvent process; Energy-Dispersive X-ray Spectroscopy
EDTA *	ethylenediamine-N,N,N',N'-tetraacetate; ethylenediamine-$N,N,N,'N'$-tetraacetic acid
EDTAN *	ethylenediaminetetraacetonitrile
EDTA Na4	EDTA tetrasodium salt
ee	enantiomeric excess
EEA *	poly(ethylene-co-acrylate); ethylene-ethylacrylate copolymer
EED	electro-explosive device
EEG	electroencephalogram
EELS	electron energy loss spectroscopy
EG *	ethylene glycol
EGDN *	ethylene glycol dinitrate
2EH *	2-ethylhexanol
EHD	electrohydrodimerization
EHEC *	ethyl hydroxyethyl cellulose
EI	electron-impact ionization
EIA	enzyme immunoassay
ELISA	enzyme-linked immunosorbant assay
EMA *	ethylene-methacrylate copolymer
EMA *	ethylene-maleic anhydride copolymer
EMD *	electrolytic manganese oxide (cells)
EMI	electromagnetic interference
EMI *	2-ethyl-4-methylimidazole
EMIA	enzyme membrane immunoassay
EMIRS	electrochemically modulated infrared spectroscopy
EMIT	enzyme-multiplied immunoassay technique
EMMA	electron microscope with microanalysis
EMTS *	ethylmercury-$para$-toluene-sulfonanilide
ENDOR	electron nuclear double resonance
EO *	ethylene oxide
EO/PO *	ethylene oxide/propylene oxide copolymer
EOR	enhanced oil recovery
EP *	ethylene-propylene copolymer (see EPM, EPR)
EP *	epoxy resin
EPC	easy processing channel
EPDM *	ethylene-propylene rubber; ethylene-propylene-nonconjugated diene terpolymer (see EPT)
EPE *	epoxy resin rubber
EPM *	ethylene propylene copolymer (see EP, EPR)
EPMA	electron probe microanalysis
EPN *	o-ethyl-p-nitrophenyl phenylphosphonothioate

16

EPR *	ethylene-propylene rubber (*see* EP, EPM)
EPR	electron paramagnetic resonance
EPS *	expanded polystyrene; polystyrene foam (*see* XPS)
EPT *	ethylene-propylene-diene terpolymer (*see* EPDM)
EPTC *	*S*-ethyl-di *N,N*-propylthiocarbamate
ER	electroreflectance
ERDA	elastic recoil detection analysis
E.S.C.	environmental stress cracking
ESCA	electron spectroscopy for chemical analysis
E.S.D.	equivalent spherical diameter
ESI	electron spray ionization
ESO *	epoxy plasticizer
ESR	electron spin resonance
ETFE *	ethylene-tetrafluoroethylene copolymer
ETU *	ethylene thiourea
EU *	polyether type of polyurethane rubber
EVA *	ethylene-vinyl acetate polymer (also E/VAC)
EVE *	ethylene-vinyl ether copolymer
EXAFS	extended X-ray adsorption fine structure spectroscopy
EXELFS	extended energy loss fine structure spectroscopy (*see* XELFS)
FAB	fast-atom bombardment
FABS	fast-atom bombardment spectroscopy
FACS	fluorescence-activated cell sorter
FAS	fast affinity chromatography
FBC	fluidized bed combustor
FBR	fast breeder reactor
FCC	fluid cracking catalyst
FE *	fluorine-containing elastomer
FEM	field emission spectroscopy
FEP *	tetrafluoroethylene-hexafluoropropylene rubber (*see* PFEP)
FEP *	fluorinated ethylene propylene
FES *	fatty ester sulfonate
FES	flame emission spectroscopy
FET	field effect transistor
FF *	furan-formaldehyde resins
FF *	furan formaldehyde copolymers
FFA *	free fatty acid
FFF *	phenyl-furfural copolymers
FFPA	free from prussic acid
FFT	fast Fourier transformation
FGAN *	fertilizer grade ammonium nitrate
FIA	fluorescent immunoassay
FID	flame ionization detector

FIM	field ion microscopy
FMN *	flavin mononucleotide
FMS	flexible manufacturing system
FOS	fuel-oxygen-scrap process
FPC *	fish protein concentrate
FPD	flame photometric detector
FPIA	fluorescence polarization immunoassay
FPM *	vinylidene fluoride-hexafluoropropylene rubber
FRM	fiber-reinforced materials
FRP	fiber-reinforced plastic
FSI *	fluorinated silicone rubber
FT	Fourier transformation
FT black	fine thermal black
FTIRS	Fourier transform infrared spectroscopy
FTNMR	Fourier transform nuclear magnetic resonance
FWA *	fluorescent whitening agents
FWHM	full width at half maximum
GABA *	γ-aminobutyric acid
GABOB *	4-amino-3-hydroxybutyric acid
GAC	gross annual cost
GC	gas chromatography
GC/MS	gas chromatography/mass spectrometry
GCEP	gas centrifuge enrichment plant
GCFBR	gas cooled fast breeder
GCR	gas cooled power reactor
GDP *	guanosine diphosphate
GHB *	γ-hydroxybutyric acid
GI *	glucose isomerase
GLC	gas–liquid chromatography
GM	Geiger–Muller
GMA *	grizzle methacrylate
GMS	galvanized mild steel
GMP *	guanosine monophosphate
GMS *	glyceryl monostearate
GP	general purpose
GPC	gel permeation chromatography
GPF black	general purpose furnace black
GR-1 *	butyl rubber (former U.S. acronym); (*see* IIR, PIBI)
GRAS	Generally Regarded As Safe
GR-N *	nitrile rubber (former U.S. acronym) (*see* NBR)
GRP *	glass reinforced plastics
GR-S*	styrene-butadiene rubber (former U.S. acronym) (*see* PBS, SBR)
GSC	gas–solid chromatography
GTP *	guanosine triphosphate
HA*	heptaldehyde-aniline condensate

HA *	heptaldehyde-aniline
HALS *	hindered-amine light stabilizers
HC*	hydrocarbon
HCB *	hexachlorobenzene
HCCH *	hexachlorocyclohexane
HCCP *	hexachlorocyclopentadiene
HCFC *	chlorofluorocarbons containing hydrogen
hCG *	human chorionic gonadotropin
hCS *	human choriosomatotropin
HDA	hydrodealkylation
HDC	hierarchial distributed control
HDLP*	high-density lipoprotein
HDPE *	high-density (linear) polyethylene
HDT *	hexamethylene diisocyanate
HEA *	2-hydroxyethyl acrylate
HEAP	Hydrogen Electric Arc Pyrolysis
HEC *	hydroxyethyl cellulose
HEDTA *	hydroxyethyl ethylenediamine triacetic acid
HEMA *	2-hydroxyethyl methacrylate
HEOD *	dieldrin
HEPA	high-efficiency particle air filters
bis-HET *	*bis*(*p*-hydroxyethyl)terephthalate
H.F.	high frequency
HFCS *	high-fructose corn syrup
HFP *	hexafluoropropylene
HFPO *	hexafluoropropylene epoxide
HFS	high-fructose syrup
hfs	hyperfine structure
HGF *	glucagon
HGMS	high-gradient magnetic separation
HHDN *	aldrin; hexachlorohexahydrodimethanonaphthalene
HHS *	high strength steel
HHV	high heating value hydrocyclone
HIP	hot isostatic processing
HIPM	high-pressure injection
HIPS *	high-impact polystyrene
HIV	retrovirus causing AIDS
HLB	hydrophilic–lipophilic balance
HMDA*	hexamethylene diamine
HMDI *	hexamethylene-1,6-diisocyanate
HMDS *	hexamethyldisilazane
HMF *	5-(hydroxymethyl)-2-furaldehyde
HMF black	high modulus furnace black
HMM *	hexamethylmelamine; hemel
HMPA *	hexamethylphosphoramide (hexametapol)
HMPT *	hexamethylphosphoric acid triamide

HMTA *	hexamethylenetetramine
HMWPE *	high molecular weight polyethylene
HMX *	cyclotetramethylenetetramine
HNM *	hexanitromannite; mannitol hexanitrate
HOMO	highest occupied molecular orbital
HOY	highly oriented yarn (PET)
HPA *	hydroxypropyl acrylate
HPC black	hard-processing channel black
HPIM	high-pressure injection molding
HPLC	high-performance liquid chromatography
HPMA *	2-hydroxy-1-methylethyl methacrylate
HPMC *	hydroxypropylmethylcellulose
HQNO *	2-n-heptyl-hydroxyquinoline-N-oxide
HREELS	high resolution electron energy loss spectroscopy
HREM	high resolution electron spectroscopy
HRT	horizontal return tubular burner
HS *	hydroxylamine sulfate
HS	high spin
HSS *	high strength steels
HTGR	high-temperature gas cooled reactor
HTST	high-temperature, short-time (pasteurizing)
HTU	height of transfer unit
HTXR	high temperature X-ray powder diffraction
HWR	heavy water reactor
IAA *	3-indoleacetic acid
IBA *	indolcbutyric acid
IBIB *	isobutyl isobutyrate
IBVE *	isobutyl vinyl ether
IC	integrated circuit
ICAP, ICP	induction coupled (argon) plasma
ICC	ignition control compound
ICP-AES	inductively coupled plasma-atomic emission spectroscopy
IDA *	iminodiacetic acid
IDF	ink donor film
IEC	ion-exchange chromatography
IES	ion selective electrodes
IETS	inelastic electron tunneling spectroscopy
IFN *	interferon
IGCC	integrated gasification combined cycle
IgG *	immunoglobulins G
IGR	insect growth regulator
IIL	integrated injection logic
IIR *	isobutene-isoprene rubber; butyl rubber (*see* GR-I, PIBI)

IJP	ink jet printing
IL-1 *	interleukin 1
IL-2 *	interleukin 2
IMMA	ion microprobe mass analysis
IMP *	inosine monophosphate; inosinic acid
IMP *	insoluble metaphosphate; Maddrell salt
IMS	inviscid melt spinning
INCP,IPC*	isopropyl *N*-phenylcarbamate
INS	ion neutralization spectroscopy
I/O	input/output
IPA *	isophthalic acid
IPA *	isopropyl alcohol
IPC *	propham; isopropyl-*N*-phenylcarbamate
IPN	interpenetrating polymer network
IPS *	impact polystyrene
IPTS	international practical temperature scale
IR	infra-red spectroscopy
IR *	isoprene rubber (British Standards Institute)
IS	immittance spectroscopy
ISE	ion-selective electrode
ISFET	ion-selective field transistor
ISG *	immune serum globulin
ISS	ion-scattering spectroscopy
IVE *	isobutyl vinyl ether
J acid *	2-amino-5-hydroxy-7-sulfonic acid naphthalene
JCPDS file	powder diffraction file (formerly ASTM file)
JH *	juvenile (insect) hormone
K acid *	2-amino-8-hydroxy-7-sulfonic acid naphthalene
K-factor	coefficient of thermal conductivity
K-value	denotes the degree of polymerization
KDCC *	potassium dichloroisocyanurate
KHN	Knoop hardness number
KTPP *	potassium tripolyphosphate
L acid *	1-hydroxy-5-sulfonic acid naphthalene
LAB *	linear alkylbenzene
LABS *	linear alkylbenzene sulfonates
LAD *	lithium aluminum deuteride
LAH *	lithium aluminum hydride
LAM	longitudinal acoustic mode (Raman)
LAS *	linear alkyl sulfonate
LATB *	lithium aluminum *tert*-butylhydride
LC	liquid chromatography
LC50	lethal concentration, 50%
LCC	liquid column chromatography
LCD	liquid crystal display
LCL	less than carload lot

LC/MS	liquid chromatography/mass spectrometry
LCP *	liquid crystal polymer
LD50	lethal dose, 50%
LDPE *	low density polyethylene
LEC	liquid encapsulated Czochralski crystal growing
LEC	ligand exchange chromatography
LED	light-emitting electrode
LEED	low energy electron diffraction
LEISS	low energy ion-scattering spectroscopy
LHDPE *	linear high-density polyethylene
LHSV	liquid hourly space velocity
LHV	low heating value; gross heating value
LIM	liquid injection molding
LIMB	limestone injection multistage burner
LIMS	laboratory information management system
LISICON *	lithium superionic conductor ($Li_{14}ZnGe_4O_{16}$)
LLC	liquid–liquid (partition) chromatography
LLDPE *	linear low-density polyethylene
LMFBR	liquid–metal fast breeder reactor
LMS	laser mass spectroscopy
LNG *	liquefied natural gas
LOX *	liquid oxygen (rocket fuel)
LPC *	leaf protein concentrate
LPG *	liquefied petroleum gas
LPG *	liquid pressurized gas
LPO	liquid phase oxidation
LRIM	liquid reaction injection molding
LRM	liquid reaction molding
LRS	laser Raman spectroscopy
LS	low spin
LSB	least significant bit
LSC	liquid–solid (adsorption) chromatography
LSD *	lysergic acid diethylamide (LSD-25)
LSI	large-scale integration
LTH *	luteotropic hormone, lactogenic hormone
LUMO	lowest unoccupied molecular orbital
LWR	light water reactor
M acid *	1-amino-5-hydroxy-7-sulfonic acid naphthalene
MAb	monoclonal antibody
MABS *	methylmethacrylate-acrylonitrile-butadiene-styrene terpolymer
MAC *	methyl allyl chloride
MAN *	methacrylonitrile
MAN *	methylammonium nitrate
MAO *	monoamine oxidase
MAP *	ammonium phosphate

MBC	minimal bactericidal concentration
MBMC *	*tert*-butyl-*meta*-cresol
MBS *	methacrylate-butadiene-styrene resins
MBT *	mercaptobenzothiazole
MBTS *	2,2'-dithiobisbenzothiazole
Mc	megacycles
MC *	methyl cellulose
MCA *	monochloroacetic acid
MCC *	microcrystalline cellulose
MCFC	molten carbonate fuel cell
MCPA *	2-methyl-4-chlorophenoxyacetic acid
MCPB *	4-(2-methyl-4-chlorophenoxy)butyric acid
MCPP *	mecoprop; 2-(2-methyl-4-chlorophenoxy)-propionic acid
MCVD	modified chemical-vapor deposition
MDA	metal deactivator
MDA *	*para, para'*-diaminodiphenylmethane, (4,4'methylene dianiline)
MDAC *	4-methyl-7-diethylaminocoumarin
MDI *	diphenylmethane-4,4'-diisocyanate (methylene diisocyanate)
MDPE *	medium-density polyethylene
MEA *	monoethanolamine
MEEP *	poly(*bis*(methoxyethoxyethoxy)phosphazene)
MEHQ *	hydroquinone monomethyl ether
MEK *	methyl ethyl ketone
MEKP *	methyl ethyl ketone peroxide
MEMC *	methoxyethylmercury chloride
MENA *	α-naphthaleneacetic acid methyl ester
MEP *	methyl ethyl pyridine
MF *	melamine-formaldehyde resin
MF	microfiltration
MFC	multifunctional concentrate
MG	metallurgical grade
M-glass	glass with high beryllia content
MH *	maleic hydrazide
MHA *	4-methylthio-2-hydroxybutyric acid
MHD	magnetohydrodynamics
MIAK *	methyl isoamyl ketone
MIBC *	methylamyl alcohol
MIBK *	methyl isobutyl ketone
MIC	minimal inhibitory concentration
MIDI *	diphenylmethane diisocyanate
MIKES	mass-analyzed ion kinetic energy spectroscopy
MIPA *	isopropanolamine
MIPC *	*o*-isopropylphenyl methylcarbamate

MIR	multiple internal reflectance
MKB *	monobasic potassium phosphate
ML	monolayer
MLA *	mixed lead alkalis
MMA *	methyl methacrylate
MMA *	monomethylamine
MMC *	metal matrix composite
MMH *	monomethyl hydrazine
MNA *	2-methyl-4-nitroaniline
MO	molecular orbital
MOCA *	4,4'-methylene-*bis*(2-chloroaniline)
MON	motor octane number
MOSFET	metaloxide semiconductor field-effect transistor without metal gate
6-MP *	6-mercaptopurine
2MP1 *	2-methyl-1-pentene
4MP1 *	4-methyl-1-pentene
MPA	multipurpose additive
3-MPA *	3-methoxypropylamine
MPC black	medium processing channel black
MPD-I *	poly-*meta*-phenylene-isophthalamide
MPF *	melamine-phenol-formaldehyde resin
MPK *	methyl propyl ketone
MRI	magnetic resonance imaging
mRNA *	messenger RNA
MS	mass spectroscopy
MSB	most significant bit
MSBR	molten salt breeder reactor
MSG *	monosodium glutamate
MSH *	melanocyte-stimulating hormone
MSI	medium scale integration
MSO*	mesityl oxide
MSP *	monosodium phosphate
MSW	municipal solid waste
MTBE *	methyl *t*-butyl ether
MT black	medium thermal black
MTD *	toluene-2,4-diamine
MTF	minimum filming temperature
MTG	methanol-to-gasoline process
MTI	maximum therapeutic index
MTZ	mass transfer zone
MVD	metal vapor deposition
MVE *	methyl vinyl ether
MVT	moisture vapor transmission
MWD	molecular weight distribution
NA	numerical aperture

NAA *	naphthalene acetic acid
NaCMC *	sodium salt of carboxymethylcellulose
NAD *	nicotinamide adenine dinucleotide
NAD$^+$ *	nicotinamide adenine dinucleotide oxidized
NADH *	nicotinamide adenine dinucleotide reduced
NADP *	nicotinamide adenine dinucleotide phosphate
NADPH *	nicotinamide adenine dinucleotide phosphate reduced
NAND	not AND (results of AND operation negated)
NaPPS *	sodium poly(styrene sulfonate)
NASICON *	sodium superionic conductor ($Na_3Zr_2PSi_2O_{12}$)
NBA *	N-bromoacetamide
NBR *	acrylonitrile-butadiene rubber
NBS	nuclear backscattering spectroscopy
NBS *	N-bromosuccinimide
NC *	nitrocellulose; cellulose nitrate (see CN)
NCR *	acrylonitrile-chloroprene rubber
NCS *	N-chlorosuccinimide
NDA	new drug application
NDGA *	nordihydroguaiaretic acid
NDOP *	n-decyl-n-octyl phthalate
NG *	nitroglycerin
NGR	nuclear gamma ray spectroscopy
NHDP *	n-hexyl-n-octyl-n-decyl phthalate
NHE	normal hydrogen electrode
NIR *	acrylonitrile isoprene rubber
NLO	nonlinear optics
NMP *	N-methyl-2-pyrrolidinone
NMR	nuclear magnetic resonance
NOBS *	nonanoyloxybenzene
NODA *	n-octyl-n-decyl adipate
NODP *	octyl decyl, or ethylhexyl decyl phthalate
NOE	nuclear Overhauser effect
NOESY	nuclear Overhauser effect spectroscopy
NOR	not OR (results of OR operation negated)
NPA *	naptalam; N-naphthylphthalamic acid, Na salt
NPDES	national pollutant discharge elimination system (EPA standard)
N-P-K	nitrogen-phosphorus-potassium in fertilizers
NPN *	n-propyl nitrate
NPO *	α-naphthylphenyloxazole
NPSH	net positive suction head
NQR	nuclear quadrupole resonance (spectroscopy)
NR*	natural rubber
NRS	nuclear reaction spectrometry
NSR *	nitrile silicone rubber

NSSC	neutral sulfite semimechanical process
NTA *	nitrilotriacetic acid
NTA *	*N,N'*-dinitroso-*N,N'*-dimethyl terephthalamide
NTAN *	nitrilotriacetonitrile
NTC	negative temperature coefficient
NTU	nephelometric turbidity unit
O.C.	oxygen consumed
OBP *	octyl benzyl phthalate
OBSH *	4,4'-oxy-*bis*(benzenesulfonyl hydrazine)
OCCP *	octachlorocyclopentene
ODPN *	β,β'-oxydipropionitrile
OER *	oil extended rubber
OI	oxygen index test
OIDP *	octyl-*i*-decyl phthalate
OMC *	oxidized microcrystalline waxes
OMPA *	octamethylpyrophosphoramide, schradan
ONB *	ortho-nitrobiphenyl
OPDN *	β,β-oxydipropionitrile
OPG *	oxypolygelatin
OPPA *	octylphenyl phosphoric acid
OPR *	propylene oxide rubber
ORP	oxydation reduction electrode
OSFET	oxide semiconductor field effect transistor
OSP *	ordinary superphosphate
OTC	over the counter drugs
OTEC	ocean thermal energy conversion
O/W	oil in water
PA *	phthalic anhydride
PA *	polyacrylate
PA *	polyamide
PAA *	polyacrylic acid
PABA *	*para*-aminobenzoic acid
PACE	pollution abatement capital expense
PAEK *	polyaryletherketone
PAF	platelet-activating factor
PAFC	phosphoric acid fuel cell
PAG *	pentaacetyl glucose
PAH *	polyaromatic hydrocarbons
PAK *	polyester alkyd resins
PAI *	polyamide-imide
PAM *	2-pyridine aldoxime methiodide
PAMS *	poly(α-methylstyrene)
PAN *	polyacrylonitrile
PAPA *	polyazelaic polyanhydride
PAPI *	polymethylene polyphenyl isocyanate
PARA *	poly(arylamide)

PAS *	*para*-aminosalicylate
PAS	photoacoustic spectroscopy
PAS *	polyarylsulfone
PAT *	polyaminotriazole
PB *	polybutene-1
PBAA *	polybutadiene acrylic acid copolymer
PBAG *	polyadipate glycol
PBAN *	polybutadiene acrylonitrile copolymers
PBD *	1,3,4-phenylbiphenyloxadiazole
PBI *	poly(benzimidazole)
PBI *	protein-bound iodine
PBMA *	poly-*n*-butyl methacrylate
PBPB *	pyridinium bromide perbromide
PBR *	butadiene-vinyl pyridine copolymer
PBS *	butadiene-styrene copolymer (*see* GR-S, SBR)
PBSMO *	3,3-*bis*(chloromethyl)oxetane polymer
PBT *	polybenzothiazole
PBT *	polybutylene terephthalate
PBTP *	polybutylene terephthalate
PC *	polycarbonate (*see* PCO)
PCB *	polychlorinated biphenyl
PCD *	poly(carbodiimide)
PCE *	perchloroethylene
P.C.E.	pyrometric cone equivalent
PCL *	polycaprolactone
PCNB *	pentachloronitrobenzene
PCO *	polycarbonate (*see* PC)
PCP *	pentachlorophenol
PCP *	phencyclidene HCl
PCS *	polycarbosilane (fiber)
PCTFE *	poly(chlorotrifluoroethylene)
PCVD *	plasma(activated) chemical-vapor deposition
PDAP *	poly(diallyl phthalate) (*see* DAP)
PDB *	*para*-dichlorobenzene
PDL	pumped dye laser
PDME	pendant-drop melt-extraction process
PDMS	plasma desorption mass spectrometry
PDMS *	poly(dimethylsiloxane)
PDP	plasma display panel
PDR	*Physician's Desk Reference*
PDU	process development unit
PE *	polyethylene
PE *	pentaerythritol
PEA *	poly(ethyl acrylate)
PEC *	chlorinated polyethylene (*see* CPE)
PECVD	plasma enhanced chemical vapor deposition

PEEK *	poly(ether ether ketone)
PEEKK *	poly(ether ether ketone ketone)
PEG *	poly(ethylene glycol)
PEI *	poly(ethyleneimine)
PEI *	poly(ether imide)
PEK *	poly(ether ketone)
PEKK *	poly(ether ketone ketone)
PEO *	poly(ethylene oxide) (*see* PEOX)
PEOX *	poly(ethylene oxide) (*see* PEO)
PEP *	ethylene-propylene copolymer (*see* EP, EPR)
PEPA *	polyether-polyamide block copolymer
PES *	polyether-sulfone
PESc *	poly(ethylene succinate)
PET *	polyethylene terephthalate
PETN *	pentaerythritol tetranitrate
PETP *	polyethylene terephthalate (British Standards Institute)
PF *	phenol-formaldehyde resins
PFA *	perfluoroalkoxy resins
PFEP *	tetrafluoroethylene-hexafluoropropylene copolymer (*see* FEP)
PG *	polypropylene glycol
PGA *	phosphoglyceric acid
PHA *	poly(β-hydroxyalkanoates)
PHB *	parahydroxy benzoic acid
PHB/HV *	copolyesters of β-hydroxybutyrate and β-hydroxy-valerate
PHE *	phenylalanine
PHR	parts per hundred parts of resin
PK *	polyketone
PI *	polyimide
PI *	*trans*-1,4-polyisoprene (British Standards Institution)
Pi *	orthophosphate
PIB *	polyisobutylene (British Standards Institution)
PIBI *	isobutene-isopropene copolymer; butyl rubber (*see* GR-I, IIR)
PIBO *	poly(isobutylene oxide)
PID	photo-ionization detector
PIDA *	phenylindane dicarboxylic acid
PIP *	synthetic *cis*-1,4-polyisoprene (*see* IR)
PIR *	poly(methyl acrylate)
PIS *	polyisobutylene
PIS	Penning ionization spectroscopy
PIT	phase inversion temperature
PIXES	particle-induced X-ray emission spectroscopy

PMA *	polymethylacrylate
PMA *	phosphomolybdic acid
PMA *	pyromellitic acid
PMAC *	polymethoxy acetal
PMAN *	polymethacrylonitrile
PMC	polymer matrix composites
PMC	fiber-reinforced plastics
PMCA *	poly(methyl-α-chloroacrylate)
PMDA *	pyromellitic dianhydride
PMDI *	polymeric methylene diphenylene isocyanate
PMHP *	p-menthane hydroperoxide
PMI *	polymethacrylimide
PMIPK *	poly (methylisopropenyl) ketone
PMMA *	poly(methylmethacrylate)
PMMI *	polypyromellitimide
PMP *	1-phenyl-3-methyl-5-pyrazolone
PMP *	poly(4-methylpentene-1)
PMR	proton magnetic resonance
PMTA *	phosphomolybdic-phosphotungstic acid mixture
PNF *	phosphonitrilic fluoroelastomer
PO *	phenoxy resins
PO *	polyolefins
PO *	poly(propylene oxide)
PO *	propylene oxide
POE *	polyoxy ethylene
POEMS *	polyoxyethylene stearate
POEOP *	polyoxyethyleneoxypropylene
POF *	dl-α-lipoic acid
POM *	polyformaldehyde
POM *	polyoxymethylene
POM *	polycyclic organic matter
POP *	poly(phenylene oxide) (see PPO)
POPDA *	polyoxypropylene diamine
POPOP *	1,4-bis[2-(5-phenyloxazolyl)]-benzene
PP *	polypropylene
PPD-T *	poly-para-phenylene terephthalamide
PPF *	plasma protein fraction
PPi *	pyrophosphate
PPI *	polymeric polyisocyanate
PPL *	poly(β-propiolactone)
PPO *	poly(phenylene oxide) (see POP, PPOX)
PPOX *	polypropylene oxide; (see POP, PPO)
PPS *	poly(p-phenylene sulfide)
PPSU *	poly(phenylene sulfone)
PPT *	poly(propylene terephthalate)
PROM	programmable read-only memory

29

PS *	polystyrene
PSB *	styrene-butadiene rubber (*see* GR-S, SBR)
PSC	prestressed cement
PSF *	polysulfone
PSO *	polysulfone
PSR	proton storage ring
PSU *	polyphenylene sulfone
PSZ	partly stabilized zirconia ceramic
PTA *	purified terephthalic acid
PTA *	phosphotungstic acid
PTC	phase transfer catalysis
PTDQ *	polymerized trimethyl dihydroxyquinoline
PTF *	polytetrafluoroethylene
PTFE *	polytetrafluoroethylene
P3FE *	polytrifluoroethylene
PTHF *	polytetrahydrofuran
PTMA *	phosphotungstic-phosphomolybdic acid mixture
PTMG *	poly(tetramethylene ether glycol)
PTMT *	poly(tetramethylene terephthalate); polybutylene terephthalate (*see* PBTP)
PTSA*	*p*-toluenesulfonamide
PTZ *	phenothiazines
PU *	polyurethane
PUR *	polyurethane
PVA *	polyvinylalcohol
PVAC *	polyvinylacetate
PVAL *	polyvinyl alcohol (*see* PVOH)
PVB *	polyvinylbutyral
PVC *	polyvinylchloride
uPVC *	unplasticized polyvinyl chloride
PVC	pigment volume concentration
PVCA *	vinyl chloride-vinyl acetate copolymer (*see* PVCAC)
PVCAC *	vinyl chloride-vinyl acetate copolymer (*see* PVCA)
PVCC *	chlorinated poly(vinyl chloride)
PVD *	polyvinyl dichloride
PVDC *	polyvinyl dichloride (used in the 1960s)
PVDC *	polyvinylidene chloride
PVDF *	polyvinylidene fluoride
PVE *	polyvinylethyl ether
PVF *	polyvinyl fluoride
PVFM *	polyvinyl formal
PVFO *	polyvinyl formal
PVI *	polyvinylisobutyl ether
PVK *	polyvinyl carbazole
PVM *	polyvinylmethyl ether

PVOH *	polyvinyl alcohol (rarely used) (*see* PVA)
PVP *	poly(*N*-vinyl-2-pyrrolidone)
PWB	printed wiring board
PWR	pressurized water nuclear reactor
PZT*	lead-zirconium titanate
QAC *	quaternary ammonium compound
QSAR	quantitative structure–activity relation
RAM	random access memory
RBE	relative biological effectiveness
RBS	Rutherford backscattering spectroscopy
RC	reinforced cement
RDA	recommended daily allowance
RDF *	refuse derived fuel
RDF	radial distribution function
RDX *	cyclotrimethylene trinitramine
RF *	resorcinol-formaldehyde resin
RFNA*	red fuming nitric acid
RHEED	reflection high energy electron diffraction
RIA	radioimmunoassay
rib *	ribose
RIE	reactive ion etching
RIM	reaction injection molding
RIMS	redrawn inviscid melt spinning, redrawn IMS
RIS	resonance ionization spectroscopy
RNA *	ribonucleic acid
RNase *	ribonuclease
RO	reverse osmosis
ROM	read-only memory
RON	research octane number
RP *	reinforced plastic
RPE	rotating platinum electrode
RP-IPP	reverse phase-ion pair partition
RR acid *	2-amino-8-hydroxy-3,6-disulfonic acid naphthalene
rRNA *	ribosomal RNA
RRS	resonance Raman spectroscopy
RS	Raman spectroscopy
RSP	salt-recovery process
RSS	ribbed smoke (sheets)
RT	room temperature
RTM	resin transfer molding
RTP *	reinforced thermoplastic
RTV	room temperature vulcanizing
RVP	Reid vapor pressure
SA *	styrene-acrylonitrile
SABRA	surface activation beneath reaction adhesives

S acid *	1-amino-8-hydroxy-4-sulfonic acid naphthalene
SAD	selected area diffraction
SAF black	super abrasion furnace black
SAIB *	sucrose acetate isobutyrate
SAM	scanning auger microprobe
SAN *	styrene-acrylonitrile copolymer
SAPO *	silicoaluminophosphates
SAR	structure–activity relation
SAS *	sodium aluminum sulfate
SAS *	sodium alkane sulfonate (paraffin sulfonate)
SASOL	modern Fischer–Tropsch process
SAXS	small angle X-ray scattering
SB *	styrene-butadiene copolymer
SBA *	secondary butyl alcohol
S.B.G.	standard battery grade (chemicals)
SBR *	styrene-butadiene rubber (see GR-S)
SCC	stress control cracking
SCE	saturated Calomel electrode
SCF	supercritical fluid
SCF	supercritical fluid chromatography
SCOT	surface coated open tubular (column)
SCP *	single cell protein
SCR *	styrene-chloroprene rubber
SCR	selective catalytic reduction
SCR I	solvent refined coal process I
SCR II	solvent refined coal process II
SDA *	specially denatured alcohol
SDD *	sodium dimethyl dithiocarbonate
SDDC *	sodium dimethyl dithiocarbamate
SDP *	4,4'-sulfonyl diphenyl
SEM	scanning electron microscopy
S-EPDM *	sulfonated ethylene-propylene-diene terpolymers
SERS	surface enhanced Raman spectroscopy
SES *	sodium 2,4-dichlorophenoxyethyl sulfate; sesone
SFE	supercritical fluid extraction
SFI	solid fat index
SFS	Saybolt Furol seconds
SFS *	sodium formaldehyde sulfoxylate
S-glass*	magnesia-alumina-silicate glass
SHC	super hybrid composites
SHE, NHE	standard hydrogen electrode
SHIPS *	super-high impact polystyrene
SHOP	Shell higher olefin process
SI	systeme internationale
SI *	silicone resins, polydimethylsiloxane
SIC	standard industrial classification

SIMS	secondary ion mass spectroscopy
SIN	simultaneous interpenetrating networks
SIP	submerged injection process (steel making)
SIPP *	sodium iron pyrophosphate
SIPS	sputter induced photoelectron spectroscopy
SIR *	styrene-isoprene rubber
SIRS	satellite infrared spectrometer
SITC	Standard Industrial Trade Classification
SLCM	structural liquid composite molding
SMA *	styrene-maleic anhydride copolymer
SMC	sheet molding compound
SMS *	styrene-α-methylstyrene copolymer
SMSI	strong metal support interaction
(SN)x *	sulfur-nitrogen polymers
SNG $^{-}$*	substitute natural gas
SNTA *	sodium nitrilo-triacetate
SOFC	solid oxide fuel cell
SPAM	scanning photoacoustic microscopy
SPEFC	solid polymer electrolyte fuel cell
SPPS	solid state peptide synthesis
SRC I & II	solvent refined coal processes
SRF black	semireinforcing black
SRIM	structural reaction injection molding
SRP *	styrene-rubber plastics (ASTM)
SRS	stimulated Raman scattering
SS	single stage
SS	suspended solids
SS acid *	1-amino-8-hydroxy-2,4-disulfonic acid naphthalene
SSD	steady state distribution
SSI	small-scale integration
STEL	short-term exposure limit
STEM	scanning transmission electron microscopy
STH *	somatotropic hormone
STM	scanning tunneling microscope
STP *	pentasodium tripolyphosphate
STP	standard temperature and pressure
STPP *	sodium tripolyphosphate
SXAPS	soft X-ray appearance potential spectroscopy
2,4,5-T *	2,4,5-trichlorophenoxyacetic acid
T *	thymine
TAC *	triallyl cyanurate
TAED *	tetraacetyl ethylene diamine
TAIC *	triallyl isocyanurate
TAME *	teriary amylmethyl ether
TBA *	tertiary butyl alcohol
TBH *	1,2,3,4,5,6-hexachlorocyclohexane; HCCH

33

TBHQ *	tertiary butyl quinone
TBMA *	tertiary butyl methacrylate
t-BOC *	poly(*tert*-butoxycarbonyloxystyrene)
TBP *	tributyl phosphate
TBT *	tetrabutyl titanate
TBTO *	*bis*(tributyltin titanate)
TC *	trichloroacetic acid
TCA	tricarboxylic acid (Krebs) cycle
TCA *	trichloroacetic acid
TCB *	tetracarboxy butane
TCBO *	trichlorobutylene oxide
TCC	tagliabue closed cup
TCD	thermal conductivity detector
TCDD *	dioxin; 2,3,7,8-tetrachlorodibenzo- *p*-dioxin
TCE *	1,1,1-trichloroethane
TCEF *	trichloroethyl phosphate
TCNE *	tetracyanoethylene
TCP *	tricresyl phosphate
TDA *	2,4 diaminotoluene
TDE *	tetrachlorodiphenylethane; DDD
TDI *	toluene diisocyanate
TDP *	4,4'-thiodiphenol
TDQP *	trimethyldihydroquinoline polymer
TDS	total dissolved solids
TEA *	triethanolamine
TEA *	triethylaluminum
TEAC *	tetraethylammonium chloride
TED	thermal emission detector
TEDP *	tetraethyl dithiopyrophosphate
TEG *	triethylene glycol
TEG *	tetraethylene glycol
TEL *	tetraethyllead
TEM *	triethylene melamine
TEM	transmission electron microscopy
TEOS *	tetraethyl orthosilicate
TEP *	triethyl phosphate
TFE *	poly(tetrafluoroethylene)
TFEO *	tetrafluoroethylene epoxide
TG	thermal gravimetry
TGA *	triglycollamic acid; nitrilotriacetic acid
TGA	thermogravimetric analysis
THAM *	*tris*(hydroxymethyl)aminomethane
THC *	Δ-9-tetrahydrocannabinol
THF *	tetrahydrofuran
THPC *	tetra-*bis*(hydroxymethyl)phosphonium chloride
THT *	tetrahydrothiophene

TIBAL *	triisobutylaluminum
TIM	thermoplastic injection molding
TIOTM *	tri-*i*-octyl trimellitate
TKP *	tripotassiumphosphate
TKPP *	tetrapotassium pyrophosphate
TLB	threshold limit value
TLC	thin layer chromatography
TLV	threshold limit values
TMA *	trimethylamine
TMA *	trimellitic anhydride
TMA	thermomechanical analysis
TMDI *	2,2,4-trimethyl-1,6-hexane diisocyanate
TMEDA *	tetramethylethylenediamine
TML *	tetramethyllead
TMS *	tetramethylsilane
TMTD *	tetramethylthiuram disulfide
TMV	tobacco mosaic virus
TNA *	tetranitroaniline
TNB *	trinitrobenzene
TNF	tumor necrosis factor
TNF *	2,4,7-trinitrofluorene
TNT *	2,4,6-trinitrotoluene
TOF *	trioctyl phosphate
TOFA *	tall oil fatty acids
TOFMS	time-of-flight mass spectrometry
TOPO *	trioctylphosphinic oxide
TOTM *	trioctyl trimellitate
tPA *	tissue plasminogen activator
TPA *	terephthalic acid
TPAK *	potassium terephthalate
TPB *	tetraphenylbutadiene
TPD	temperature programmed desorption
TPDE	temperature programmed decomposition
TPE *	thermoplastic elastomer
TPG *	triphenylguanidine
TPN *	triphosphopyridine nucleotide
TPO *	thermoplastic polyolefin (rubber)
TPP *	triphenyl phosphate
TPR *	1,5-*trans*-poly(pentenamer)
TPR	temperature programmed reduction (or reaction)
TPS*	toughened polystyrene (British Standards Institute)
TPT *	triphenyltetrazolium chloride
TPT *	tetraisopropyl titanate
TPU *	thermoplastic polyurethane
TPX *	polymethylpentene
T.R.	temperature rise

TR	thio rubber (British Standards Institute)
TRF *	thyrotropin-releasing factor
TRF *	thyroliberin
TRL	transistor-resistor logic
tRNA *	transfer RNA
TSA *	toluenesulfonic acid
TSC	thermal stress cracking
TSH *	thyrotropic hormone
TSP *	trisodium phosphate
TSP *	triple superphosphate
TSPA *	triethylenethiophosphoramide
TSPP *	tetrasodium pyrophosphate
TSR	technically specified rubbers
TTB *	2,2'6,6'-tetrabromo-3,3'5,5'-tetramethyl-4,4'-dihydroxydiphenyl
TTC *	tetrazoliumchloride
TTD *	tetraethylthiuram disulfide
TTL	transistor-transistor logic
TTS	total suspended solids
UART	universal asynchronous receiver transmitter
UDMH *	unsymmetrical dimethyl hydrazine
UDP *	uridine diphosphate
UDPG *	uridine diphosphate glucose
UF *	urea formaldehyde resins
UF	ultrafiltration
UFC	unidirectional fiber composite
UHMPE *	ultra-high molecular weight polyethylene
UHMW-PE *	ultra-high molecular weight polyethylene
UHV	ultra high vacuum
UMP *	uridine monophosphate; uridylic acid
UP *	unsaturated polyesters (British Standards Institute)
UPS	ultraviolet photoelectron spectroscopy
USAN	U.S. Adopted Name (for pharmaceuticals)
USP	United States Pharmacopeia
USRDA	U.S. recommended daily allowances (nutrition)
UTP *	uridine triphosphate
UV	ultra violet spectral range
VA *	vinyl acetate
VAD	vapor-axial deposition
VAM *	vinyl acetate monomer
VC *	vinyl chloride plastic
VC *	vinylidene chloride
V/C	volume to capacity ratio
VC/E *	vinyl chloride-ethylene copolymer
VC/E/VA *	vinyl chloride-ethylene-vinylacetate copolymer
VCH *	vinyl cyclohexane

VCM *	vinyl chloride monomer
VC/MA *	vinyl chloride-methyl acrylate copolymer
VC/MMA *	vinyl chloride-methylmethacrylate copolymer
VC/OA *	vinyl chloride-octyl acrylate
vcp	vacuum condensing point
VC/VAC *	vinyl chloride-vinyl acetate copolymer
VC/VDC *	vinyl chloride-vinylidene chloride
VDC *	vinylidene dichloride
VDM	vacuum fluorescent displays
VF *	vinyl fluoride
VF *	vulcan fiber
VFM *	vinyl fluoride monomer
VF2 *	polyvinylidene fluoride
VGA	variable graphic array
VGCF*	vapor phase grown carbon fiber
VHSIC	very high speed integrated circuit
VI	viscosity index
VLE	vapor–liquid equilibrium
VLSI	very large-scale integration
VOC	volatile organic compounds
VPC	vapor phase chromatography
VPO	vapor-phase oxidation
VTVM	vacuum-tube voltmeter
WCOT	wall coated open tubular (column)
WDS	wavelength dispersive X-ray spectroscopy
WEP *	water extended polyesters
WFNA *	white fuming nitric acid
WHSV	weight hourly space velocity
W/O	water in oil
WPA *	wet process phosphoric acid
WVT	water vapor transmission
XAES	X-ray auger electron spectroscopy
XANES	X-ray absorption near edge spectroscopy
XAPFS	extended X-ray appearance potential fine structure spectroscopy
XELFS	extended electron energy loss fine structure spectroscopy (EXELFS)
XES	X-ray energy spectrometry
XIS *	isomerization process of xylenes to *p*-xylene
XLPE*	cross-linked polyethylene
XPD	X-ray photoelectron diffraction
XPS *	expanded or expandable polyethylene (*see* EPS)
XPS	X-ray photoelectron spectroscopy
XRD	X-ray diffraction
XRDF	X-ray radial distance function
XRF	X-ray fluorescence

YAG	yttrium-aluminum-garnet laser
ZDP *	zinc dithiophosphate
ZMA *	zinc metaarsenite
ZMBT *	2-mercaptobenzothiazole
ZSM-5 *	a zeolite catalyst

Acronyms for Organizations

AAAS	American Association for the Advancement of Science
AATCC	American Association of Textile Chemists and Colorists
ACHEMA	Ausstellungs-Tagung für Chemisches Apparatewesen
ACS	American Chemical Society
ACS	American Cancer Society
ACGIH	American Conference of Governmental Industrial Hygienists, Inc.
AGA	American Gas Associated
AIC	American Institute of Chemists
AIChE	American Institute of Chemical Engineers
AIME	American Institute of Mining, Metallurgical and Petroleum Engineers
AIP	American Institute of Physics
AISI	American Iron and Steel Institute
ANSI	American National Standards Institute
AOCS	American Oil Chemists Society
APHA	American Public Health Association
API	American Petroleum Institute
ASM	American Society for Metals
ASME	American Society of Mechanical Engineers
ASTM	American Society for Testing and Materials
BATF	Bureau of Alcohol,Tobacco and Firearms of the U.S. Department of the Treasury
BRITE	Basic Research in Industrial Technologies in Europe
BSI	British Standards Institute
CAA	Clean Air Act
CAER	Community Awareness and Emergency Response (CMA)
CDC	Center for Disease Control
CEFIC	European Council of Chemical Manufacturers Associations
CEQ	Council on Environmental Quality
CERCLA	Comprehensive Environmental Response, Compensation and Liability Act

CERN	European Nuclear Research Center (Geneva)
CFR	Code of Federal Regulations
CHEMRAWN	Chemical Research Applied to World Needs
CMA	Chemical Manufacturing Association
CMRA	Chemical Marketing Research Association
CPSC	Consumer Product Safety Commission
CSIRO	Commonwealth Science and Industry Research Organization (Australia)
CWM	Chemical Waste Management (EPA)
DECHEMA	Deutsche Gesellschaft für chemisches Apparatewesen
DIN	Deutsche Industrie Normen
DIN	Deutsches Institut für Normung
DOA	Department of Agriculture
DOD	Department of Defense
DOE	Department of Energy
DOT	Department of Transportation
ECMRA	European Chemical Market Research Association
EEC	European Economic Community
EMBL	European Molecular Biology Laboratory
EPA	Environmental Protection Agency (US)
ESA	European Space Agency
ESF	European Science Foundation
ESPRIT	European Strategic Program for Research and Development in Information Technology
EVAF	European Association for Industrial Market Research
FAO	Federal Energy Administration
FCC	Food Chemicals Codex
FDA	Food and Drug Administration
FEA	Federal Energy Administration
FEMA	Flavor Extracts Manufacturers' Association
FEPCA	Federal Environmental Pesticides Control Act
FHSA	Federal Hazardous Substances Act
FIFRA	Federal Insecticide, Fungicide, and Rodenticide Act
FPC	Federal Power Commission
FR	Federal Register
FRB	Federal Reserve Board
FTC	Federal Trade Commission
GPI	Glass Packing Institute
IATA	International Air Transport Association

ICC	Interstate Commerce Commission
IMRA	Industrial Marketing Research Association (Britain)
IP	Institute of Petroleum (U.K.)
IPM	Integrated Pest Management
IRLG	Interagency Regulatory Liaison Group
ISO	International Organization for Standardization
IUPAC	International Union of Pure and Applied Chemistry
JET	Joint European Torus
MCA	Chemical Manufacturers' Association
MITI	Ministry of International Trade and Industry (Tokyo)
NAAQS	National Ambient Air Quality Standard
NAS	National Academy of Science
NASA	National Air and Space Administration
NBS	National Bureau of Standards, formerly NTIS
NCI	National Cancer Institute
NFPA	National Fire Protection Association
NIDA	National Institute on Drug Abuse
NIH	National Institutes of Health
NIOSH	National Institute for Occupational Safety and Health
NIST	National Institute of Science and Technology, formerly NBS
NPRM	Notice of Proposed Rulemaking
NRC	Nuclear Regulatory Commission
NRL	Naval Research Laboratory
NSF	National Science Foundation
NTIS	National Technical Information Service
OECD	Organization for Economic Cooperation and Development
OSHA	Occupational Safety and Health Administration
OSTP	Office of Science and Technology Policy
OTA	Office of Technology Assessment
PIA	Plastics Institute of America
PMA	Pharmaceutical Manufacturers Association
PMN	Premanufacture Notification
PTO	Patent and Trademark Office
RACE	Research and Development in Advanced Communication Technologies for Europe
RC	Regulatory Council

RCRA	Resource Conservation and Recovery Act
RPAR	Rebuttable Presumption Against Reregistration
SAE	Society of Automotive Engineers
SCI	Society of Chemical Industry
SERC	Science and Engineering Research Council (United Kingdom)
SERI	Solar Energy Research Institute
SIC	Standard Industry Classification
SOCMA	Synthetic Organic Chemists Manufacturing Association
SPE	Society of Plastics Engineers
TAPPI	Technical Association of the Pulp and Paper Industry
TSCA	Toxic Substances Control Act
TVA	Tennessee Valley Authority
UL	Underwriters Laboratory
USDA	United States Department of Agriculture
U.S.P.	United States Pharmacopeia
USRDA	United States Recommended Daily Allowance

WHO
World Health Organization
Chemical Emergency Numbers

Canada	CANUTEC	Emergency 613 996 6666
		Information 613 992 4624
United States	CHEMTREC	Emergency 800 424 9300
		Information 800 282 8200
United Kingdom and Ireland	Chemical Emergency Agency Service	0235-83 48 00

Encyclopedic Dictionary of Chemical Technology

A

A acid: 1,7-dihydroxy-naphthalene-3,6-disulfonic acid; dye intermediate

abaca: manila hemp; strong vegetable fiber

abherents: inert substances that prevent adhesion of materials to themselves or other materials, e.g., lubricants, dusting agents

abietic acid: chief ingredient of pine tree rosin; terpenoid; use: in varnishes, paints, paper sizing, compounding with resins and rubbers

ABIETIC ACID

ablative agents: materials used for dissipation of high heat by removal of mass; commonly used in space vehicles for reentry

abrasives: hard granules or powder used for cleaning, smoothing, polishing, or shaping surfaces; natural abrasives include garnet, diamond dust, corundum, silica, rouge, pumice, and feldspar; synthetic abrasives include boron carbides, silicon carbides, and fused alumina

absorbed dose: in nuclear technology, the energy transferred to a material during irradiation; measured in gray (Gy) where 1 Gy = 1 J/kg; formerly measured in rad (1 rad = 0.01 Gy or 100 erg/g)

absorbed power density: rate of absorption of energy per milliliter

absorption: the penetration of one substance into another; compare with adsorption

absorption chromatography: *see* liquid–solid chromatography

absorption spectra: set of spectral lines or bands missing from an otherwise continuous spectrum; electromagnetic radiation is absorbed by materials; the wavelength of the absorbed radiation depends on the molecular and/or atomic structure of the material, e.g., visible light (λ = 4000–7000 Å) is absorbed by electrons, infrared radiation ($\lambda \geq 7000$ Å) is absorbed by covalent bonds that are made to vibrate and rotate due to the absorbed energy, ultraviolet radiation ($\lambda \leq 4000$ Å) is absorbed by electrons that are held tighter than those absorbing visible light

ABS resin: *see* acrylonitrile-butadiene-styrene copolymer

acacia gum: exudate from stems of *Acacia Senegal; see* arabic gum

acaricides: pesticides effective against ticks and mites; e.g., amitraz, dicofol

acaroid: natural resin derived from *Xanthorrhoea* trees; closely related to balsams; use: in varnishes, metal lacquers, coatings, sizing, binders, and for production of picric acid

accelerators: in polymer and rubber technology, substances that reduce the time needed for polymerization or cure; in radiation technology, machines (e.g., betatrons, cyclotrons, linear accelerators) that accelerate particles to high energy

acceptor: in chemical science, an atom that accepts a pair of electrons during bond formation; a Lewis acid

acceptor stem: in genetic engineering, the

tRNA structure that attaches to an amino acid that then is transported and incorporated into a growing peptide or protein

acclimation: in genetic engineering, adjustments of an individual organism to local environmental changes

acclimatization: adjustments of a species over several generations to significant environmental changes

accroides: *see* resins

acenaphthene: produced from aromatic coal tar fraction; limited use: gas phase oxidation to phthalic anhydride and naphthalic anhydride

ACENAPHTHENE

acephate: *see* orthene naphthalene-1,8-dicarboxylic acid; hazardous chemical waste (EPA)

acetaldehyde: CH_3CHO; ethanal; produced by catalytic ($CuCl_2 \cdot PdCl_2$) oxidation (air or O_2) of ethylene (Wacker–Hoechst process), by noncatalytic oxidation of C3/C4 hydrocarbons (gas phase at 425–460°C and 7–20 bar or liquid phase in the presence of air or oxygen) (Celanese), by catalytic dehydrogenation of ethanol (Cu catalysts activated with Co, Cr, or Zn, 270–300°C in the absence of air or catalytic Ag at 450–550°C and 3 bar in the presence of air); use: precursor for numerous products including acetic acid, acetic anhydride, n-butanol, chloral, ketene, ethyl acetate, crotonaldehyde, 2-ethylhexanol, trimethylolpropane, pentaerythritol, peracetic acid; *see* paraldehyde

acetaldol: $CH_3CH(OH)CH_2CHO$; produced from acetaldehyde at room temperature in the presence of NaOH; use: feedstock for the production of 1,3-butanediol and crotonaldehyde

acetal resins: linear macromolecules of the type $-O-CH_2-O-CH_2-$; hard, rigid, strong polymers; use: engineering plastics; *see* poly(oxymethylene)

acetals: in general, products resulting from the reaction of a carbonyl group with 2 moles of alcohol; in particular, acetal is the trivial name for $CH_3CH(OC_2H_5)_2$; use: solvent and organic intermediate

acetamide: CH_3CONH_2; produced from acetic acid and ammonium carbonate; use: solvent, liquid flux for soldering, softening agent for leather coatings and glue, plasticizer, antifreeze, chemical intermediate

acetaminophen: p-$CH_3CONHC_6H_4OH$; produced by acetylation of p-aminophenol with glacial acetic acid and acetic anhydride; use: analgesic (Tylenol)

acetanilide: *N*-phenyl acetamide; produced by acetylation of aniline with glacial acetic acid; use: intermediate in the pharmaceutical and dye industries, rubber accelerator, antiseptic, stabilizer

acetarsone: chemotherapeutic agent effective against protozoa

ACETARSONE

acetate dyes: *see* disperse dyes

acetate fibers: *see* cellulose acetate

acetate film: *see* cellulose acetate

acetic acid: CH_3COOH, HAc, HOAc; vinegar contains 2–10% acetic acid; glacial acetic acid contains 99.5–99.7% acetic acid; produced by liquid phase oxidation of n-butane (Celanese, Hüls, UCC), of n-butenes (Bayer, Hüls), of light gasoline (BP, British Distillers); by catalytic oxidation of acetaldehyde using oxygen (UCC, FMC, Daicel Japan, British Celanese), or using air (Hoechst, Rhône-Poulenc); catalytic carbonylation of methanol (BASF, Borden, Shell, Monsanto); from synthesis gas using metallic rhodium on alumina or silica (UCC); fermentation of ethanol has been phased out; widely used in chemical industry for manufacture of vinyl acetate monomer, cellulose acetate, butyl and isopropyl acetates, acetic anhydride, acetanilide, acetylchloride, acetamide, and chloroacetic acid; as solvent for the commercial production of terephthalic acid from p-xylene; methyl, ethyl, isopropyl, and butyl acetates serve as solvents for paints and resins; as pesticide, acidizer for oil wells; Na, Pb, Al, and Zn acetates as auxiliary agents in textile dye and leather industries

acetic anhydride: $(CH_3CO)_2O$, Ac_2O; produced by (1) catalyzed (Mn, Cu, or Cr acetates) oxidation (air or O_2) of acetaldehyde; (2) reaction of acetic acid with ketene; and (3) reaction of methylacetate with syngas in the presence of rhodium catalyst; use: acetylation, particularly for manufacture of cellulose and vinyl acetates, chloroacetic acid, acetylsalicylic acid (aspirin)

acetin: $C_3H_5(OH)_2OOCCH_3$; glyceryl monoacetate; produced by acetylation of glycerin with acetic anhydride; use: in the manufacture of dynamite, solvent for basic dyes and tannins; diacetin is the diacetate of glycerol; the triacetate (triacetin) is used as plasticizer

acetoacetanilide: $C_6H_5NHCOCH_2COCH_3$; produced by reaction of ethyl acetoacetate and aniline or from aniline and ketene; use: for organic synthesis

acetoacetic acid: CH_3COCH_2COOH; its ester, ethyl acetoacetate (acetoacetic ester), is used in organic synthesis

Acetobacter: genus of bacteria used for manufacture of acetic acid (vinegar) by fermentation of glucose

acetone: CH_3COCH_3; dimethylketone; produced by liquid phase catalytic ($PdCl_2$) oxidation (O_2 or air) of propene at 110–120°C and 10–14 bar (Wacker–Hoechst process); by catalytic oxidation of isopropanol (e.g., with Ag or Cu at 400–600°C, or with ZnO at 300–400°C, or with Cu-Zn [brass] at 500°C and 3 bar); as coproduct in phenol manufacture from cumene (Hock process) and during oxidation of crude oil distillates; from fermentation of starch using *Bacillus macerans*; use: chemical intermediate for products (e.g., bisphenol A, isopropanol, methyl isobutyl ketone, methyl isobutyl carbinol, hexylene glycol, methacrylates, mesityl oxide, ketene), solvent (incorporated in high-low solvent blends as the low boiling component), drying agent, extractant

acetone cyanohydrin process: ACN process; *see* methacrylic acid

acetonitrile: CH_3CN; methyl cyanide; produced as by-product in oxidation of propene to acrylonitrile; use: solvent, organic intermediate

acetonylacetone: $(CH_3COCH_2)_2$; produced as by-product of hydration of acetylene to acetaldehyde; use: solvent, organic intermediate

acetophenetidine: $p-CH_3CONHC_6H_4O$ C_2H_5; produced by reaction of p-acetaminophenol and ethylbromide; use: analgesic

acetophenone: $C_6H_5COCH_3$; produced as by-product from oxidation of cumene (Hock) to phenol and from the peroxide oxidation of ethylbenzene to styrene (ARCO); use: solvent, organic intermediate

acetostearin: acetylated glyceryl monostearate; use: protective coating for food

acetoxylation: reaction of an olefin or aromatic compound with acetic acid in the presence of air leading to an acetate e.g., manufacture of ethylene glycol from ethylene by way of the diacetate followed by hydrolysis; the reaction can be stopped at the monoacetate

acetylation: formation of esters or amides by reaction of alcohols or amines with acetic anhydride or acetyl chloride; *see* acylation

acetyl chloride: CH_3COCl, AcCl; produced by reaction of acetic anhydride and HCl, or acetic acid with PCl_3; used as acetylating and chlorinating agent, catalyst for esterification and chlorination, rubber antiscorch agent

acetylcholine: $CH_3COOCH_2CH_2N(CH_3)_3OH$, ACh; natural product; neurotransmitter at junction between muscles and nerves, between nerve endings in the autonomic nervous system and within the brain; believed to be implicated in Alzheimer's disease

acetylcholine acetate: intermediate in the citric acid cycle, the biological pathway leading to citric acid from glucose

acetylcholinesterase: AChE; enzyme for hydrolysis of acetylcholine; in nerve–muscle interaction acetylcholine must be hydrolyzed after it is excreted from the nerve to prevent overstimulation of the muscle. Carbamate esters that inhibit AChE activity are used as insecticides

acetylene: $HC{\equiv}CH$; industrial gas; prepared by cracking of hydrocarbons using an electric arc (e.g., Hoechst–Hüls' HEAP [*h*ydrogen *e*lectric *a*rc *p*rocess] process, other plasma processes at UCC, DuPont, USSR); by indirect cracking of hydrocarbons using two regenerator ovens operating alternately hot and cold (Wulff process modified by UCC, also Kureha process using superheated steam as heat transfer agent); by cracking process using heat from partial combustion of feed: (e.g., Sachsse–Bartholome process for light petroleum, submerged flame process (BASF, Sisas, SBA, Montecatini, Hoechst's HTP [*h*igh *t*emperature *p*yrolysis]), from CaC_2; use: synthesis notably to prepare industrial products (e.g., vinyl chloride [q.v.]

and vinyl acetate monomers, 1,4-butanediol (BASF, DuPont, GAF), and propargyl alcohol); for welding and cutting metals as the oxygen–acetylene flame produces the highest temperature (estimated to be 3300°C) of any combustible gas, for flame hardening, for shaping synthetic sapphires and rubies

$$HC{\equiv}CH$$

ACETYLENE

acetylene black: carbon obtained from thermal decomposition of acetylene; use: dry cells, lubricants; *see* carbon blacks

acetylides: salt-like metal derivatives of acetylene (e.g., CaC_2); acetylide salts are explosive; *see* carbides

acetylsalicylic acid: o–$CH_3COOC_6H_4COOH$; aspirin; manufactured in a continuous process from salicylic acid and acetic anhydride; use: analgesic, antipyretic, appears to reduce blood clotting; *see* FLOW DIAGRAM FOR PRODUCTION OF ASPIRIN; arachidonic acid

Acheson process: manufacture of silicon carbide by reaction of SiO_2 and C; *see* polycarbosilane; silicon carbide; ceramics

acicular growth: crystallization in the form of needles

acid-azo pigments: salts obtained from acidic azo dyes by reaction with alkali or alkaline earth ions; these pigments include lithol rubine, Persian orange, red lake, and tartrazine; use: printing inks, paints, rubber products

acid dope: *see* pyroxylin

acid dyes: anionic dyes, negative dyes; water-soluble dyes that contain hydrophilic groups (e.g., -COOH, $-SO_3H$, or -OH) that attach the dyes to fibers containing basic groups (e.g., cotton, wool, and polyamides [nylons]); acid dyes, include azo dyes, anthraquinones, azines, triarylmethanes, chrome, and nitro dyes

acid number: milligrams of KOH needed to neutralize free acid in a 1 *g* sample; in poly-

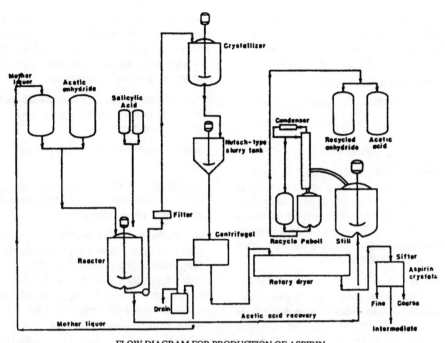

FLOW DIAGRAM FOR PRODUCTION OF ASPIRIN

[Excerpted by special permission from *Chem. Eng.* 60 (6):116–120 (1953; copyright 1953 by McGraw-Hill, Inc. Copyright *Riegel's Handbook of Industrial Chemistry*, 8th ed., Van Nostrand-Reinhold Co., Inc.]

mer chemistry, used to determine the number average molecular weight of polymers containing terminal acid groups

acid rain: rain water with low pH due to dissolved SO_3, SO_2, and NO_x; acid rain damages stone statuaries and buildings, increases acidity in lakes and rivers, disturbs plant and fish life; acid rain has been implicated in forest damage; the sulfur and nitrogen oxides appear to originate in volcanic eruptions and from human activities such as industrial facilities and automobile exhaust; the gases are carried into the atmosphere by prevailing winds

acids: substances that liberate H^+ in water solutions and that react with bases to form salts; common industrial acids include strong mineral acids (HCl, HNO_3, H_2SO_4, H_3PO_4), the weak carbonic acid, H_2CO_3, and organic acids (carboxylic acids RCOOH), such as acetic and propionic acid, dicarboxylic acids (e.g., phthalic acid), amino acids, saturated and unsaturated fatty acids

acidulant: acids occurring in or added to fruit and vegetables; use: preservatives, chelating agents or flavor enhancers; examples: citric and fumaric acids

acidulation: acidification

Acinetobacter: genus of bacteria used in industry e.g., for degradation of chlorinated biphenyls and of polybutadiene, denitrification of activated sludge and production of single cell protein (SCP) from ethyl alcohol

aconitic acid: H(COOH)C=C(COOH)CH_2COOH; produced by sulfuric acid dehydration of citric acid; intermediate in the biological uptake of CO_2; use: antioxidant, intermediate for plasticizers and wetting agents, for organic synthesis

acoustic collectors: sound with a frequency of 1000–10,000 Hz causes flocculation of fumes and mists; devices that produce such sound can be used as particle collectors

acridine: *see* acridine dyes

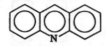

ACRIDINE

acridine dyes: basic dyes based on acridine; examples include acridine yellow (produced by fusion of toluene-2,4-diamine with glycerin and oxalic acid followed by oxidation with $FeCl_3$); acridine orange is used to detect nucleic acids: it fluoresces green with DNA and yellow with RNA

acrolein: CH_2=CHCHO; produced by catalytic gas phase oxidation of propylene (Shell, British Distillers, Montecatini use Cu_2O or Cu as catalysts, SOHIO, Nippon Shokubai, and Ugine-Kuhlmann use mixed oxides of Bi, Sb, Sn, Fe, Co, and W), formerly produced by gas phase reaction of acetaldehyde with formaldehyde using Na silicate/SiO_2 as catalyst (Degussa process); use: manufacture of glycerol, methionine, allyl alcohol, 2-picolines, acrylic acid, and acrylonitrile; hazardous chemical waste (EPA)

acrylamide: CH_2=CHCONH$_2$; produced from acrylonitrile by hydrolysis using sulfuric acid or a copper-based catalyst; use: preparation of acrylamide polymers and to modify other polymers; moderate neurotoxin

acrylamide polymers: *see* polyacrylamide

acrylates: *see* acrylic acid

acrylic acid: CH_2=CHCOOH; produced primarily by air oxidation of propylene via acrolein; also by addition of HCN to ethylene oxide followed by hydrolysis (cyanohydrin process); catalytic carbonylation of acetylene in the presence of water (Reppe process) as well as hydrolysis of acrylonitrile

have been phased out; use: raw material for esters, the most important of which are ethyl, butyl, 2-ethylhexyl, and methyl acrylates; acrylic acid polymers comprise most of the commercial water-soluble resins

acrylic elastomers: *see* acrylonitrile-butadiene copolymer; acrylonitrile-butadiene-styrene copolymer

acrylic ester polymers: *see* polyacrylic esters

acrylic fibers: fibers spun from copolymers that are made from at least 85% acrylonitrile monomer; *see* poly(acrylonitrile)

acrylonitrile: ACN or AN; CH_2=CHCN; produced by catalytic gas phase oxidation of propylene in the presence of NH_3, air, and water, typically at >400°C and 1.5 bar; catalysts include $Bi_2O_3 \cdot MoO_3$ (SOHIO), MoO_3 fixed bed (BP Distillers), Te-, Cc-, and Mo-oxide (Montedison), Mo/V or Bi fixed bed (SNAM Progetti); use: in manufacture of polymers and copolymers (e.g., acrylic fibers, nitrile rubber), synthetic intermediate (e.g., for acrylamide, succinonitrile); hazardous chemical waste (EPA)

acrylonitrile-butadiene copolymer: NBR; rubber; polarity of nitrile group causes interchain cohesion and insolubility in hydrocarbon solvents; *see also* styrene-acrylonitrile; styrene-butadiene; acrylonitrile-butadiene-styrene copolymers

acrylonitrile-butadiene-styrene copolymers: ABS copolymers; rubber and thermoplastic composites obtained by grafting acrylonitrile and styrene copolymer onto a polybutadiene matrix; the product contains 5–30% polybutadiene; higher polybutadiene content increases impact strength and lowers tensile strength, stiffness, hardness, and thermal properties; in suspension and bulk polymerizations lightly cross-linked polybutadiene is dissolved in styrene and acrylonitrile, the system is run to ~30% conversion at 80–100°C, then the viscous material is dispersed in water with a dispersing agent and the reaction completed at 100–170°C; various types of ABS resins can be blended to give desired impact strength, melt flow, tensile strength; use: electrical and electronic equipment, pipes, fittings, automotive parts, packaging, luggage, furniture

ACTH: *see* adrenocorticotropic hormone

Actinomycetes: genus of bacteria having fungus-like filaments; used industrially to produce antibiotics (e.g., streptomycin, actinomycin), enzymes, and other biologicals

actinomycin: antibiotic; isolated from *Streptomyces*; experiences some drug resistance

activated alumina: γ-Al_2O_3; produced by heating alumina to drive off water; highly porous aluminum oxide used as adsorbant, drying agent, and catalyst

activated carbon: highly porous amorphous carbon; produced by destructive distillation of wood followed by heating to 900°C in the presence of steam and CO_2; use: adsorbant, decolorizing, and purification agent

activated sludge: in waste disposal, aerated sludge containing bacteria that destroy organic matter; used during secondary waste treatment (activated sludge process) after the solids have been removed by screening; the activated sludge and waste water are kept in an aeration tank for several hours, then the waste water ("mixed liquor") is removed to another sedimentation tank and the activated sludge is returned to the aeration tank to digest a new batch of waste water

activation energy: energy needed to start a reaction; catalysts lower the activation energy of a process

activators: in rubber technology, substances that promote vulcanization; common activators include metal oxides (most commonly ZnO) and fatty acids such as stearic (C18) and lauric (C12) acid

active oxygen method: AOM; technique used to predict the shelf-life of fats and oils; oil is heated to 97.8°C while air is blown through the sample; the AOM value equals the number of hours needed to reach a peroxide value of 100 mEq/kg

actuator: transducer; device that translates the control signal from one type or level of energy (or power) to another; actuators can be manual, hydraulic, pneumatic, or electric

acylases: enzymes that catalyze the incorporation of –(R)C=O into a molecule; the reaction proceeds via an enzyme–acyl [enzyme–(R)C=O] intermediate

acyls: compounds containing the –C=O group, i.e., a carboxyl (–COOH) group from which the –OH has been lost

adamantane: $C_{10}H_{16}$; *sym*-tricyclodecane; obtained from crude oil; hydrocarbon with diamond-like carbon skeleton; use: antiviral agent

ADAMANTANE

ADAM-EVA process: proposed energy transport system by KFA Jülich; heat supplied from a nuclear reactor converts CH_4 to CO/H_2 (synthesis gas) in the EVA reformer; the syngas is supplied to energy consumers and again converted to methane in the ADAM methanator from which it is recycled to the EVA reformer

Adam's catalyst: PtO_2; made by fusing H_2PtCl_6 with $NaNO_3$ at 500°C; catalyst is the fine insoluble residue left after washing with water; use: promotes reductions in the presence of H_2

addition polymerization: chain polymerization; in this process the polymerization of a monomer is initiated by a free radical and continues (propagation reaction) until two radicals combine or another chain is started (termination reaction); in contrast to condensation polymerization, this reaction is fast and exothermic

addition polymers: macromolecules made from unsaturated monomers (e.g., butadiene, formaldehyde, perfluoroethylene, vinyl, and vinylidene compounds); *see* polymerization

additives: materials added to formulations to improve the properties of the compounded products; e.g., antiknock compounds such as methyl-*t*-butyl ether (MTBE) in gasoline, viscosity index improvers in lubricating oils, enzymes in detergents, antiozonants in rubber, metal scavengers in purification processes

additives, plastics: materials added to stabilize the products, to facilitate processing and to improve properties; the most widely used additives are (1) plasticizers, not only to improve the processability of plastics but also their flexibility; generally such external plasticizers lower the glass transition temperature (T_g) and the elastic modulus without changing the chemical structure of the polymer; copolymerization with monomers that lower the T_g (internal plasticizers) avoid extraction of plasticizer from the final product, however, they are limited by narrow use ranges; most important plasticizers are phthalates and phosphates; (2) antioxidants to inhibit autoxidation of monomers and polymers; they serve to stabilize polymers at elevated temperatures; the most important antioxidants are sterically hindered phenols, secondary aromatic amines, thioethers, phosphites, and phosphonates; (3) metal scavengers whose action is related to the action of antioxidants; catalyst metal cations remaining in the polymer lead to free radical products and thus to decomposition of the polymer; metal deactivators sequester the cations and are generally nitrogen compounds such as amides, hydrazones and hydrazides; (4) light stabilizers include free radical scavengers because light promotes the production of radicals; energy absorbed by chromophores in the plastic can be removed by quenching agents such as nickel complex ions; stabilizers are selected depending on the final use of the plastic product (outdoor materials need U.V. light absorbers such as derivatives of 2,4-dihydroxybenzophenone); (5) lubricants serve as processing aids to permit viscous polymers to pass through narrow slots and to facilitate the removal of products from molds; the most common lubricants are fatty acids and their derivatives; (6) fillers serve as extenders or reinforcements; they increase density, stiffness, hardness, and heat deflection while decreasing shrinkage and cost; fillers include inorganic and organic materials ranging from powders (talc, wood flour) to particles (sand, nut shells) to fibers (glass, graphite whiskers); (7) flame retardants serve by shielding the surface of the polymer through charring or by influencing the pyrolysis and the combustion mechanism of the polymer; common fire retardants are polyhalogen compounds, phosphorus compounds, and antimony trioxide; (8) antistatic agents disperse the surface accumulation of static electricity by increasing surface conductivity; internal agents are surfactants that are added before or during molding; external antistatic agents are generally limited to fiber applications.

Other common additives are pigments, whitening agents, blowing agents, accelerators, impact modifiers, and biostabilizers (pesticides)

adduct: addition product; also two or more molecules associated with each other without covalent or ionic bonding, e.g., hydrates $CaSO_4 \cdot 5H_2O$

adenine: 6-aminopurine; one of the four bases present in DNA and in RNA

ADENINE

adenosine: adenine riboside; building block of RNA, also of ATP, ADP

ADENOSINE

adenosine triphosphate: ATP; adenosine–PO_3H–PO_3H–PO_4H_2; contains a high energy bond (7300 cal/mole) that is used to store energy in organisms including the human body; energy is released during the reaction of ATP $\rightarrow$ ADP + $P_{inorganic}$ (ADP, adenosine diphosphate)

adhesives: materials that bond surfaces of different substances through chemical bonding or van der Waals forces; to achieve adhesion, surfaces must be close together; strong adhesion occurs when adhesives and substrates have similar solubility parameters; adhesives may be liquid or tacky semisolids, natural or synthetic, organic or inorganic, waterborne, solventborne, or solventless; one distinguishes between pressure-sensitive and nonpressure sensitive systems; properties of adhesives are designed through formulation (e.g., incorporation of metal results in conductive adhesives). *Natural adhesives:* water solutions of starch, dextrins, or caseins are used for paper, inexpensive plywood, corrugated cardboard; glues derived from animal protein are used for paper, gummed labels, as well as for sizing of textiles; bituminous adhesives derived from petroleum asphalt, coal, and pitch are used as roofing adhesives; rubber and latex are applied in the automotive industry and for pressure-adhesive tapes; polyphenolic proteins from mussels show promise for industrial applications; sodium silicate (water glass) is used for ceramics. *Synthetic adhesives:* include acrylics, cyanoacrylates, epoxies, synthetic rubbers, phenolic resins, polyolefins, polyamides, polyesters, styrene-butadiene polymers, silicones, urethanes, and vinyl polymers and copolymers; they are used for metals, wood, glass, and ceramics; they often supplant other forms of joining; markets that use adhesives extensively include construction, transportation, textiles (nonrigid bonding), furniture, and appliances, consumer goods, tapes, and packaging; also used in medicine for surgical adhesives

adiabatic extrusion: *see* autothermal extrusion

adiabatic processes: processes in which there is a transfer of matter but not of heat, e.g., adsorption, compression are often adiabatic

adipic acid: $HOOC(CH_2)_4COOH$; produced by catalytic (Cu or V salts) oxidation (HNO_3 or air) of cyclohexane; use: precursor for hexamethylenediamine, for manufacture of nylon 66, polyurethanes, plasticizers

adiponitrile: $NC(CH_2)_4CN$; ADN; produced by the catalytic dehydrative amination of adipic acid in the liquid or gas phase; by indirect hydrocyanation of 1,3-butadiene via the 1,4-dichlorobutene intermediate, by direct catalytic hydrocyanation of 1,3-butadiene with HCN, or by electrohydrodimerization of acrylonitrile (EHD process); use: to prepare hexamethylene diamine (HMDA), a raw material for nylon 66

adjuvant: additive that enhances the activity of a primary ingredient, e.g., clay in pharmaceutical formulations

Adkins catalyst: copper–chromium oxide catalyst; prepared by reacting $Cu(NO_3)_2$ and Cr_2O_3 with ammonia, followed by washing and roasting; use: for high-pressure hydrogenations

adrenaline: *see* epinephrine

adrenocorticotropic hormone: ACTH; stimulant for the secretion of adrenal hormones, such as cortisone, that inhibits inflammations

adriamycin: doxorubicin; antibiotic and anticancer drug; produced by fermentation from *Streptomyces* species; acts by disrupting the synthesis of RNA

ADRIAMYCIN

adsorbates: substances that adhere to a surface

adsorbents: solids having surfaces to which substances, mainly gases and liquids, adhere; to be effective adsorbents must have large surface areas; typical adsorbents are activated carbon and alumina, ion-exchange resins, silica gel, magnesia, acid-treated clay, and Fuller's earth; used for purification and isolation of products; *see* zeolites

adsorption: adhesion of substances to the surface of solids; as adsorption involves no phase change it is used for energy-efficient purifications and separations; use: removal of odor, color, and toxic gases; water and process stream purification by ion exchange; drying of gases and petroleum fractions; and gas phase separations of low molecular weight hydrocarbons

adsorption chromatography: *see* liquid–solid chromatography

advanced waste treatment: AWT; *see* tertiary waste water treatment

advancing rolls: in continuous fiber production, rollers that permit yarn to remain in an area long enough so that it can be subjected to operations

aeration: process of passing air through a liquid; aeration tanks are used during secondary waste water and sludge treatment

aerobic: refers to the requirement of air; aerobic bacteria live only in the presence of air

aerogel: dispersion of gas in a solid; produced from gels by using air to displace the solvent above its critical temperature; aerogels are characterized by large pore volume and low densities (<0.01 g/mL); use: catalysts, insulation, detector for high energy particle

aerosol: suspension of a liquid or solid in gas; natural aerosols include fog, smog, and smoke; aerosols can be produced by passing a pressurized gas through a liquid or solid; use: to dispense personal materials (e.g., perfumes, shaving cream, and deodorants), household materials, paint, pesticides, coatings, finishes, and foods; aerosol cans contain the material to be dispersed, a propellant gas such as air, hydrocarbons, CO_2 and N_2O, and a solvent for the gas; chlorofluorocarbon propellants have been discontinued in the United States because of their destructive effect on the atmospheric ozone layer and their use is being curtailed worldwide

affination: *see* sugar refining

affinity chromatography: used to isolate and purify minute quantities of biological materials; the stationary phase consists of immobilized ligands to which materials in the mobile phase bind reversibly; use of highly specific enzymes permits the isolation of biological materials from mixtures in which they may be present in quantities of parts per million; porous supports permit large protein molecules to diffuse through the matrix; the most commonly used matrix is beaded agarose (derived from agar), which is suitable for separation of molecules having molecular weights between 1 million and 1 billion; the agarose is modified to increase its stability to heat and solvents; other supports include cellulose, cross-linked dextran, and cross-linked polyacrylamide; to prevent steric hindrance ligands are separated from the matrix by short alkyl chains (spacer arms); applications of affinity chromatography include isolation and purification of biological receptors, nucleic acids, hormones, plasma membrane proteins, and viruses; potential medical applications include the removal of undesirable substances from the body; *see* chromatography, immobilization

affinity complexation: *see* affinity sorption

affinity sorption: technique that uses the specific reversible association between a biomacromolecule and another molecule; affinity complexation; *see* affinity chromatography

affinity syrup: *see* refiner's blackstrap

aflatoxins: a group of naturally occurring toxic, carcinogenic fungal metabolites found in mold-infected grain, fruit, vegetables, and peanuts; FDA regulates their permissible level in food products

AFLATOXIN B$_1$

after-stretching: in textile technology, stretching of fibers after spinning

agar: agar-agar; a gelatinous polysaccharide derived from seaweed; forms a firm gel at concentrations as low as 1%; consists primar-

ily of agarose that is composed of an alternating (1 → 3)-linked β-D-galactopyranose and (1 → 4)-linked 3,6-anhydro-α-L-galactopyranose structure; use: culture medium in microbiology, gelling agent in foods, in pharmaceuticals and impression materials

agarose: *see* agar

Agent Orange: herbicide and defoliant containing a mixture of 2,4,5-trichlorophenoxy acetic acid (2,4,5-T) and 2,4-dichlorophenoxy acetic acid (2,4-D) as well as traces of dioxin (2,3,7,8 tetrachlorodibenzo-*p*-dioxin; TCDD); used extensively during the Vietnamese conflict

age resistors: in rubber technology, the collective name for additives that prevent elastomers from becoming sticky, brittle, or developing cracks; these additives include antioxidants and antiozonates

agglomeration: size enlargement of fine particles; achieved by flocculation of colloidal particles, compaction of powders by application of pressure, extrusion through a die, fusion of smaller particles using heat or a binder, or mechanical agitation using a rotating drum or disk or an airstream in the presence of a binder; aggregation increases bulk density thus reducing storage space requirements; when agglomeration is used for formulation (e.g., heavy-duty detergents or fertilizers) mixers are combined with agglomerating equipment; use: preparation of free-flowing bulk materials, tablets, briquettes, and pellets; *see* agglutination

agglutination: agglomeration of particles in the presence of specific proteins; the term is usually restricted to antibody–antigen interaction

aggregation: *see* agglomeration

aglycone: *see* glycoside

agonists: diet supplements, replacement drugs; chemicals that the body needs but cannot synthesize itself, e.g., vitamin C

agricultural chemicals: agrochemicals; chemicals used to promote the growth and health of cultivated plants and domestic animals (e.g., fertilizers, hormones, and soil conditioners) and to control the distribution of competing plants and animals, i.e., pesticides that include such broad classes as herbicides, insectides, and fungicides

agrobacterium: family of microorganisms affecting agricultural crops

agrochemicals: *see* agricultural chemicals

air: major industrial source of oxygen, nitrogen, and rare gases; air can be enriched with oxygen by passing pressurized air through a membrane (the vented gas is nitrogen enriched); use: coolant, power source, propellant in aerosol cans, blowing agent, source of individual constituent gases by fractional distillation of liquefied air; oxygen-enriched air is used for waste water treatment, oxygen bleaching, enhanced combustion, and in fermentation systems

air-felting process: *see* dry-forming process

air liquefaction: cooled compressed air is liquefied by expansion through a valve (Joule–Thompson expansion cycle); cooling is achieved by first expanding the air in an engine that does useful work; use: for separation of oxygen and nitrogen and recovery of argon

Air–Liquide process: Linde process; low temperature separation of synthesis gas (CO + H_2); most of the CO_2 is removed by scrubbing with ethanolamine and the remaining CO_2 and H_2O are removed with molecular sieves; CH_4 and CO are then separated by cooling the $CO/CH_4/H_2$ mixture to $-180°C$ at 40 bar, followed by fractional distillation that removes CO at the head of the column; also *see* Cosorb process

air pollution: air pollutants consist of natural and man-made gases or particulates; natural air pollutants include windblown dust and pollen, ash, and gases from volcanic eruptions and forest fires, ozone from lightening, terpenes from vegetation, gases from biological decomposition and radioactivity; major man-made pollutants arise from transportation, industrial, and residential sources, and include nitrogen and sulfur oxides (NO_X, SO_X), CO and CO_2, organic vapors, chlorinated hydrocarbons, and particulates; NO_X and SO_X result from combustion of fossil fuels and are blamed for acid rain; reduction in emission of these oxides can be achieved either by cleaning the fuel before combustion or by precipitating the pollutants from the flue gases; pollution patterns depend on long-range and short-range transport of pollutants; efforts to promulgate regulations for emission control are underway in the United States and western Europe; excessive production of CO_2 and CH_4 is feared to result in global heating leading to melting of the ice caps and changes in vegetation patterns; *see* greenhouse effect

akardites: stabilizers and gelatinizers for gun powders; akardite I (diphenylurea), akardite II (methyldiphenylurea), akardite III (ethyldiphenylurea)

alachlor: $2,6-(C_2H_5)_2,1-N(COCH_2Cl)CH_2 OCH_3-C_6H_3$; pre-emergence herbicide; used for corn, cotton, and soy beans

alanine: $CH_3CH(NH_2)COOH$; natural amino acid; use: nutritional supplement

Alberger process: production of NaCl by evaporation of a slurry from superheated brine; also *see* Grainer process

albumin: protein obtained from blood, egg white, and milk; use: emulsifying agent, adhesive, antidote for mercury poisoning, for clarification of wines

albuterol: *see* salbutamol

Alcoa process: aluminum production by electrolysis of aluminum chloride

alcohols: organic substances having one

(monohydric) or more (polyhydric) hydroxyl groups; lower alcohols are water soluble; use: solvents and organic intermediates; industrially important alcohols include methanol, ethanol, isopropanol, propanol, butanol, glycol, butanediol, amyl alcohol, glycerol, pentaerythritol, ethylene glycol, allyl alcohol; the term alcohol is commonly used for grain alcohol (ethanol); denatured alcohol is ethanol that is made unsuitable for consumption by addition of methanol, butanol, and/or bittering agents such as alkaloids; *see* Alfol process; hydroformylation

alcoholate: alkoxide; metal derivative of an alcohol, e.g., sodium methoxide (CH_4ONa), potassium propoxide (C_3H_7OK)

alcohol ethoxylate: alkyl ethoxylate; low-sudsing nonionic surfactant prepared by treating a long chain alcohol such as lauryl alcohol (C12) with ethylene oxide: CH_3 $(CH_2)_{10}CH_2OH + 8 \ C_2H_4O \rightarrow CH_3 \ (CH_2)_{10}$ $CH_2(OCH_2CH_2)_8OH$

alcoholysis: reaction with alcohols analogous to reaction with water (hydrolysis); e.g., alcoholysis of an amide leads to an ester: $ROH +$ $H_2C=CHCONH_2 \rightarrow H_2C=CHCOOR$

aldehyde resin: *see* polyoxymethylene

aldehydes: class of compounds containing the –CHO group; although they can be prepared by oxidation of alcohols, modern industrial manufacture involves catalytic oxidation or hydroformylation of olefins (*see* Oxo process); use: important reducing agents, chemical intermediates, and raw materials for polymer manufacture include formaldehyde, paraldehyde, acetaldehyde, crotonic aldehyde, propionaldehyde, acrolein

aldicarb: temik; $CH_3S(CH_3)_2CH=NOCO$ $NHCH_3$; pesticide (insecticide, acaricide, nematocide)

aldolyzation: *see* aldol condensation

aldol condensation: acid- or base-catalyzed condensation of aldehydes yielding β-hy-

droxy aldehydes; although in the past the reaction was widely used industrially it has been largely replaced by other catalytic processes

Aldox process: *see* 2-ethylhexanol

Alfen process: also known as Ziegler process; production of unbranched α-olefins containing an even number of carbon atoms from ethylene using $TiCl_3·Et_3Al$ (Ziegler) catalyst; growth of ethylene chain occurs by successive addition of ethylene to Et_3Al to form $Al[(CH_2CH_2)_n]_3$; the oligomers are released with simultaneous catalyst regeneration by heating to 200–300°C at 50 bar, the chain length is controlled by process conditions; the triethylaluminum is recycled; *see* Ziegler process; Alfol process

Alfin catalyst: polymerization catalysts produced from the reaction of sodium, isopropanol, and olefins, for example, to give high molecular weight *trans*-polybutadienes

Alfol process: Ziegler growth reaction; synthesis of primary, unbranched alcohols having an even number of carbon atoms by successive addition of ethylene to triethylaluminum; chain growth to $Al[(CH_2CH_2)_nEt]_3$ occurs at 120°C and 100–140 bar; the trialkyl aluminum compounds are oxidized with dry air to the corresponding alkoxides, which are saponified to aluminum hydroxide and the alcohol

algae: water plants ranging from single cell organisms to large kelp; use: food, source of agar, potential importance as biomass

algicide: herbicide active against algae

algin, alginate: polysaccharide derived from kelp; exists as insoluble salts of alginic acid, which is a high molecular weight polymer of D-mannuronic acids and L-guluronic acids [isomers of $HOOC(CHOH)_4CHO$]; alginates are hydrophilic colloids that form gels and films; use: emulsifiers, water-retainers, swelling agents, stabilizers, and thickeners in food products, paints, rubber, and pharmaceuticals; also stiffeners and binders in cotton finishing; alginate fibers are produced by extruding sodium alginate into acid; they are made into nonflammable yarn with low wet strength; use: production of lace-like embroidery or thin materials by weaving alginate fibers together with cotton or wool and then dissolving the alginate

aliphatics: straight chain, branched, and cyclic hydrocarbons and their derivatives; the compounds may contain single, double, or triple carbon–carbon bonds; aliphatics are distinguished from aromatics by the absence of benzene or fused ring systems

alizarin: 1,2-dihydroxyanthraquinone; originally isolated from madder root; now produced by oxidation of anthraquinone; use: red-orange dye, also as starting material for other anthraquinone dyes

alkali cellulose: *see* soda cellulose

alkali metals: group I metals: Li, Na, K, Rb, Cs; they are highly reactive, strong reducing agents, react with water to give basic solutions

alkaline cells: *see* electric cells

alkaline earth metals: group II metals: Be, Mg, Ca, Sr, Ba, Ra; reactive metals but less reactive than alkali metals, good reducing agents; in compounds they are present as divalent cations; beryllium forms an amphoteric hydroxide, whereas the other elements in group II form basic hydroxides

alkaline manganese dioxide battery: electric battery having a manganese dioxide anode, a zinc cathode, and aqueous KOH electrolyte; *see* batteries

alkali silicates: only sodium and potassium silicates are industrially important; anhydrous silicate glass consists of SiO_2/alkali oxide with a ratio ≥ 2; it is produced by reaction of pure quartz sand with alkali carbonates at 1300–1500°C; the melted product

flows into casting molds from the furnace; dissolving the anhydrous silicate glass in water at 5 bar and 150°C leads to water glass solutions; the only industrially important alkali silicate of composition Si_2/alkali oxide ≤ 1 is sodium metasilicate; use: in detergent, adhesive, paint, enamel, ceramic, cement, and foundry industries; also used for manufacture of catalysts and zeolites, for ore flotation, water treatment, and as silica sols and gels and as fillers

alkaloids: nitrogen-containing bases derived from plants; these materials are usually optically active and are characterized by their bitter taste and specific physiological activity; many are used as medicinals; a partial list of alkaloids includes the stimulants caffeine (theine) and cocaine, the pain-relieving opiates codeine and morphine, nicotine (which is found in tobacco), the heart stimulant atropine, which is also used to dilate the pupil of the eye, the nasal decongestant epinephrine, the hallucinogen peyote derived from cacti, the ergot alkaloids used to induce contractions of the uterus, the tranquilizing *Rauwolfia* alkaloids, the antimalarial quinine, and the highly poisonous strychnine; structural modification of alkaloids has resulted in many products with useful medicinal properties

alkane sulfonates: high sudsing anionic surfactants; produced from α-olefins by reaction with SO_3; sodium salts are used as detergents

alkanol amides: amides carrying one or more hydroxyl groups; prepared by reaction of alkanol amines with fatty acids; use: nonionic surfactants, foam stabilizers, emulsifiers, dispersants and wetting agents for aqueous–nonaqueous systems

alkanol amines: nonvolatile, water-soluble amines containing one or more hydroxyl groups (e.g., ethanol amine, tripropanol amine); prepared from ammonia and alkylene oxide or by reduction of the appropriate nitro-alcohol; use: anionic and nonionic surfactants, chemical intermediates, weak bases, cross-linking agents for urethanes, chelating agents, corrosion inhibitors

Alkar process: gas phase ethylation of benzene using BF_3/Al_2O_3 at 290°C and 60–65 bar; the process has been developed by UOP (now part of Allied Signal) and displays high selectivity for ethylbenzene

Alkazid process: purification of synthesis gas (i.e., removal of H_2S, COS, CO_2) using alkali salts of amino acids such as N-methylaminopropionic acid or dialcohol amines

alkoxides: *see* alcoholates

alkyd resins: thermosetting polymers made by esterification of polybasic acids or acid anhydrides with polybasic alcohols; the resins have good adhesion to metals, good dielectric strength, they are transparent, flexible, and tough; properties depend on starting materials: maleic anhydride increases the melting point, phthalic anhydride promotes hardness and stability, long chain acids impart toughness and flexibility; use: coatings, plastic molding compounds, adhesives and plasticizers; also used in paints, binders, enamels

alkyl: group remaining when a hydrogen atom is removed from an aliphatic hydrocarbon; e.g., methyl–CH_3, *tert*-butyl–$C(CH_3)_3$; usually denoted by R–

alkyl aluminum halides: $R_n AlX_{3-n}$; use: cocatalyst for Ziegler–Natta catalysts; stereospecificity of catalyzed polymerizations decreases with increasing size of R

alkylaromatics: aromatic compounds that carry alkyl substituents on the rings, e.g., toluene, cumene

alkylaryl sulfonate: generic term for anionic surfactants used as laundry detergents; prepared by sulfonation of "alkylates" (alkylbenzenes) followed by neutralization with base; sulfonation is carried out in a continuous process either by using 20–25% oleum

$(SO_3·H_2SO_4)$ or air–SO_3; alkylates with C10 to C15 side chains have maximum detergency, those with C10 to C12 side chains are used for light-duty liquid systems, with C12 to C14 side chains for heavy-duty detergents

alkylate gasoline: *see* alkylates

alkylates: alkylate gasoline; in the petroleum industry, "alkylates" refers to highly branched alkanes produced during gasoline manufacture; *see* alkylation; for the detergent industry; *see* alkylaryl sulfonate

alkylating agents: *see* anticancer drugs

alkylation: in the petroleum industry refers to the formation of highly branched alkanes that boil in the gasoline range; isobutane is reacted with propylene and butylene using sulfuric acid (50°F) or hydrogen fluoride (100°F) at pressures high enough to keep the reaction mixture liquid; alkylates have octane ratings in the low to middle nineties; in organic synthesis, the introduction of an alkyl group into a molecule

alkylbenzene sulfonate: ABS; a high sudsing detergent; poorly biodegradable; *see* alkylaryl sulfonate

alkylbenzenes: benzene rings carrying one or more alkyl groups; the industrially important alkylbenzenes toluene ($C_6H_5CH_3$), xylene ($C_6H_4(CH_3)_2$), and ethylbenzene ($C_2H_5 C_6H_5$) are isolated from petroleum refining and reforming; linear alkylbenzenes (LAB) with C10 to C13 side chains are used for production of detergents

alkylene-: radical that indicates unsaturation, e.g., ethylene or a cyclic derivative, e.g., ethylene oxide, C_2H_4O; the IUPAC convention uses the suffix -ene, i.e., ethene and ethene oxide; alkylene oxides of mono-, di-, and triglycerides are used as antifoaming agents

alkyl ethoxylate: *see* alcohol ethoxylate

alkyl ethoxylate sulfate: high-sudsing anionic detergent suitable for use in hard water

alkyl glyceryl sulfonate: AGS; high-sudsing anionic surfactant

alkyl hydroperoxides: ROOH; produced by oxidation of alkanes or by alkylation of hydrogen peroxide in acid medium; lower peroxides are unstable and subject to explosion; use: intermediate for boric acid oxidation (Bashkirov oxidation) of alkanes to alcohols; oxidizing agent in the oxirane process

alkyllithium compounds: RLi; derived from Grignard reagents (RMgX); use: catalysts for polymerization of olefins

alkylphenols: RC_6H_4OH; many anionic surfactants are polyethoxyethylenes and polyethoxypropylene derivatives of alkylphenols; alkylphenol disulfides serve as vulcanizing agents for rubber

alkyl sulfates: high-sudsing anionic surfactants; used in cosmetic products and all-purpose detergents

allanite: mineral; black, lustrous silicate of variable composition; important source of Ce

allantoin: isolated from plants, produced by reaction of glyoxalic or dichloracetic acid and urea; end product in purine metabolism; intermediate in uric acid metabolism leading to urea; use: medicinal, for topical treatment of wounds

ALLANTOIN

allele: in genetics, alternate form of a gene, usually arising from mutation

allelopathic substances: plant material that adversely affects other plants, e.g., some alkaloids

allergens: substances that induce hypersensitivity (allergy), may be evidenced by skin rashes, breathing difficulties, malaise, or hayfever

allethrin: pyrethroid insecticide; used against flies

ALLETHRIN

allophanate: RN(CONHR)COOR'; produced by reaction of isocyanate (RNCO) with alcohol leading to RNHCOOR'; reaction with a second mole of isocyanate RCNO leads to the allophanate; used to protect sensitive and tertiary alcohols

allopurinol: acts by blocking the receptor for xanthine oxidase, the enzyme needed for the production of uric acid; uric acid accumulation in joints cause the pain in gout

ALLOPURINOL

allosteric enzymes: enzymes having more than one active region; allosteric inhibition is caused by conformational change due to enzymatic binding

allotropy: existence of one substance in more than one form, e.g., carbon exists in various solid forms: graphite, diamond, buckminsterfullerene (buckyballs), and several amorphous structures

alloy: mixture of two or more metals; alloys have different properties, e.g., strength, corrosion resistance, or hardness than the metal alone; examples of alloys are brass, bronze, 14-carat gold, amalgam, and steels; *see* polyblend

allyl acetate: $CH_2=CHCH_2OOCCH_3$; produced by reaction of propylene, acetic acid and oxygen (acetoxylation) using Pd catalysts at 150–250°C; used for production of allyl alcohol

allyl alcohol: $CH_2=CHCH_2OH$; produced under pressure (13–14 bar) by alkaline hydrolysis of allyl chloride, by catalytic (Li_3PO_4 or Cr_2O_3) isomerization of propylene oxide, by gas phase reduction of acrolein with isopropanol or isobutanol using a metal oxide catalyst (MnO-ZnO), by acid hydrolysis of allyl acetate; raw material for the manufacture of glycerol and epoxy resins

allyl chloride: AC; $CH_2=CHCH_2Cl$; produced by gas phase addition of chlorine to excess propylene at 500°C; used for production of allyl alcohol

allyl compounds: characterized by the structure $RCH=CH-CH_2-$; the hydrogen on the carbon α to the double bond is highly reactive; use: chemical raw materials

allylic rearrangement: *cis* or *trans* isomerization of the allylic double bond due to the abstraction of the alpha hydrogen: e.g., reaction of *cis* or *trans* mixture of 3,4-dichloro-2-butene with HCN at 80°C leads to 95% 1,4-dicyano-*cis*-butene

allylic resins: thermosetting plastics that are generally clear with good electric properties

alpha cellulose: cellulose component of wood and paper pulp

alpha emitter: radioactive species that spontaneously emits alpha particles (helium nuclei)

alpha helix: *see* helix

alpha particle: charged particle consisting of 2 protons and 2 neutrons; helium nucleus

alpha-naphthyl acetic acid: ANA; plant growth regulator

alprazolam: tranquilizer; diazepam derivative

alternating copolymers: copolymers in which the monomers alternate regularly along the chain: ABABAB...; for example, copolymers of maleic anhydride with styrene or stilbene, copolymers of olefins (propylene, butene, isobutene) and sulfur dioxide; *see* polysulfone, copolymerization

alum: aluminum sulfate, $Al_2(SO_4)_3$; produced by heating alumina (Al_2O_3) or bauxite with sulfuric acid at 105–110°C for 15–20 hrs; use: primarily for coagulating and coating pulp fibers to produce a hard paper surface; also as coagulant in water purification, mordant (in dyeing), in medicine, in leather manufacture; the Na salt is used as leavening acid in baking

alumina: aluminum oxide, Al_2O_3; produced from clay deposits; source for aluminum metal; use: absorbant, abrasive (corundum), drying agent (activated alumina), ceramics, refractories, gems, and pigments; used as catalyst carrier and as catalyst (alkylations, cyclizations [e.g., butadiene and methylamine to pyridine]), dehydrations (e.g., 1,6 dihydroxy hexane to furan), dehydrosulfurization of petroleum residue, dimerization of alkylstyrenes, hydrocracking of hydrocarbons (e.g., Fischer–Tropsch synthesis, hydrolysis, isomerizations, oxidation [e.g., alkyl aromatics to phenols] and reductions [e.g., NO_x to ammonia])

aluminium: British spelling of aluminum

aluminized trinitrotoluene: *see* tritonal

aluminum acetate: red liquor; in the textile industry used as mordant for dyeing and for water proofing

aluminum: Al; element; the most important nonferrous metal; light weight (sp.g. = 2.7) with high thermal and electrical conductivity; oxidation in air results in a transparent Al_2O_3 coat that protects the metal against corrosion; production: Hall–Heroult process: electrolysis of bauxite (aluminum oxide) in cryolite (Na_3AlF_6) at 940–980°C; addition of LiF to the melt improves energy efficiency; electrolysis from $AlCl_3$ (Alcoa process) promises to be better environmentally; use: in alloys (for construction, transportation, consumer goods) to reduce weight, improve corrosion resistance and hardness; aluminum powder is added to explosives to improve their efficiency

aluminum alloys: the most common elements used in alloying aluminum are Fe, Si, Cr, Cu, Mn, Ni, Zn, Ti, Va, Na, Zr, and Ga; the Aluminum Association system for aluminum alloys uses numbers for the alloying elements: Cu (2), Mn (3), Si (4), Mg (5), Mg + Si (6), and Zn (7); the 1000 series contains 99+% aluminum, the 2000 series contains 2–6% Cu and its mechanical properties approximate mild steel; the 3000 series contains ≤1.2% Mn and is used for housings, cooking utensils, sheet metal, storage tanks, and cryogenic equipment; the 4000 series, which contain Si, has reduced melting points and is used for welding wires and as casting alloys; the 5000 series contains 1–5% Mg, has a lower density than pure aluminum and high resistance to marine atmospheres; the 6000 series, which contains up to 1.3% each of Mg and Si, has an excellent corrosion resistance and is used for construction, building, and decorative application where corrosion resistance is most important; the relatively new 7000 series contains 1–7.5% Zn and 2.5–3.3% Mg, it is the strongest of the alloys and is widely used in the aerospace industries

aluminum compounds: natural products include the gem stones emerald, beryl, garnet, and the abrasive corundum (Al_2O_3), and the soft minerals mica and kaolin (silicates); aluminum chloride is a Lewis acid and is used as catalyst for the Friedel–Crafts reaction and for reforming and polymerizing hydrocarbons; finely divided solid aluminum hydroxide is used as a suspension agent; aluminum nitride ceramics conduct heat better than other ceramics and are used as substrates and housing for integrated circuits; aluminum phosphates are mined as a source of phosphate; aluminum triethyl (ATE) serves as a polymerization catalyst for olefins; also *see* alum, alumina

aluminum silicate: $Al_2O_3 \cdot nSiO_2$, widely distributed in soil and rocks: clay is hydrated aluminum silicate; used for ceramics, glass making, refractories, fillers, coatings, water softener, catalyst supports; the mineral kyanite is used as refractory; the calcium aluminum silicate, zoisite is used as gemstone (tourmaline, tanzanite); synthetic sodium alumina silicate is a zeolite and is used as a molecular sieve; alumina-silica fibers are low-cost, high-temperature resistant reinforcements for composites; the higher cost continuous alumina-boria-silica fibers are suitable for pultrusion reinforcements

amalgam process: manufacture of sodium hyposulfite by treating sodium sulfite with NaHg (5% Na) at pH 5–7; the amalgam is derived from the cathode of a mercury-chlorine cell, mercury is released during the reaction and recycled

amalgams: mercury alloys; use: dental fillings (Hg + Ag·Sn), electrodes, reducing agent

amatols: pourable high explosive consisting of variable proportions of $TNT:NH_4NO_3$; the 20:80 mixture is used to fill grenades with the help of an extruder

amber: fossil resin from extinct pines; *see* resins

ambient: referring to the temperature and pressure conditions of the environment; i.e., ambient pressure is usually 1 atmosphere, ambient temperature is room temperature

American Society for Testing and Materials: *see* ASTM

amethyst: purple-colored quartz

amides: organic compounds having the $-CONR_2$ group (R can be a hydrogen, alkyl, or aryl group); produced by treatment of acids with anhydrous ammonia in the presence of a catalyst; simple amides can be converted to substituted amides by treatment with primary or secondary amines; use: water repellents, lubricants, starting material for synthesis; *see* polyamides

amiloride: diuretic that does not lead to the excretion of potassium

AMILORIDE

amination: process of introducing an amino group, $-NH_2$, into an organic compound

amine oxides: $R_3N \rightarrow O$, $Ar_3N \rightarrow O$ are used as high sudsing, amphoteric detergents

amines: derivatives of ammonia in which one, two, or three hydrogen atoms are replaced by alkyl or aryl groups; like ammonia, amines are bases and form quaternary salts; one distinguishes between primary amines: $-NH_2$, secondary amines: $-NH-$, tertiary amines $-N<$; compounds may have more than one amino group, i.e., di-, tri- . . . polyamines; use: weak bases, solvents, as chemical intermediates; amines of long chained fatty acids are used also as fabric softeners, flotation and rubber-mold releasing agents, and petroleum additives; the simplest aromatic amine is aniline; some important polyamines are 2,4-diaminotoluene (TDA), which is used to manufacture toluenediisocyanate (TDI); 4,4'-methylene dianiline (MDA) is used to manufacture methylene di-paraphenylene diisocyanate (MDI); *see* polyurethanes

amino acids: L-alpha amino acids are the building blocks of proteins; they are used as food additives; essential amino acids that cannot be synthesized in the human body and must be taken as food are marked by an asterisk in the table of amino acids

aminoacylase: bacterial enzyme used for L-amino acid production

p-aminobenzoic acid: PABA; essential microbial metabolite mimicked by sulfa drugs; ingredient in sunscreen lotions

ω-aminocaproic acid: $HOOC(CH_2)_5NH_2$; *see* polyamides

aminoform: hexamine; *see* hexamethylenetetramine

aminoglycosides: broad-spectrum, water-soluble antibiotics that contain a cyclic poly-hydroxy unit (a sugar) bearing amino groups (aminocyclitol unit); the first aminoglycoside discovered was streptomycin

1,6-aminohexanol: produced by hydrogenation of caprolactone at 250°C and 280 bar over a copper catalyst; used to manufacture hexamethylenediamine (HMDA), which is a raw material for nylon 6,6

aminoplast resins: *see* amino resins

amino resins: *see* urea-formaldehyde and melamine-formaldehyde resins

ω-aminoundecanoic acid: produced from undecenoic acid (C11) by addition of HBr in the presence of peroxide followed by reaction with ammonia; precursor of nylon 11

amitriptyline: antidepressant drug

AMITRIPTYLINE

ammines: coordination compounds of metal ions and ammonia, e.g., $Cu(NH_3)_4^+$

ammonia: NH_3; colorless water-soluble weakly basic gas; produced from synthesis gas by reaction of nitrogen and hydrogen (Haber–Bosch process) at 2000–10,000 psi and 400–600°C over an iron catalyst; various processes such as those used by BASF,

63

amino acids

Name	Abbreviation	Symbol	Formula
alanine	Ala	A	$CH_3CH(NH_2)COOH$
arginine	Arg	R	$HN{=}C(NH_2)NH(CH_2)_3CH(NH_2)COOH$
asparagine	Asn	N	$H_2NCOCH_2CH(NH_2)COOH$
aspartic acid	Asp	D	$HOOCCH_2CH(NH_2)COOH$
	Asx		
cysteine	Cys	C	$HSCH_2CH(NH_2)COOH$
cystine	Cys$_2$		$(\text{-}SCH_2CH(NH_2)COOH)_2$
glutamic acid	Glu	E	$HOOC(CH_2)_2CH(NH_2)COOH$
glutamine	Gln	Q	$H_2NCO(CH_2)_2CH(NH_2)COOH$
	Gex		
glycine	Gly	G	H_2NCH_2COOH
histidine	His	H	

HISTIDINE

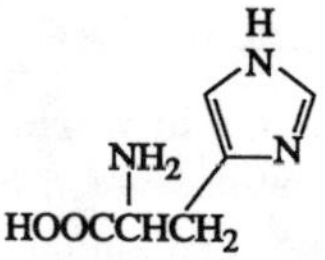

Name	Abbreviation	Symbol	Formula
isoleucine*	Ileu	I	$CH_3CH_2CH(CH_3)CH(NH_2)COOH$
leucine*	Leu	L	$i\text{-}PrCH_2CH(NH_2)COOH$
lysine*	Lys	K	$H_2N(CH_2)_4CH(NH_2)COOH$
methionine	Met	M	$CH_3S(CH_2)_2CH(NH_2)COOH$
phenylalanine*	Phe	F	$C_6H_5CH_2CH(NH_2)COOH$
proline	Pro	P	

PROLINE

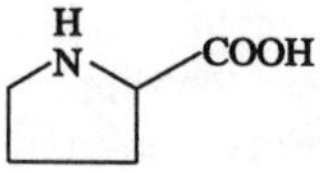

Name	Abbreviation	Symbol	Formula
serine	Ser	S	$HOCH_2CH(NH_2)COOH$
threonine*	Thr	T	$CH_3CH(OH)CH(NH_2)COOH$
tyrosine	Tyr	Y	$p\text{-}HOC_6H_4CH_2CH(NH_2)COOH$
tryptophan	Trp	W	

TRYPTOPHAN

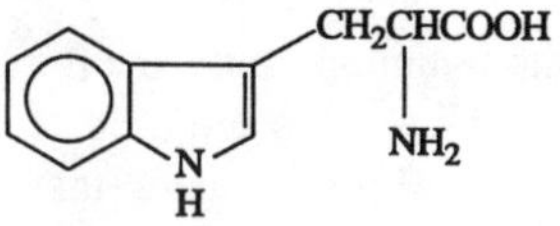

Name	Abbreviation	Symbol	Formula
valine*	Val	V	$i\text{-}PrCH(NH_2)COOH$

Casale, ICI, Kellogg, Topsoe, TVA, and Uhde differ by modifications in pressure, temperature, catalyst, means of feedstock and product purification, recycling, by-product utilization, and heat recovery; use: fertilizer and in general chemical processing: for the manufacture of urea, nitric acid. and explosives (e.g., nitrates, azides); in the fiber and plastics industries for the production of melamine, urea–formaldehyde, nylons, acrylonitriles, and aramids; in the rubber industry for production of polyurethanes and blowing agents for foam rubber; in aqueous solution, ammonia is used for saponification of fats and oils, for etching aluminum, as cleaning agent and as deodorant; gaseous ammonia serves as a refrigerant and for nitriding of steel; *see* AMMONIA SYNTHESIS

ammonia dynamites: explosive in which a small amount of nitroglycerin is used to sensitize ammonium nitrate to explode

ammonia gelatin: secondary explosives consisting of nitrogelatin in which some of the nitroglycerin has been replaced with ammonium nitrate.

ammonia-soda process: *see* Solvay process

ammonium dihydrogen phosphate: *see* ammonium phosphates

ammonium nitrate: AN; raw material for manufacture of industrial explosives; constituent of rocket propellants; ammonium nitrate explosives consist of mixtures of AN and wood, oil, or coal

ammonium nitrate–fuel oil: ANFOL; prilled explosive mixture that requires a booster to explode

ammonium perchlorate: APC; prepared by neutralizing ammonia with perchloric acid; explosive; used in composite propellants

ammonium phosphates: the industrially significant ammonium dihydrogen phosphate ($NH_4H_2PO_4$) and diammonium hydrogen phosphate ($[NH_4]_2HPO_4$) are produced by reaction of ammonia and pure phosphoric acid; ammonium polyphosphate ($[NH_4PO_3]_n$) is produced by reaction of phosphoric acid and urea; used as fertilizers

ammonium salts: widely used as fertilizers; other use: the nitrate is an industrial explosive and is used for manufacture of nitrous oxide, the chloride is the electrolyte in dry cells, the bicarbonate is a leavening agent in the food industry, the sulfate is part of fire-retardant formulations

ammonolysis: reaction of ammonia analo-

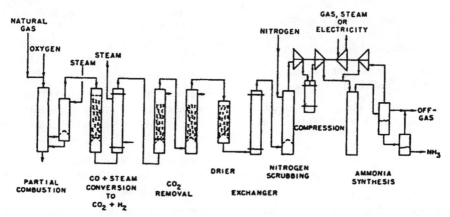

AMMONIA SYNTHESIS VIA PARTIAL COMBUSTION OF NATURAL GAS

[Copyright *Riegel's Handbook of Industrial Chemistry*, 8th ed., Van Nostrand-Reinhold Co., Inc.]

gous to the reaction of water in hydrolysis; reaction with halides or alcohols leads to amines (e.g., chlorobenzene to aniline, phenol to aniline), with esters to amides (e.g., methyl formate to formamide)

ammoxidation: oxidative reaction with ammonia in the presence of a catalyst, e.g., propylene to acrylonitrile: $CH_2=CHCH_3 + NH_3 + 1.5 O_2 \rightarrow H_2C=CHCN$

amorphous metal alloys: materials produced by rapid quenching from the melt; they have high hardness, strength, and corrosion resistance, good electric and magnetic properties, superior fatigue and abrasion resistance and catalytic capability

amorphous polymer: noncrystalline polymer; polymers that are irregular in their chemical configuration or structure (bulky side groups, nonlinear units, nonuniform conformations or randomness in the copolymer); thus cannot form ordered arrangements as in liquid-crystalline or crystalline systems (materials with one-, two-, or three-dimensional order); amorphous polymers also form on rapid cooling (quenching) from the melt (thus preventing crystallization in the range between the melting temperature and the glass transition temperature) or by rapid solvent evaporation and cooling; examples: normal melt spinning of poly(ethylene terephthalate) (PET) and dry spinning of cellulose triacetate solutions lead to amorphous fibers

amorphous silicon: noncrystalline silicon; commercially used in photovoltaic cells

amphetamine: 1-phenyl-2-aminopropane; central nervous system stimulant, appetite suppressant

amphipathic substances: compounds that have hydrophobic and hydrophilic properties, e.g., soap

ampholytic substances: compounds that can behave both as acid or base after being dissolved; used as solvents and detergents

amphoteric substances: compounds with both acidic or basic properties

ampicillin: antibiotic; penicillin derivative with a $C_6H_5CH(NH)CONH-$ side chain on the β-lactam ring; *see penicillin*

amyl alcohols: saturated aliphatic alcohols containing five carbon atoms; pentanols; produced by hydroformylation of butenes; industrially important compounds include *n*-amyl alcohol (1-pentanol), *sec*-amyl alcohol (2-pentanol), *tert*-amyl alcohol (2-methyl-2-butanol), *iso*-amyl alcohol (3-methyl-1-butanol); use: solvents and plasticizers

amylases: group of enzymes that hydrolyze starch to sugars; use: production of sugar syrups, solubilizing starch in the brewing industry, modification of starch viscosity in the paper industry, laundry products for presoak to loosen stains

amylenes: common name for pentenes; α-*n*-amylene (1-pentene), β-*n*-amylene (2-pentene)

amylodextrin: amylopectin; highly branched polyglucose component of starch

amyloglucosidase: glucoamylase; amylase that decomposes starch by splitting off one glucose unit at a time

amylopectin; the outer relatively water-insoluble layer of starch granules; a branched polyglucose; stains violet with iodine; use: thickening agent and coating for paper and textiles; *see* starch

amylose: the inner relatively water-soluble portion of starch granules; a linear polyglucose; stains blue with iodine; use: adhesive for coatings and sizings; *see* starch

amyl propionate: produced by esterification of propionic acid; use: solvent for resins and cellulose derivatives

anaerobic: the term stands for "in the absence of air"; some adhesives cure only in the absence of air; anaerobic microorganisms are used to decompose organic matter in landfills

analeptics: compounds capable of restoring depressed brain functions; stimulants

analgesics: pain relief agents that do not reduce consciousness, e.g., morphine family, salicylates (aspirin), acetaminophen, *p*-aminophenols, meperidine, methadone, ibuprofen, phenylbutazone, propoxyphene, and pyrazolones

analytical electron spectroscopy: AEM; surface analytical method permitting near-atom resolution of structures; bombardment with high-energy electrons leads to the emission of high-energy electrons and X-rays; permits elucidation of elemental composition

analytical methods: methods of identifying materials by their chemical or physical characteristics; one can distinguish between methods that destroy the sample (e.g., wet chemistry or combustion) or nondestructive physical investigations (e.g., spectroscopic analyses)

anaphylactic reactions: in medicine, response to the presence of histamine; *see* antihistamines

anatase: crystalline TiO_2; powdered mineral is used as white chalking-type (i.e., nonreacting) pigment; *see* titanium dioxide

androgens: male sex hormones responsible for the development of secondary male sexual characteristics such as facial and body hair

Andrussow process: *see* hydrocyanic acid

anesthetics: compounds that relieve pain, induce sleep, and provide sedation; the first inhalation anesthetic used for surgery was laughing gas (N_2O), which was introduced in 1842; during the next 100 years there followed diethyl ether, chloroform, cyclopropane, vinyl ether, 1,1,2-trichloroethylene, isopropenyl vinyl ether, and propyl methyl ether; the first modern nonflammable fluo-

rine containing anesthetic, halothane ($CF_3CHBrCl$), was introduced in 1956; other fluorine-containing inhalation anesthetics include fluroxene ($CF_3CH_2OCH=CH_2$), methoxyflurane ($CH_3OCF_2CHCl_2$), enflurane ($CHFCl\ CF_2OCHF_2$), isoflurane ($CF_3CHClO\ CHF_2$), sevoflurane ($CF_3)_2CHO\ CH_2F$ and I-653 ($CF_3\ CHFOCHF_2$); local anesthetics are applied in solution, topically, or by injection

anhydrite: $CaSO_4$ ore, gypsum; use: white paint, paint extender by coprecipitation with TiO_2, in cement production

anidex: generic acrylate elastomer; use: soil release finishes

aniline: $C_6H_5NH_2$; produced by reduction of nitrobenzene using hydrogenation catalysts (e.g., Cu, Ni, or Cr) that are activated by sulfur; by aminolysis of chlorobenzene using aqueous ammonia under pressure; by aminolysis of phenol using metal oxide catalysts; use: chemical intermediate to manufacture secondary products including isocyanates, cyclohexylamine, acetanilide, 4,4'diaminodiphenylmethane; as an intermediate for antioxidants, rubber chemicals, polymers agricultural chemicals, and dyes

aniline black: polymeric azine dye for cotton; the material to be dyed is soaked with aniline hydrochloride, sodium chlorate, ammonium vanadate, and potassium ferrocyanide and aged at 60°C; *see* azine dyes

aniline-formaldehyde resins: thermoplastics made by condensing formaldehyde and aniline in an acid solution; used for molded and laminated insulating materials

aniline inks: fast drying inks for printing on plastics

anionic clays: *see* clays

anionic dyes: *see* acid dyes

anionic polymerization: *see* ionic polymerization

anionic surfactants: usually sodium salts of alkylaryl sulfonates

anisole: $C_6H_5OCH_3$; phenyl methylether; use: solvent for plastics and recrystallizations; heat transfer agent because of its wide liquid range ($-37°$ to $154°C$)

anisotropy: departure from uniformity; some materials have molecular or crystal structures that make them inherently anisotropic (e.g., calcite); others (polymers) are isotropic but when deformed become anisotropic; the anisotropy is determined by measuring the difference in the refractive indices parallel and perpendicular to the deformation direction; molecular chains in anisotropic fibers are aligned in the fiber direction and, therefore, are weaker and less stiff than normal to this direction; *see* isotropy

annealing: heat treatment of materials (e.g., metals, plastics, fibers, films, glass) to stabilize their structure by removing stresses introduced during processing; metals and glasses are annealed by heating and slow cooling; fiber, yarn, and fabric annealing proceeds at stepwise increasing temperatures; the final treatment provides stability in films and "ease of care" characteristics in fabrics

annular liquid chromatography: separation is achieved in a bed that rotates slowly around its axis; solutions are fed continuously to the top end of the column, effluents are withdrawn at the bottom; the technique is applicable to continuous processing and can be scaled up for industrial use

anode: the electrode in batteries and electrolytic cells at which oxidation occurs; e.g., in the electrolysis of NaCl chlorine gas is liberated at the anode

anodex: unvulcanized and prevulcanized latex compounds

anodic protection: anticorrosion methodology in which an inert cathode is placed close to the surface to be protected; a crude potentiostat is used to maintain the potential in the passive region

anolyte: electrolyte at or near the anode of an electrolytic cell; the anolyte is oxidized during the reaction; *see* photoelectrochemical cells

anoxic environment: *see* anaerobic

ANSI: American National Standards Institute

antagonism: action in opposition; e.g., some optical isomers have opposite biological actions; thus, it is important to use only one of the enantiomers; the opposite of synergism

antagonists: substances that inhibit the action of biologically active substances, such as enzymes, drugs; antagonists bind with specific receptor sites but do not have the activity characteristic of the substances they displace

anthracene: aromatic hydrocarbon isolated from coal; used for manufacture of anthraquinone; considered a hazardous chemical waste (EPA)

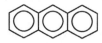

ANTHRACENE

anthracite: hard coal containing 86–97% carbon; contains small amounts of volatile matter and ash; burns with little smoke and has a heating value of 15,000 Btu/lb; forms of anthracite are ranked as follows (in decreasing order of carbon content): meta-anthracite, anthracite and semi-anthracite; ash and volatile content less than 14%

anthraquinone: produced by catalytic (Cr_2O_3, Fe_2O_3-V_2O_5) oxidation of anthracene, by Diels–Alder condensation between phthalic anhydride and benzene (Bayer process), by dimerization of styrene by way of an indane intermediate (BASF process); use: manufacture of hydrogen peroxide by catalytic (Ni or Pd) reduction to anthrahydroquinone followed by rapid autooxidation with release of H_2O_2; aids delignification when added to alkali liquors during wood pulp production; dye intermediate

ANTHRAQUINONE

anthraquinone dyes and pigments: colored compounds based on anthraquinone and related polycyclic quinones; the chromophore in these compounds are C=O groups; manufactured by chemical transformation of the parent quinones or their derivatives; anthraquinone dyes are light-fast and include acid, mordant, disperse, and vat dyes; anthraquinone pigments include alizarin, indanthrene, madder lake

antianxiety agents: tranquilizers, anxiolytics, neuroleptics; agents that allay anxiety without affecting consciousness, such as meprobamate, and the benzodiezepams (chlordiazepoxide, diazepam [Valium]) and phenothiazines (chlorpromazine)

antibacterial agents: bacteriostatic agents that interfere with the bacterial metabolism; term is usually applied to sulfur drugs that interfere with bacterial growth (bacteriostats) but do not kill the bacteria (bacteriocides)

antibiotics: compounds that kill bacteria (bacteriocides) by inhibiting growth or metabolism, e.g., chloramphenicol, erythromycin, penicillins, streptomycin, tetracyclins

antiblocking agent: material used to reduce adhesion; it is usually composed of finely divided infusible material

antibody: globular protein produced in the body in response to a foreign material (anti-

gen); the antibody bonds to the antigen in an attempt to neutralize it; "antibody–antigen interaction"; *see* agglutination

anticaking agents: materials added to powders or granules to prevent compaction and permit uninterrupted flow; in detergents compaction arises in formulations with high sodium carbonate content or strongly anionic character; anticaking agents include sodium benzoate, tricalcium phosphate, silicon dioxide, and microcrystalline cellulose

anticancer drugs: antineoplastic drugs; agents that control the uninhibited cell growth typical of cancer by affecting the DNA in the cancer cell; the drug classes include alkylating agents (e.g., chlorambucil, nitrosoureas) that bind with DNA, antimetabolites (e.g., methotrexate, 5-fluorouracil, cyclophosphamide) that disrupt the cancer cell metabolism, nucleophilic antibiotics (e.g., bleomycin, adriamycin) that attack DNA, platinum complexes (e.g., cisplatin) that intercalate into the DNA helix, alkaloids (e.g., vinblastine) that interfere with DNA synthesis, steroids that are active against sex-linked cancers; all anticancer drugs are highly toxic because they also attack healthy cells

anticarcinogens: *see* anticancer drugs

antichlor: in photography, sodium bisulfite

anticoagulants: substances that prevent spontaneous coagulation of latex, e.g., ammonia, sodium sulfate, and formaldehyde; in medicine, substances used to prevent coagulation of blood, e.g., heparin

anticodon: in genetic engineering, the tRNA structure that matches the mRNA codon by pair bonding between nucleotide bases; the anticodon consists of a sequence of three bases that matches the triplet on mRNA; *see* codon, messenger RNA, ribonucleic acid, transfer RNA

anticonvulsants: in medicine, agents used for the treatment of epilepsy, e.g., carbamazepine, phenobarbital, phenytoin (diphenylhydantoin), valproic acid

antidegradation clause: in pollution control, a legal provision that prohibits deterioration of air and water quality in regions that have pollution levels that are lower than those permitted by law

antidepressants: in medicine, agents that include monoamine oxidase (MAO) inhibitors (e.g., phenelzine) and tricyclic agents (e.g., amitriptyline)

antiferromagnetism: solid state property due to the opposing alignment of the spins of the orbital electrons; such materials are not magnetic

antifoaming agent: additive that reduces surface tension, e.g., 2-octanol, silicon fluids, sulfonated oils, silicone coated silica in hydrocarbon oil dispersion, fatty alcohol–fatty acid mixtures in hydrocarbons, alkylene oxides of mono-, di-, and triglycerides

antifogging agent: mild wetting agents that lower the surface tension of water, e.g., polyoxyethylene esters with oleic acid, sorbitol esters

antifouling agents: organometallic compounds used in marine paint, e.g., *bis*(tributyltin)oxide or fluoride, metallic naphthenates (Cu)

antigen: in medicine, a substance alien to the body that causes the formation of antibodies

antihistamines: agents that act against allergic and anaphylactic reactions induced by histamine; H_1 blockers, e.g., chlorpheniramine, diphenhydramine, tripelennamine

antihypertensive agents: agents that lower blood pressure; they act in several different ways: on the nervous system (e.g., clonidine, methyldopa), by dilating blood vessels (e.g., prazosin), by blocking neurotransmitters activated from the adrenal gland adrenergic nervous system (e.g., guanethidine), or by

inhibiting the angiotensin converting enzyme that is activated by the kidney (e.g., captopril)

antiinflammatory agents: group of agents based on steroids secreted by the adrenal cortex, e.g., cortisone and prednisone; noncorticoid agents include ibuprofen, indomethacin, and phenylbutazone

antiknocks: antiknock agents; *see* octane number

antimetabolites: *see* anticancer drugs

antimony oxide: Sb_2O_3; white pigment also used as fire retardant

antineoplastic agents: *see* anticancer drugs

antioxidant: substance that inhibits oxidation; use: to prevent autoxidation of monomers, stabilize and protect polymers, prevent or retard rancidity of fats and oils, prevent spoilage of foods and vitamins; antioxidants approved for foods include sterically hindered substituted phenols (e.g., butylated hydroxy toluene [BHT], butylated hydroxy anisole [BHA]), citric and phosphoric acids

antiozonants: in rubber technology, protective agents such as paraffin waxes (unbranched C20 to C53 hydrocarbons), secondary aromatic amines, furan derivatives and quinoline

antipyretic: fever-reducing agents such as aspirin

antiredeposition agent: dispersants; in laundry detergents, a material that prevents soil from resettling on fabrics after it has been removed by washing; the agents act by being adsorbed on the soil as well as on the fabric; the sodium salt of carboxymethylcellulose (NaCMC) is the most widely used agent

antiscurvy vitamin: *see* ascorbic acid

antistatic agents: substances that impart

some electrical conductivity to finished products, e.g., quarternary ammonium salts, polyethylene glycols; used in industries such as polymer, textile and detergent

antitussive: cough-repressant agent

antiulcer agents: antihistamines that prevent secretion of stomach acid; H_2 blockers, e.g., cimetidine, ranitidine; *see* histamine

anxiolytic drugs: *see* antianxiety agents

apatite: group of phosphate ores; used for fertilizer production; *see* calcium sulfate

API gravity: American Petroleum Institute gravity; a measure of density expressed in °API: °API = (141.5/s.g. at 60°F)−131.5; thus an oil with specific gravity of one is a 10°API oil, the lower the density the higher the API value; used to indicate the gasoline and kerosene content

aquaculture: controlled maintainance of freshwater, estuarine, and marine environment to promote growth of plant animal species

aquifer: underground stream of water or a water-bearing stratum of permeable rock, gravel, or sand, e.g., the Ogallala aquifer that stretches from South Dakota to Texas

Ar: symbol for aryl group; symbol for the gas argon

arabic gum: exudate of *Acacia arabica*; a complex polycarbohydrate containing Ca, Mg, and K salts of arabinose, galactose, mannomethylose, and glucoronic acid; use: adhesives, textile coatings, emulsifying agent for drugs, and cosmetics, binder for tablets

arabinose: nonfermentable pentose (C5 sugar) isolated from wood hydrolysates

arachidonic acid: $CH_3(CH_2)_4(CH=CHCH_2)_4$ CH_2CH_2COOH; an "essential" fatty acid, i.e., it is not synthesized by the human body; the biological starting material for pain-transmitting (prostaglandins) and blood-clotting

agents; aspirin prevents the reaction of arachidonic acid that leads to a cascade that results in these bioactive substances

aramids: low-specific gravity, high-performance aromatic nylons; two commercial aramid fibers are available: poly-*meta*-phenylene-isophthalamide (MPD-I) (Nomex), which is produced from *meta*-phenylenediamine and isophthaloyl chloride and is dry spun into fibers, and poly-*para*-phenylene terephthalamide (PPD-T) (Kevlar, Dupont; Twaron, Akzo), which is produced from *para*-phenylenediamine and terephthaloyl chloride and is dry-wet spun from a liquid crystalline solution in 100% H_2SO_4; aramids produce composites with high modulus, fatigue resistance, heat resistance, and good electric properties; used for hot gas filters, protective clothing; *see also* PBI

arc process: nitrogen fixation by rapidly passing air through an electric arc to give nitrogen oxides

area mining: used for coal mining in flat or rolling countryside; land reclamation follows removal of coal

Arge process: modification of Fischer–Tropsch synthesis of hydrocarbons using fixed-bed catalyst

argon: Ar; element; the gas is obtained from air and from natural gas in ammonia plants; use: in incandescent lamps, as inert gas shield especially for production of Ti and Zr metals and manufacture of semiconductor devices

aromatic copolyesters: high-melting aromatic esters based primarily on *para*-disubstituted monomers such as terephthalic acid, *para*-hydroxybenzoic acid, hydroxyquinone, 2,6 dicarboxy-, or 2-hydroxy-6-carboxy-naphthalene and similar monomers; these polymers have liquid crystalline behavior in a certain temperature range (thermotropic character); because there are no known spinning solvents for such materials (their polymer melting points are above their range of decomposition), copolymers of the above are used to lower their melting points to a useful melting range (~300–320°C); another option to lower melting is the use of these monomer components, modified by side chain substitution with phenyl or other groups or the use of meta structures such as isophthalic acid; applications: high melting matrix materials for composites, high modulus industrial fibers, films

aromatics: unsaturated cyclic organic compounds hydrocarbons and their derivatives; benzene, C_6H_6, is the parent compound; aromatics are produced by cracking petroleum; the aromatic hydrocarbons benzene, toluene, xylene (BTX), and ethylbenzene are derived from reforming of coal and naphthas; separation of aromatic hydrocarbons from aliphatics is achieved by azeotropic or extractive distillation and by liquid–liquid extraction; because of their high octane numbers aromatics are used for blending in gasoline; used as feed stock for organic chemicals; *see*

SYNTHETIC ORGANIC CHEMICALS, *p*-xylene for its isolation from C8 mixtures

arsenic: As; soft element with metallic luster; use: chemical raw material, for doping silicon to make *n*-Si semiconductors; toxic

arsenic acid: $(H_3AsO_4)_2 \cdot H_2O$, arsenic trioxide; produced by oxidation of arsenic with HNO_3 and HCl; use: rodent poison, insecticide and weed killer; for manufacture of pigments, pharmaceuticals, war gases, to decolorize glass; added to outdoor wood products to protect them from insect damage

arsenicals: biocides containing arsenic, e.g., cacodylic acid; *see* arsenic acid

artificial silk: *see* rayon

asbestos: the only natural mineral fiber; fibrous silicates of magnesium and calcium, containing also iron and aluminum; most of the asbestos is found in Russia (Ural mountains), Canada (Quebec), and South Africa; and exists as the white mineral chrysotile $(3MgO \cdot 2SiO_2 \cdot 2H_2O)$; carcinogen

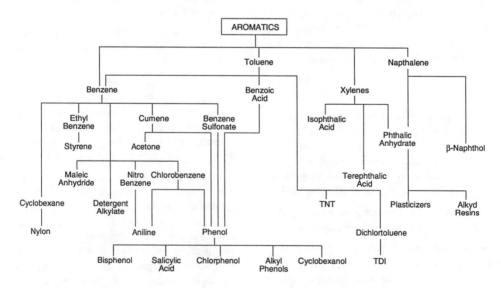

SYNTHETIC ORGANIC CHEMICALS DERIVED FROM AROMATIC COMPOUNDS

[Copyright *Riegel's Handbook of Industrial Chemistry*, 8th ed., Van Nostrand-Reinhold Co., Inc.]

ascorbic acid: vitamin C; antiscurvy vitamin; occurs in plants, primarily citrus fruit and tomatoes; destroyed by cooking; produced industrially from glucose using *Acetobacter suboxidans* or members of the mold family *Aspergillus*; use: as nutritional supplement, as antioxidant; the optical isomer, isoascorbic acid, is biologically inactive but is used as antioxidant

$$CH_2OH$$
$$HCOH$$

ASCORBIC ACID

-ase: suffix meaning enzyme

ash: sodium carbonate; *see* soda ash

A-site: in biological cells the position on ribosomes where protein synthesis occurs

aspect ratio: in fiber technology, the ratio of length to diameter

Aspergillus: family of molds (fungi) grown on agar or liquid media or on agricultural, industrial or domestic wastes; use: for industrial fermentation processes (*A. niger, A. oxyzae,* and *A. terreus* oxidize sugars to citric acid), leaching of iron from quartz sands, clays, and kaolin, reducing the toxic tannin content of tannery effluents and removing colored products from liquid molasses wastes

asphalt: black semisoft mixture of hydrocarbons; obtained from natural deposits that unlike coal, are derived from animal origins and consist of bitumen, sand, and water; also obtained from residue left after removing tar tailings in the distillation of petroleum; use: adhesive, coating (roads, roofs, floors, wood, paper), sealant, substrate for single-cell protein production, rubber and paint additive

asphaltene: component of bitumen or asphalt that is soluble in carbon disulfide; contain aromatic heterocyclic compounds

aspirin: *see* acetylsalicylic acid

assay: quantitative analytical method

A-stage: early stage in curing of thermosetting resin; "resol"; *see* phenol-formaldehyde resin

ASTM: American Society for Testing and Materials; organization chartered to develop standards on characteristics and performance of materials, products, systems, and services

ASTM distillation: a test to measure the volume percent of petroleum distilled at various temperatures; often expressed as the temperature at which a given volume percent vaporizes; a measure to characterize petroleum feedstock

atactic: pertaining to a random molecular arrangement; in atactic polymers side groups are randomly arranged along the polymer backbone; for types of molecular arrangement along polymer chains *see* tacticity; also *see* isotactic polymers

ATACTIC

ATE: aluminum triethyl, triethyl aluminum; polymerization catalyst for olefins

atomic absorption spectroscopy: AAS; analytical method that utilizes the absorption of electromagnetic radiation by individual atoms; the sample is usually in a solution that is first nebulized in a spray chamber and the solvent is evaporated; the dried sample is then vaporized and finally atomized using a high temperature flame or a furnace that is electrically heated; the wavelength of light absorbed while passing through the atomized sample depends on the nature of the atoms; the intensity of absorption of the atom-specific light beam depends on the atom concentration; in a double beam instrument the differences of light absorption in the presence or absence of sample are measured; *see also* atomic fluorescence spectroscopy, flame emission spectroscopy

atomic emission spectroscopy: AES; quantitative or qualitative analytical method that utilizes the emission of light from "excited" atoms; energy needed to excite the atoms in the vaporized sample are supplied by an electric arc or spark, a laser or an argon plasma; the emission spectra that are characteristic for each element are measured photographically or photoelectrically

atomic fluorescence spectroscopy: AFS; analytical method similar to atomic absorption spectroscopy; the instrumentation differs in that the incoming beam is at right angles to the axis of the spectrometer; after excitement by the flame the atoms in the sample emit a characteristic fluorescence that is proportional to the exciting radiation flux; used for trace metal analysis; *see* atomic absorption spectroscopy

atrazine: preemergence herbicide used primarily for corn

$(CH_3)_2CHNH$ N Cl
N N
$HNCH_2CH_3$

ATRAZINE

atropine: *see* alkaloids

attenuation: in radiation technology, the reduction of beam intensity by interaction with some medium

atto: Systéme Internationale (S.I.) prefix standing for $\times 10^{-18}$

attrition mills: *see* mills

Auger electron spectroscopy: AES; technique for examining elemental composition of solid surfaces with beams of medium-energy electrons; lateral resolution is in nanometers, surface depth penetration is a few atom layers

auger mining: technology used to reach coal in stripped areas by placing large augers onto the surface mine floor and boring horizontally into mine face

autoacceleration: in some vinyl polymerizations, an increase in molecular weight as polymerization reaches completion

autocatalytic degradation: process in which the rate increases after the onset of degradation

autoclave: high-pressure vessel used for heating and sterilizing

autogenous extrusion: *see* autothermal extrusion

automotive gasoline: ASTM designation D 439; gasoline fraction having a maximum boiling temperature of 374°F

autosome: in biological cells, a nonsex chromosome

autothermal extrusion: autogeneous extrusion, adiabatic extrusion; process in which a polymer is heated entirely by friction of the extrusion screw

autotrophic: self-nourishing, i.e., organisms that can form organic matter from inorganic material

auxins: group of plant hormones; their growth stimulating action is called auxotrophic

auxochrome: in the dye industry, a color enhancer

auxotrophic: growth promoting; *see* auxins

available chlorine: in bleach, the oxidizing power equivalent to that of free chlorine

average molecular weight of polymers: value that falls between the weight average and the number average molecular weight; roughly equal to viscosity molecular weight

Avgas: contraction for aviation gasoline

aviation gasoline: ASTM designation D

910; fuel having a maximum boiling temperature 275°F

aviation turbine fuel: ASTM designation D 1655; fuel with maximum boiling temperature of 470°F

azelaic acid: $HOOC(CH_2)_7COOH$; use: for manufacture of nylon 6,9, lacquers, alkyd salts

azeotrope: *see* azeotropic mixture

azeotropic copolymer: copolymer that has the same relative number of monomer units as the mixture of monomers from which it is derived

azeotropic mixture: solution of two or more liquids (azeotropes) whose composition remains unchanged on distillation

azides: salts of hydrazoic acid (N_3H); sodium azide is produced by reaction of $NaNH_2$ and N_2O; use: explosives, inflation of car airbags

3'-azidothymidine: AZT, zidovudine; drug licensed for treatment of acquired immunodeficiency syndrome (AIDS); antiviral drug

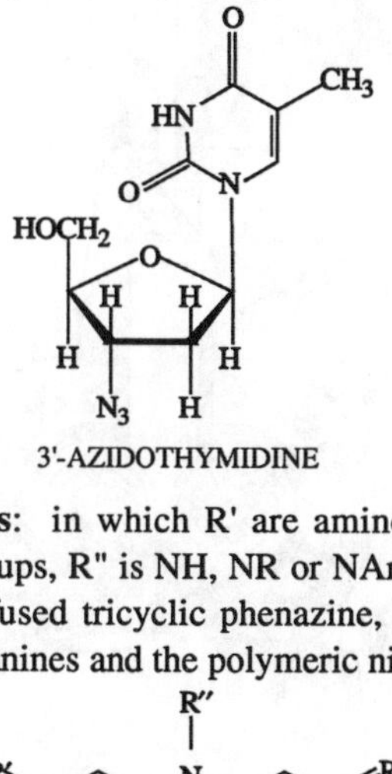

3'-AZIDOTHYMIDINE

azine dyes: in which R' are amino or hydroxyl groups, R" is NH, NR or NAr derived from the fused tricyclic phenazine, dyes include safranines and the polymeric nigrosines

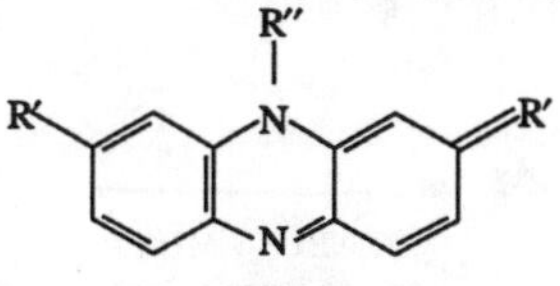

AZINE DYES

azlon: fibers made from regenerated protein

1,1'-azobisformamide: $H_2NCONNCONH_2$; ABFA; synthesized from hydrazine and urea; use: blowing agent

2,2'-azobisisobutyronitrile: $(CH_3)_2C(CN)$ $NNC(CN)(CH_3)_2$; AIBN; free radical catalyst; use: for low-temperature polymerization initiator

azo dyes: dyes characterized by the chromophore Ar–N=N–Ar'; produced by batch processes starting with diazotization of primary aromatic amines followed by coupling with aromatic amines or phenols; azo dyes acid, mordant, and disperse dyes

azoic dyes: water-insoluble azo dyes that are produced in situ on the fiber

azo pigments: azo dyes that are insoluble in the medium in which they are used

AZT: *see* 3'-azidothymidine

B

bac: latex from modified butadiene copolymer; used to improve bonding between rubber and textiles

bacillus thuringiensis: biological insect control agent; acts by infecting insect larva

bacitracin: antibiotic polypeptide produced by *Bacillus subtilis* and *B. licheniformis*

bacteriocide: agent that kills bacteria; *see* bacteriostat

bacteriophage: phage; virus that infects and multiplies within bacteria; λ bacteriophage is commonly used as carrier in recombinant DNA experiments

bacteriostat: agent that stops bacterial growth but does not kill the bacteria; *see* bacteriocide

baffle: device inserted in a flow channel to restrict flow of material or divert its path

bagasse: in sugar manufacture the residue left after extraction from sugar cane or sugar beet; also the waste remaining after fiber has been removed from fiber plants such as sisal; use: biomass for fuel, reinforcement in laminates and molding powders, raw material for pulp and paper

bag filter: *see* baghouse

baghouse: dust-collecting device consisting of an array of large fabric bags; dust-laden air is sucked into the bags and then drawn out through the sides; the dust is allowed to drop periodically into hoppers below the bags

bag molding: method to manufacture reinforced plastic laminates; a fibrous sheet impregnated with resin is placed on a hard form and a flexible bag on top of the sheet applies uniform pressure and heat to cure the laminate

bag paper: paper grade that consists of 100% unbleached softwood treated by Kraft process; for white bags bleaching is required

bakelite: plastic first commercialized by Leo Baekeland; *see* phenol–formaldehyde resins

balanced dope: in explosives technology, active absorbants in dynamite that protect against the shock-sensitive nitroglycerin; the absorbants consist of carbonaceous materials (e.g., wood pulp, starch, meals), oxidizers (e.g., nitrates), and antacids (e.g., $CaCO_3$ or ZnO); the mixtures (dopes) are oxygen balanced to provide maximum strength and minimum fumes

balata: natural balata is a thermoplastic similar to gutta-percha; synthetic balata is *trans*-1,4-polyisoprene

baling: bundling up and compacting material before shipping; recycling baled postconsumer plastics leads to discolored, low-value products

ballas: *see* diamond

Ballestra process: continuous air-SO_3 sulfonation of alkylbenzene, fatty alcohols, and α-olefins for the production of detergents; *see* alkylaryl sulfonate

ball mills: *see* mills

balsam: viscous solution of resins in volatile oils; *see* oleoresins

bamboo: in areas where soft and hard woods are not available this plant can be a raw material for the pulp and paper industry

Banbury mixer: mixing occurs in a closed chamber with two spiral blades rotating in opposite directions; the material to be mixed is kept under pressure with a pneumatic ram; at higher temperatures and pressures and in the presence of catalysts the mixer can be used for reclaiming rubber

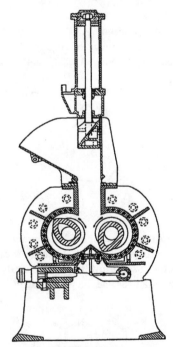

BANBURY MIXER
[Farrel Co. Copyright *Perry's Engineer's Handbook*, 6th ed. (1984), McGraw-Hill, Inc.]

baratols: pourable TNT mixtures containing 10–20% $Ba(NO_3)_2$

barbiturates: a group of sedative–hypnotic drugs

$R_1 = R_2 = C_2H_5$ veronal, barbital
$R_1 = C_2H_5, R_2 = C_6H_5$ phenobarbital

BARBITURATES

Barcol hardness: value obtained by measuring resistance to penetration of a steel point; *see* hardness

barium perchlorate: $Ba(ClO_4)_2$; use: with barium chlorate ($BaClO_3$) as oxidizer in propellant formulations; to produce green color in fireworks

barium process: desugaring of beet sugar molasses by precipitation of barium saccharate

barium sulfate: $BaSO_4$; white salt; use: "blanc fixe" powder and paste, densifying agent in medical X-ray radiology, ingredient of the pigment lithopone, suspension agent in suspension polymerization

barium sulfide: BaS; *see* lithopone

barrier plastics: thermoplastics with low permeability to liquids or gases; use: food wrapping, beverage bottles; nonporous "breathable" barrier films and coatings permit transmission of water vapor and are used to improve comfort of textiles

barrier resins: *see* nitrile barrier resins

barrier sheet: in laminates the layer between the core and the outer layer

baryte: *see* barium sulfate

basagran: *see* bentazon

Bashkirov oxidation: oxidation of *n*-paraffins to secondary alcohols without chain cleavage; air oxidation takes place at ambient pressure and 140–180°C with 0.1% (wt) $KMnO_4$ and 4–5% boric acid; boric acid serves to protect the initially formed –OH groups from further oxidation by borate formation; industrially important process for manufacture of cyclododecanol, which is a raw material for nylon 12 manufacture; *see* alkyl hydroperoxides

basic dyes: cationic dyes; these dyes contain amino groups that become protonated under acid conditions of the dye bath; dyes form salt linkages with anionic groups in the textile fibers and have a high color value; the newer light-fast dyes are made by attaching a cationic tail to azo- or anthraquinone chromophores; the tail is not conjugated with the chromophore

basic slag: by-product of the steel industry, source of P_2O_5

batch autoclave splitting: production of fatty acids for the manufacture of soap; steam hydrolysis of glycerides is carried out in a copper or stainless steel vessel with or without catalyst

batch process: the manufacture of products in individual batches in contrast to continuous processing

bate: in leather manufacture, material used during the removal of lime from animal skins before tanning; bating results in unswollen collagen fibers thus making the skins soft and flaccid; natural bate is obtained from dung, modern bate is made from pancreatic enzymes such as trypsin

BATF: Bureau of Alcohol, Tobacco and Firearms of the United States Department of the Interior

batteries: energy storage devices that convert chemical energy into electric energy (electricity); functional components of batteries are electrodes, electrolyte, and separator; the electrolyte may be aqueous, molten, or solid; alkaline cells use aqueous NaOH or KOH as electrolyte; the separator maintains spacing between the electrodes and thus prevents short circuits; primary batteries are discharged only once and then either discarded, recycled, or their spent electrode is replaced; secondary batteries (storage batteries) are recharged by reversing their potential using an external source; the dimensions of batteries depend on their use: small button cells power watches, calculators, hearing aids, and cameras; cylindrical cells power flashlights; larger systems are used for electric vehicle propulsion, solar energy storage, industry, and electric utility energy storage (the batteries are charged at night during low demand time and discharged during peak demand); common specifications for battery systems are cost ($/kWh), life time, energy density (Wh/L), power density, and energy efficiency. The standard 1.5 V flashlight battery (energy density 2.7 Wh/in.^{3}) is based on $Zn-MnO_2$ chemistry using aqueous $NH_4Cl-ZnCl_2$ as electrolyte (Leclanché system); the lead-acid battery is based on the oxidation of Pb and reduction of PbO_2 to aqueous $PbSO_4$; advanced lead acid batteries are used in electric vehicles and in industry; the Zn-HgO cell sets the standard for advanced button cells; other advanced batteries with high energy densities are now coming on the market; some of these work at elevated temperatures (e.g., sodium–sulfur at 300–350°C, lithium–iron sulfide at 425–500°C), others work near room temperature (e.g., iron–chromium, iron–nickel oxide, nickel–air, nickel–iron, nickel–zinc, zinc–air, zinc–bromine, zinc–chlorine, zinc–nickel oxide). New Li batteries ($Li-CuO$, $Li-SO_2$, $Li-SOCl_2$, $Li-SO_2Cl_2$) have energy densities (8–18 Wh/in.^{3}) and do not use aqueous systems; $Li-CF_x$ is a commercial primary button cell; silver-zinc battery has the highest current density but because of its cost is limited to military applications

battery acid: sulfuric acid used in lead-acid batteries; *see* battery

battery separator: *see* batteries

batu: East India tree resin; use: alcoholic and oleoresinous varnishes

Baumé: units of specific gravity, Bé; at 60°F, for liquids heavier than water, the specific gravity is $145/(145-n)$, for liquids lighter than water, the specific gravity is $140/(130 + n)$ where n is the °Bé reading on the Baumé scale

Baum jig: in separating coal and impurities suspended in water, a device using intermittent air pressure to cause pulsations through the water that moves the lighter coal to the surface, where it overflows; the impurities are discharged at the bottom

bauxite: aluminum ore, $Al_2O_3\cdot2H_2O$; *see* alumina

B-black powder: *see* black powder

BBP: *see* butyl benzyl phthalate

B cells: B lymphocytes; white blood cells that (in mammals) are formed in the bone marrow; members of the immune system that are responsible for the production of antibodies; *see* T cells

beads: in tire technology, circles of wire that reinforce the tire where it touches the rim; the wires must be removed before shredding during recycling

beam: in fabric manufacture, yarns are wound from bobbins onto cylinders, called beams, to be used as warp in weaving of fabrics; the process is called beaming

beater-addition process: in the manufacture of paper-based laminates, resins are added to the pulp in the beater and are then precipitated on the fiber by addition of alum or acid; the resulting paper–resin combination is called resin-filled paper

beck: during textile dyeing, machines move

the fabric through a dye bath; when the fabric is relaxed it is called beck; when the fabric is moved under tension it is called jig; *see* dyeing

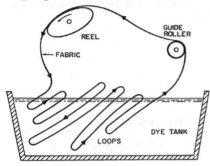

DYE BECK
An endless band of fabric may be formed by sewing one end to another with fabric movement occurring as shown. Although the goods are in the tank during most of the cycle, the turning reel, by constantly moving the material, assures uniform distribution of the dye. Dye becks also process fibers in rope form. [Courtesy D. M. Considine, P.E.]

beet sugar: derived from the sugar beet that grows in the temperate zone; sugar is extracted from washed beet slices by countercurrent water diffusion, and is then concentrated, purified, and crystallized

Beken mixer: Duplex–Beken kneader, Bramley–Beken mixer; a mixer that uses two overlapping blades, one of which turns twice as fast as the other

belted-bias tire: a tire in which the body plies are positioned as in bias tires and the outer tread stabilizer is used as in radial tires; *see* tires

beneficiation: in mineral extraction, a process that concentrates the ore from the waste products (gangue) before further processing; separation steps take advantage of differences in visual appearance, specific gravity and of wetting, magnetic, and electric properties; operations include calcination, clarification, filtration, milling, screening and thickening

benefin: dinitroaniline herbicide

benomyl: systemic fungicide that enters the plant so that it can destroy fungi removed from the area of application

BENOMYL

bentazon: bentazone; postemergence herbicide

BENTAZON

benthos: plants or animals that live on the bottom of a body of water

bentonite: natural aluminum silicate that has a high water absorption; use: stabilizer, gelling agent and filter, suspension agent for polymerization

benzene: simplest aromatic compound; produced together with other aromatics by (1) coking hard coal at 1000–1400°C; the coke-oven gas extract contains about 65% benzene together with alkylbenzene, (2) catalytic (metal oxides catalysts such as Al_2O_3, $CoO–MoO_3$) cracking of gasoline under pressure, (3) steam reforming of naphtha leads to BTX, (4) disproportion of toluene to benzene and xylene, (5) catalytic (metal oxide catalysts such as Cr_2O_3, Mo_2O_3) dehydroalkylation of toluene (e.g., Hydeal [UOP], Detol [Houdry–Air Products], and Bextol [Shell, BASF] processes), (6) thermal dehydroalkylation of toluene (Gulf, Atlantic Richfield, Mitsubishi), and (7) separation from other aromatics by distillation.

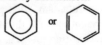

BENZENE

Major derivatives include aniline, hexachlorobenzene, ethyl benzene, cumene, phenols, cyclohexane, sulfonates; major products include plastics and resins; benzene serves as chemical feedstock; hazardous chemical waste (EPA)

γ-benzenehexachloride: $C_6H_6Cl_6$; BHC; 1,2,3,4,5,6-hexachlorocyclohexane; produced by batch or continuous addition of chlorine to benzene; use: only the γ isomer is an insecticide primarily used for treatment of seeds; hazardous chemical waste (EPA); do not confuse with hexachlorobenzene

benzenestearosulfonic acid: *see* fat splitting

benzenesulfonic acid: $C_6H_5SO_3H$; produced by reaction of benzene with sulfuric acid; used as intermediate in phenol synthesis

1,2,4,5-benzenetetracarboxylic acid: produced by catalytic (V_2O_5) oxidation with HNO_3 of 1,2,4,5-tetramethylbenzene (durene); the anhydride (pyromellitic anhydride) is used in the production of polyamides, polyimides and plasticizers

1,2,4-benzenetricarboxylic acid: produced by catalytic (V_2O_5) oxidation with HNO_3 of 1.2.4-trimethylbenzene (pseudocumene); the anhydride, trimellitic anhydride is used for production of polyamides, polyimides and plasticizers

benzidine: p–$H_2NC_6H_4$–p–$C_6H_4NH_2$; formerly used as raw material for azodyes; a carcinogen banned in the United States; hazardous chemical waste (EPA)

benzodiazepines: group of anxiety-reducing drugs; *see* diazepam

benzoic acid: C_6H_5COOH, BzOH; produced by catalytic (e.g., Co salts) oxidation of toluene; use: production of phenol, production of ε-caprolactam, food preservative (maximum amount is 0.1% due to toxicity), antiseptic, for flavoring tobacco, as chemical intermediate

benzol: name for gasoline in some countries

benzo(a)pyrene: BaP; carcinogenic polynuclear aromatic hydrocarbon; present in cigarette smoke; hazardous chemical waste (EPA)

BENZO(a)PYRENE

1,2-benzopyrone: *see* coumarin

benzoyl chloride: C_6H_5OCl; *see* benzoyl peroxide

benzoyl peroxide: $(C_6H_5CO)_2O_2$; prepared by reaction of benzoyl chloride and sodium peroxide; use: polymerization initiator, bleaching agent for fats and flour, cross-linking agent, "drying" agent for unsaturated oils; explosive that must be wetted for transportation

Bergbau process: *see* hydrazine

Bergius–Rheinau process: wood hydrolysis using dry chips and 40–45 wt% HCl; the acid is recycled and fortified by gaseous HCl from salt-sulfuric acid retorts; the hydrolysate that contains the oligosaccharide is concentrated under vacuum and spray-dried; the intermediary sugars are hydrolyzed for further use

beryllia: BeO; beryllium oxide; produced from mineral deposits; because of its high melting point and chemical stability BeO is used in ceramics, in particular in printed circuits

beryllium: Be; metallic element; produced by reduction of BeF_2 with magnesium; a light metal used in alloys for aerospace application due to their high elasticity/weight ratio, tensile stress and thermal conductivity, as moderator in nuclear reactors due to its high ability to absorb neutrons

beta blocker: drug that competes with the neurotransmitters epinephrine and norepinephrine for their receptor site and thus protects the heart against overstimulation; *see* propranolol

beta-carotene: isoprenoid precursor of Vitamin A; gives orange color to carrots

beta-cellulose: soluble fraction obtained from treating cellulose with caustic soda; *see* hemicellulose

betaine: alkaloid

betaines: $-OOCCH_2N^+(CH_3)_3$; group of amphoteric ions

beta particle: the electron

Bextol process: *see* benzene

BHA: *see* butylated hydroxyanisole

BHC: 1,2,3,4,5,6 hexachlorocyclohexane; lindane; insecticide

BHT: *see* butylated hydroxytoluene

bias tires: *see* tires

bicarb: *see* sodium bicarbonate

bicomponent fibers: bilateral fibers; fibers formed by extruding two polymers side-by-side from spinnerettes; in the resulting fibers the two components affect each other's physical behavior much like two joined metal strips affect each other; differential swelling and shrinking of the fibers causes bicomponent fibers to crimp or form spiral distortions; examples are fibers made from two different acrylic copolymers; polyethylene–polypropylene bicomponent fibers in nonwoven materials use PE for thermal bonding, nylon bicomponent fibers were used for crimped yarns, but recent texturing improvements have reduced their need

bidentate: *see* ligand

bigas process: methane production from coal; a coal slurry and steam are fed separately into the first gasifier stage where the mixture is contacted by hot gas from the oxygen-fed lower stage; the resulting product, consisting of CH_4, CO, CO_2, H_2, and char is quenched with water, undergoes a CO/H_2 shift reaction to produce more H_2 followed by methanation of the remaining CO and H_2; the process was designed to treat all types of coal without pretreatment but difficulties in handling the solids remain; *see* substitute natural gas

bilateral fibers: *see* bicomponent fibers

bimetallic catalysts: in petroleum refining, platinum catalysts with promoters (rhenium, germanium, tin) improve hydroforming reactions and reduce hydrocracking and coke formation; *see* catalysts

binders: materials that hold solid particles together; such as the resinous constituents of paint coatings; produced from natural drying oils or from synthetics such as acrylics, alkyds, cellulosics, epoxies, melamines, phenolics, polyvinyl acetates, polyurethanes, silicones, ureas, or vinyls; also asphalt or bitumen for paving

bioassay: analysis using living organisms

biochemical oxygen demand: BOD; in waste water treatment, a measure of oxygen consumed by biological processes that break down organic chemicals, e.g., BOD_5 is the amount of oxygen consumed in 5 days

biocides: term generally limited to antimicrobial and antifungal agents used as antibiotics, antifoulants, disinfectants, preservatives, and antiseptics; agents against slimes and mildew; use: for sanitation and in foods, water supply, swimming pools, and cosmetics; often incorporated in paint, polymer and rubber formulations, in sizings for textiles and paper

biodegradable plastics: materials subject to decomposition by microorganisms, such as copolymers of natural and synthetic polymers, that are produced by graft polymerization of starch or cellulose with polystyrene; polymerization using microorganisms leads to biodegradable poly(β-hydroxyalkanoates) (PHA) such as ICI's commercial batch fermentation of *Alcaligenes eutrophus,* that yields copolyesters of β-hydroxybutyrate and β-hydroxyvalerate (PHB/HV); biodegradability can be promoted by incorporation of groups that are labile to heat, oxidation, or light; biodegradable plastics are subject to degradation only outside of most landfills; use: in medicine for sutures, tapes, and temporary prostheses

biodegradation: breakdown of materials due to action of living organisms, e.g., natural polymers such as starch, synthetic biodegradable polymers, and detergents

biodisk process: secondary treatment process of waste water in which agitation is carried out by banks of rotating disks

biologicals: substances biologically active such as antibiotics, aminoacids, nucleosides, and nucleotides, vitamins and growth factors, steroids, tetracyclins, and enzymes

biomass: plant material used for energy production and for chemical feedstock; includes living plants, agricultural and urban wastes

biomaterials: in medicine: metals, ceramics, carbon, polymers, and natural substances in contact with biological fluids; use: therapeutics, diagnostics (e.g., protheses, catheters)

biosensors: highly sensitive and selective detectors using enzymes and other biological materials

biotechnology: the commercial exploitation of recombinant DNA (genetic engineering) and monoclonal antibody technologies

biotin: a growth factor

bipolymer: polymer derived from two different monomers; *see* bicomponent fibers

birefringence: double refraction; occurs in substances whose refractive indices differ with direction (e.g., nicol prism, calcite); in oriented fibers their birefringence indicates the degree of overall orientation in the fiber and correlates with mechanical properties

bis-: prefix denoting two; used for compound and complex radicals and functional derivatives

***bis*(benzonitrile)dichloropalladium (II)**: $PdCl_2(C_6H_5CN)_2$; homogeneous catalyst; has been used for carbonylation (e.g., RC_6H_4OH to $(RC_6H_4O)_2CO$, $H_2C=CH$ CH_2OH to $H_2C=CHCH_2COOH$); dimerization of acrylonitrile to $NCCH_2CH=CH$ CH_2CN; for hydration of nitriles to amides; for reduction of nitrobenzene to aniline; for isomerization of allyl acetates and of 2,3-dihydrofurans to 2,5-dihydrofurans

***bis*(cyclopentadienyl)titanium dichloride**: $(C_5H_5)_2TiCl_2$; homogeneous catalyst; has been used for the reductive coupling of $HOCH_2CH_2C\equiv CH$ to $CH_3CH_2CH=CHCH_2$ CH_2OH; for dehalogenation of alkyl halides in the presence of sodium borohydride; for the dimerization of alkenes; for the Fischer–Tropsch process; and for the reduction of 1,2-alkenes to alkanes in the presence of lithium aluminumhydride

bisphenol A: $(p\text{-}HOC_6H_5)_2C(CH_3)_2$; produced by (continuous or batch) condensation of excess phenol with acetone in the presence of sulfuric acid; in the Hooker process dry HCl and CH_3SH are used as promoter; the UCC process uses a heterogeneous catalyst; use: precursor for thermoplastics

bisphenol A polycarbonate: $H[p\text{-}OC_6H_4$ $C(C\{CH_3\}_2)p\text{-}C_6H_4OCO]_nCl$; made from bisphenol A and phosgene in pyridine and CH_2Cl_2; use: plastic for molding materials

***bis*(p-hydroxyethyl)terephthalate**: (*bis*-HET); monomer for a high molecular weight polyester (polyethylene terephthalate [PET])

***bis*(tributyltin) fluoride oxide**: *see* antifouling agents

bisulfate pulping: batch process involving

the digestion of spruce or fir wood with alkaline solutions of HSO_3^- at 140–150°C for 4–5 hr; the products are strong, light-colored pulps and lignosulfonates; most of the waste liquor is concentrated and burned to recover the cooking chemicals

bitrinitroethylnitramine: $(NO_2)_3CCH_2NH(NO_2)CH_2C(NO_2)_3$; BTNENA; high explosive

bitrinitroethylurea: $(NO_2)_3CCH_2NHCON-HCH_2C(NO_2)_3$; BTNEU; high explosive

bittering agents: substances, such as alkaloids, that are added to make ethyl alcohol unfit for consumption; denaturants

bitumen: semisolid and solid mixture of hydrocarbons occurring naturally and as residue after distillation of petroleum; use: as process aid, softener, and tackifier in rubber compounding, fuel

bituminous coal: soft coal containing 69–86% carbon; coal of rank II

biuret: $NH_2CONHCONH_2 \cdot H_2O$; prepared from urea by heating; forms during polymerization of urethanes; use: analytical reagent

black lead: *see* graphite

black liquor: dark liquor recovered from alkaline wood pulping; the liquor contains lignin and some carbohydrates; it is concentrated and burned for process energy

black iron oxide: Fe_3O_4; used as pigment

black news ink: inks based on carbon black pigment

black orlon: *see* ladder polymers

black powder: explosive consisting of potassium nitrate, charcoal and sulfur; standard composition: 75% potassium nitrate, 10% sulfur, and 15% charcoal; other compositions contain 64–74% potassium nitrate; B-black powders have corresponding compositions but also contain sodium nitrate; speed of burning increases with potassium nitrate content; manufactured by batch process; sensitive to impact, friction, and sparks; use: primarily for fuses, igniters, blasting powder

blackstrap molasses: residual mother liquor from cane sugar production; contains 70 wt% sucrose and reducing sugars; use: cattle feed, fermentation to ethanol, citric acid and acetic acid

blanc fixe: barium sulfate, used to impart X-ray opacity and to increase specific gravity of materials

blast finishing: process of removing flash from objects by impinging particles like steel balls, plastic pellets, crushed apricot pits, or walnut shells on them

blast furnace: vertical coke-fired furnace for smelting iron ore and other ores; the exit gas from the furnace has high CO content

blasting gelatin: explosive consisting of nitroglycerin + nitrocotton in a 92:8 ratio; used as comparison explosive to determine relative weight strength (strength of explosive material/unit weight)

blasting powder: *see* black powder

bleach activators: materials added to detergent formulations to increase the activity of bleaching agents; examples: tetraacetyl ethylene diamine (TAED), pentaacetyl glucose (PAG), diacetyl dioxohexahydrotriazine (DADHT), and nonanoyloxybenzene (NOBS)

bleaches: materials that whiten and remove stains from fabrics, paper, and pulp by oxidation or reduction; the most commonly used household bleaches are aqueous hypochlorite (OCl^-) and sodium perborate; household detergents may contain bleach as well as bleach activators; in paper (Kraft pulp) and textile manufacture sodium chlorite $(NaClO_2)$, ClO_2 and hydrogen peroxide are used; *see* optical bleach

bleeding: diffusion of color into a surrounding surface; also termed: blooming, bronzing, crocking, or migrating

blending: in polymer technology, the mixing of two or more polymers with or without the help of a compatibilizing agent

blending resin: extender resin; resin used to change properties and to reduce costs of primary resin

blends: homogeneous mixture of two or more materials as in polymer and fiber blends or in rubber formulations

blister packaging: preformed blisters from thermoplastic films such as cellulosics, polystyrene, and vinyls

blisters: imperfections in finished metals, plastics, rubber due to small air pockets

block polymer or copolymer: polymers composed of separated sections of two or more different polymer units, such as AB or ABA where A and B are polymer blocks; examples: a flexible section of polyoxybutane or an aliphatic polyester coupled to stiff aromatic polyurethane sections (*see* Spandex fibers, urethane block copolymers); blocks of polystyrene and polybutadiene; in these block systems, the phases do not mix and the mechanical and structural characteristics depend on the relative concentration and size of the blocks, ranging from rubbers to very stiff materials that, however, are thermplastic at elevated temperatures and can be reformed; block copolymers usually have higher impact strength than each homopolymer or a mixture of the polymers; *see* polymer block

bloom: in polymer technology, undesireable effect on the surface of plastics or rubber caused by the migration of additives to the surface; in metallurgy, discoloration of a metal mold; a piece of steel made from an ingot; in textile technology, glossy finish applied to wool or cotton

blooming: *see* bleeding

blowing agent: foaming agent; substance capable of imparting a cellular structure, i.e., foam; gases or substances that produce gas upon decomposition; examples: baking powder for food production and low boiling organics for polyurethane foam; chemical blowing agents are stable at normal storage and processing conditions and evolve a gas controllably at a specific decomposition temperature; most modern blowing agents are organic nitrogen compounds: azobisformamide (ABFA) has general application, dinitrosodimethyl terephthalamide (NTA) is used with plasticized PVC, dinitrosopentamethylene tetramine (DNPT) produces foam rubber, oxy-*bis*(benzenesulfonylhydrazide) is used for foamed polyethylene insulation of wire or cable because its residue has less effect on the electrical properties of the insulation; *see* unicel

blow molding: formation of bottles and other hollow articles by melt extruding a polymer tubing into an open mold and then expanding it into the mold shape by compressed air or steam

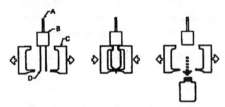

BLOW MOLDING
(A) AIR LINE; (B) DIE; (C) MOLD; (D) PARISON.
[Copyright Kirk-Othmer. *Encyclopedia of Chemical Technology,* 3d ed., John Wiley & Sons, Ltd.]

bluing: blue pigment, usually ultramarine blue that counteracts the yellowing of fabrics; *see* optical bleach

blush: whitish discoloration on a freshly applied solution coating or lacquer; blush is caused by cooling the atmosphere below its dew point because of fast evaporation of solvent; this temperature drop causes moisture to condense on the wet surface; the term is also used for chalky areas in plastics produced by stress

boart: *see* diamond

body: general term denoting all-over consistency of liquids; in textile technology, fabric quality such as drape and hand

boghead coal: bituminous coal that contains large amounts of fossil algae

Bohemian glass: *see* hard glsss

bolstar: $p-CH_3SC_6H_4OP(S)(OC_2H_5)S$ $(CH_2)_2CH_3$; insecticide

bolus alba: *see* kaolin

bonded fabric: web of fibers held together by an adhesive

bonding resins: agents used for bonding aggregates such as sand, asbestos, and paper; resinous adhesives for plywood; *see* isocyanates

BON pigments: red and maroon pigments based on β-oxynaphthoic acid; these pigments are resistant to bleeding, widely used in plastics and rubber industry

bone phosphate of lime: BPL; $Ca_3(PO_4)_2$; tricalcium phosphate; phosphate content in rock is often expressed as equivalent tricalcium phosphate or grade of BPL; high grade phosphate rock has 66 BPL–78 BPL

boom powder: ignition mixture that produces incandescent particles in fireworks; consists of iron oxide, titanium, zirconium in a cellulose nitrate binder

boort: *see* diamond

booster: *see* builder

borates: boric acid esters; trimethyl, tri-n-butyl and tri-*p*-cresyl borates are used as flame retardants

borax: $Na_2B_4O_7$; anhydrous, also exists as penta- and deca-hydrates; use: for ceramics, as builder for detergents, as herbicide

Bordeaux mixture: basic copper sulfates; old insecticide

boria: B_2O_3; boric oxide; *see* boric acid

boric acid: H_3BO_3; produced by treating borax with sulfuric acid; use: in glass and glass fibers, flux for metallurgy, fire-retardant for textiles, antiseptic

boron carbide: hard abrasive; B_6C is harder than B_4C

boron fibers: low-density, high-strength, stiff fibers used for aerospace applications

borosilicate glass: *see* glass

boss: in molding, a protuberance provided to an article to facilitate handling during manufacture

bottoms: material in the lowest part of a column or tank

BR: butadiene rubber

BR: butyl ricinoleate

Brabender plastograph: instrument used to determine flow properties, mixing and extrusion effects, and temperature/viscosity relationships of plastics and rubber; used to determine effects of additives during experimental compounding

Bramley–Beken mixer: *see* Beken mixer

branched polymer: nonlinear macromolecule; produced by free radical transfer (e.g., high-pressure ethylene polymerization), polymerization of difunctional monomers having unequal reactivity of the two vinyl groups (which may also lead to cross-linking), reaction of amino acids with three-functional carboxylic acids, alcohols and amines

brass: copper–zinc alloys having zinc content ≤40 wt%

braze: chlorinated rubber + hypochlorinated rubber, bonding agent for rubber/metal

brazing alloys: *see* solder

break elongation: *see* elongation

break length: the theoretical length of yarn

at which it would break under its own weight; this value is used mostly in Europe

breaker strips: in tires, protective strips in the tire centers

break tenacity: *see* tenacity

breathable film: plastic film that is at least somewhat permeable to gas

breeder: *see* nuclear breeder

Brewer's yeast: *see* yeast

brewing: production of ale and beer by fermentation of starch to alcohol

brightener: *see* brightening agent

brightening agents: optical brighteners (e.g., coumarins, naphthylimide, diaminostilbene disulfonates) that absorb UV rays and emit visible violet-blue light that overcomes yellow casts and helps to enhance clarity and brightness; use: primarily on fabrics; not suitable for materials that contain UV absorbers

brightness: the measure of reflectance of a pressed powder cake compared with standard MgO at a wavelength of 474 nm

bright rayon: clear yarn product of rayon manufacture; addition of opacifier (TiO_2 pigment) converts it to matte product

brine: impure NaCl solution

Brinell hardness: measure of hardness determined by pressing a 10-mm steel ball into the specimen under constant load

briquetting: formation of small bricks having any shape (briquet) using molds, rollers, or extruders

brisance: shattering power of high explosives; gelatin dynamites have the highest brisance

British thermal unit: Btu; the energy needed to raise the temperature of 1 lb H_2O by 1°F

brittleness temperature: in polymer technology, the temperature at which polymers rupture on impact under specified conditions; related to the glass transition temperature

Brix scale: density scale for sucrose solutions; degrees brix equal percent sugar in solution

broadgoods: woven materials with widths over 18in

bromacil: herbicide

BROMACIL

bromides: metal salts of bromine; use: KBr in photography, high density $CaBr_2$-$ZnBr_2$ solution for deep well oil drilling

bromine: Br_2; element; produced from hot bromide-containing brine by mixing with chlorine and steam; use: in manufacture of organic bromides such as the pesticide ethylene dibromide (EDB), the fumigant methylbromide and of tetrabromobisphenol A, which is used as modifier for polymers

bronze: copper-tin alloys; also any copper alloy that does not contain Zn or Ni

bronzing: *see* bleeding

Brookfield viscometer: rotating spindle viscometer to determine viscosity of thixotropic liquids; by rotating spindle at varying speeds one can determine the degree of thixotropy

brown sugars: specialty dry sugars containing sucrose, water, invert sugar, and some nonsugars; obtained from low-purity process liquor or by addition of impure syrup or molasses to refined granular sugar

BS: British Standard

BSI: British Standards Institution

B-stage: intermediate stage of curing of thermosetting resins; *see* A-stage and C-stage

Btu: *see* British thermal unit

BTX: abbreviation for a mixture of benzene, toluene, and xylene

buckminsterfullerenes: bucky balls, fullerenes; an allotropic form of carbon consisting of 60 or more carbons with 5- or 6-membered rings arranged in a soccerball-like shape; the hollow spheres can serve as hosts for atoms, molecules, or ions

buckyballs: *see* buckminsterfullerenes

builder: in the detergent industry, a material that upgrades the cleaning efficiency of the surfactant; builders include chelating agents, water softeners, buffers, and antiflocculants; examples of builders: sodium tripolyphosphate (TPP), tetrasodium pyrophosphate (TSPP), silicates, sodium carbonate, sodium citrate, nitrilotriacetate (NTA), and zeolites; built soaps and built detergents are formulated with a builder; unbuilt detergents are made for light duty

built soap: *see* builder

bulk density: the apparent density of powder or granules

bulk polymerization: mass polymerization; polymerization in the absence of solvents; *see* thermal polymerization, condensation polymerization

bulky yarn: textured yarns made by crimping or false twisting to produce twisted yarns and fabrics with increased air space and thus lower skin–fabric contact

bullet proof glass: *see* safety glass

buna: synthetic rubber developed in Germany in 1926–1930; name stems from *buta*diene-*na*trium because metallic sodium was originally used as catalyst to effect polymerization; buna rubbers are identified by numbers that represent about 10^{-3} of the

molecular weight: buna 32 is a low viscosity softener, buna 85 is used for hard rubber linings and has good resistance to chemicals, buna 115 has good low temperature characteristics; buna S and buna SS are copolymers of butadiene and styrene, and buna *N* is a copolymer of butadiene and acrylonitrile

butadiene: $CH_2=CHCH=CH_2$; production methods include cracking of hydrocarbons (naphtha is mostly used), dehydrogenation of *n*-butylene over a fixed bed catalyst at ~1100°F, dehydrogenation of *n*-butane (Houdry process), catalytic conversion of ethanol (Ostromislenski process); use: raw material for styrene butadiene rubber, polybutadiene rubber

butadiene rubber: BR, produced by emulsion or solution polymerization of butadiene; BR has poor milling and extrusion properties, thus it is usually blended with other elastomers

butadiene-styrene copolymers: SBR, production by emulsion or solution polymerization; widely used as elastomer

1,4-butanediol: $HOCH_2CH_2CH_2CH_2OH$; manufactured from 1,3-butadiene in the presence of acetic acid by air oxidation (Mitsubishi process) with tetrahydrofuran as coproduct; by stepwise hydrogenation (Ni-Re and Ni-Co-ThO$_2$/SiO$_2$ catalysts) of maleic anhydride via butyrolactone and tetrahydrofuran (Mitsubishi and Kao); from propylene oxide via allyl alcohol and CO/H$_2$ (Kuraray process); use: raw material for chemical synthesis leading to engineering plastics, elastomers, elastic fibers, artificial leather, solvents, pharmaceuticals, adhesives

butanols: butyl alcohols; (1) butanol is produced by hydroformylation of propylene followed by hydrogenation; by aldol condensation of acetaldehyde to crotonaldehyde, followed by hydrogenation; by reaction of propylene with CO and water in the presence of a pentacarbonyliron catalyst (Reppe process); (2) butanol is produced by hydration of *n*-butene in the presence of sulfuric acid, followed by hydrolysis; *tert*-butanol is produced by hydration of isobutene in the presence of sulfuric acid, followed by hydrolysis; butanols are important chemical starting materials and solvents

butenes: butylenes, C_4H_8; obtained as by-product from motor fuel production and from petroleum cracking processes; extraction of the C4 mixture separates the butenes and butanes (raffinate) from 1,3-butadiene; the bulky shape of isobutene [$CH_2=C(CH_3)_2$] permits its isolation from the raffinate by use of a molecular sieve (Olefin–Siv process); the higher reactivity of isobutene than that of the other two butenes is the basis of its separation by selective industrial acid hydrolysis to tertiary butanol [$(CH_3)_3COH$] followed by dehydration to isobutene; separation of 1-butene from the 2-butenes can be achieved by distillation; use: starting materials for industrial syntheses to products such as butanols, methyl-*tert*-butyl ether, acetic acid, and methacrylic acid

button cells: *see* batteries

butyl alcohols: *see* butanols

butylate: thiocarbamate herbicide

butylated hydroxyanisole: BHA; antioxidant for oils and fats

BHA

butylated resins: resins containing the radical –C_4H_9

butylated hydroxytoluene: BHT; 2,6-di-*tert*-butyl-4-methylphenol; use: antioxidant in food, animal and vegetable oils, petroleum, plastics, rubbers, soaps

butyl benzyl phthalate: BBP; *o*–C_4H_9OO $CC_6H_4COOCH_2C_6H_5$; plasticizer for cellulosics and vinyl resins; good gelling agent for PVC

butylenes: *see* butenes

butyl rubber: BR; copolymer of isobutylene with 1.5-4.5% isoprene or butadiene; this copolymer has good resistance to heat, O_2 and O_3 and low permeability to air; use: inner tubes, hoses for steam and chemicals, conveyor belts for hot materials, protective coverings against acids, cable insulations

2-butyne-1,4-diol: $HOCH_2C\equiv CCH_2OH$; manufactured from acetylene and aqueous formaldehyde in the presence of copper acetylide catalyst (Reppe process); intermediate in 1,4-butanediol synthesis

butyraldehyde: $CH_3CH_2CH_2CHO$; produced by hydroformylation of propene or aldol condensation of acetaldehyde; use: for organic synthesis (e.g., 2-ethylhexanol), manufacture of poly(vinylbutyral)

butyrophenone: $C_6H_5COCH_2CH_2CH_3$; butyrophenone derivatives form a class of tranquilizers

BUX: *m*–$CH_3NHC(O)OC_6H_4CH(CH_3)CH_2$ CH_2CH_3; soil insecticide

C

cacodylic acid: $(CH_3)_2As(O)OH$; arsenical pesticide

cadmium: Cd; metallic element; isolated from zinc ores; use: as component in solders and brazing alloys; cadmium pigments have brilliant colors and are stable at high temperatures so that they can be incorporated into plastics at high processing temperatures; colors range from yellow (CdS) to red (Cd[S,Se]); cadmium salts are toxic but because the pigments are very insoluble their oral LD_{50} is >10 g/kg rat

caffeine: central nervous system stimulant; present in coffee (110 mg/cup), tea (24 mg/cup [3-minute brew]), soft drinks (20–46 mg/12 fl oz can), milk chocolate (24 mg/4 oz bar), in drug formulations (pain relievers, cold medicines, diuretics) (30–200 mg); adverse effects arise with doses >1000mg; *see* alkaloids

calcination: prolonged heating at high temperature to remove moisture and increase hardness, e.g., magnesium carbonate to MgO with elimination of CO_2

calcium: Ca; metallic element; produced by reduction of calcium oxide with aluminum at 1200°C; use: reducing agent, in maintainance-free batteries; presence of Ca^{2+} results in "hard water" in which soaps precipitate and causes the deposit of scale in boilers and pipes; because Ca^{2+} ions regulate biological cell membrane potentials and serve as cellular messengers, their activity plays an important role in the design of drugs

calcium carbide: CaC_2; produced by reaction of pure calcium oxide with coke in an electric arc-reducing furnace; use: for manufacture of acetylene (C_2H_2), calcium cyanamide ($CaCN_2$), and calcium hydride (CaH_2); for desulfurization of iron during steel manufacture

calcium carbonate: $CaCO_3$; produced by mining the minerals aragonite, calcite, chalk, limestone, lithographic stone, marble, marl, travertine, whiting; use: in the construction and cement industry, for manufacture of quicklime and fertilizer, for desulfurizing flue gases, as filler and additive for paints, plastics, plastisols, and paper

calcium chloride: $CaCl_2$; produced as by-product in the Solvay process and in the chlorhydrin process for manufacture of propylene oxide, isolated from waste HCl by reaction with limestone; use: drying agent, for road de-icing, as additive to concrete and to oil drilling mud

calcium cyanamide: $CaCN_2$; produced by heating calcium carbide under a N_2 atmosphere; use: fertilizer, pesticide, raw material for synthesis of nitrogen compounds

calcium fluoride: CaF_2; fluorspar; obtained by mining; use: for production of fluorine

calcium hydride: CaH_2; produced from calcium carbide and hydrogen under pressure; potential use: supplying hydrogen to vehicular engines

calcium hydroxide: $Ca(OH)_2$; slaked lime, lime hydrate; produced by slow addition of water to calcium oxide (lime slaking); wet slaking produces slurries used for mortar manufacture, dry slaking uses twice the stoichiometric amount of water in a heated vessel resulting in dry calcium hydroxide particles; *see* calcium oxide

calcium hypochlorite: $Ca(OCl)_2$; produced by chlorination of calcium hydroxide; used as bleach

calcium oxide: CaO; quicklime; produced by calcining lime stone at 1000–1200°C; use: for removing phosphorus and sulfur from molten metal, for manufacture of calcium compounds, for making slaked lime (calcium hydroxide), for desulfurizing flue gas; widely used in the chemicals, glass, sugar, paper, and construction industries

calcium phosphates: $CaHPO_4$, $CaH_4(PO_4)_2$; manufactured by reaction of lime (CaO) with phosphoric acid; use: in fertilizers and animal feed, as baking powder in the food industry, as cleansing agent in toothpastes

calcium sulfate: $CaSO_4$; gypsum, anhydrite; mined from natural sources, produced as by-product of phosphoric acid manufacture by reaction of apatite ($Ca_3(PO_4)_2$) with sulfuric acid; use: as filler and pigment extender; *see* cement

calandria pans: heating elements in a vacuum evaporator

calender: machine used to produce smooth, uniform sheeting and to impregnate or coat fabrics; machine consists of two counter-rotating heated rollers

calender grain: internal orientation resulting from calendering

caliche: *see* iodine

calorie: energy unit: 1 calorie is needed to raise the temperature of 1 mL of water from 14.5–15.5°C

cam wound package: package of yarn that carries zero twist; for twisted yarn *see* ring spinning

Canada balsam: a natural resin

Candida: a group of yeast species (e.g., *C. acidothermophilum, C. ethanothermophilum, C. utilis*) used to prepare single cell protein (SCP) from ethanol. *C. brumptii* are used to produce citric acid from glucose; *C. hydrocarbofumarica* are used to make α-ketoglutaric acid from paraffin

candle filter: in wet spinning the final filter before the polymer solution emerges from the spinerette

cannabis: Indian hemp; marijuana, pot; contains mood changing chemicals; a controlled substance in the United States

CANUTEC: Canadian Chemical Emergency Service, Telephone numbers (1) Information (613) 992 4626, (2) Emergency (613) 996 666

capillary electrophoresis: instrumentation capable of rapid and highly efficient separation of samples that are placed inside capillary tubes that are filled with a gel or buffer; protein separations with efficiencies greater than a million plates have been achieved

capping: in elastomeric copolymers the formation of hard groups between soft segments

by reacting the prepolymer with short-chain glycols or diamines (*see* urethane block polymers), in acetals like POM to stabilize with end groups

caprolactam: (1) from cyclohexanone and hydroxylamine (Raschig or DSM HPO processes) to cyclohexanone oxime followed by a Beckmann rearrangement from the oxime with oleum (H_2SO_4) to the lactam; (2) by reduction of benzoic acid followed by reaction with nitrosylsulfuric acid ($NOHSO_4$) (Snia Viscosa process); (3) by peracetic acid oxidation of cyclohexanone to lactone followed by reaction with ammonia (UCC process); primarily used for manufacture of nylon 6

CAPROLACTAM

captan: fungicide; produced from tetrahydrophthalimide; acts by disrupting the fungal cell wall and by inhibiting enzymes

CAPTAN

captofol: systemic fungicide

captopril: blood pressure controlling drug; acts by inhibiting the production of angiotensin-converting-enzyme (ACE) that is essential to the body's ability to increase blood pressure

CAPTOPRIL

carbamate pesticides: group of agents having the general structure R'OC(O)*N*(R")R'";

these pesticides interfere with the enzyme cholinesterase and thus disturb the nerve impulses of the insects

carbamazepine: anticonvulsant; drug of choice for partial seizures

CARBAMAZEPINE

carbamic acid esters: NH_2COOR; *see* carbamate pesticides; isocyanates

carbaryl: produced by reaction of 1-naphthol and methyl isocyanate; a broad-spectrum pesticide for agricultural and garden use

CARBARYL

carbides: binary carbon compounds such as SiC, boron carbide (B_4C), and Al_4C_3; these nonoxide ceramics are characterized by hardness and strength and are used in cement and metal composites

carbidopa: used in medication for Parkinson's disease; *see* levodopa-carbidopa mixture

carbofuran: carbamate insecticide

CARBOFURAN

carbohydrases: amylases, enzymes that hydrolyze starch; *see* enzymes

carbohydrates: *see* saccharides

carbolic acid: *see* phenol

carbon: C; element; *see* diamond; graphite; buckminsterfullerene

carbonaceous: containing carbon

carbonado: *see* diamond

carbonate-apatite: a phosphate mineral

carbonatite: alkaline rock associated with apatite

carbon blacks: amorphous powdered carbon produced by incomplete combustion of hydrocarbon gases or oils; generic term includes acetylene black, bone blacks, furnace black, lamp blacks, channel or impingement, and thermal blacks; use: pigment, filler for plastics and rubber, for manufacture of hard materials and refractories, as oxidation–reduction catalyst

carbon dioxide: CO_2; colorless gas; produced from the combustion of coke (ammonia plants), by calcination of limestone, as by-product in syntheses involving CO or as by-product of fermentation; use: as gas for refrigeration, fire extinguishers, in enhanced oil production, beverages, chemical manufacture, fumigant for stored grain; solid CO_2 ("dry ice") is used as refrigerant

carbon disulfide: CS_2; produced by reaction of charcoal with liquid sulfur at 720–750°C; use: as reagent and solvent in the chemical industry; for manufacture of viscose fiber, cellophane, CCl_4, rubber chemicals, corrosion inhibitors

carbon-fiber-reinforced plastics: CFRP; *see* composites

carbon fibers: obtained from polyacrylonitrile (PAN) fibers or extruded pitch by heating under controlled conditions to between 1200 and 3000°C; when grown in the vapor phase whiskers are produced; use: for manufacture of high-strength or high-modulus, heat-resistant fibers

carbonization: heating of carbon-rich materials in the absence of air; carbonization of product gases during gasification of coal can cause damage to structures due to reaction of

91

carbon with metals; ease of carbonization of polymers increases their fire retardance; *see* destructive distillation

carbon monoxide shift: water shift reaction, shift conversion; the catalytic conversion of CO to CO_2 and H_2 in the presence of steam

carbon steel: iron alloy containing carbon; low carbon steels (mild steels) contain ≤0.30% carbon; medium carbon steels contain 0.31-0.55% carbon and high carbon steels contain ≤1% carbon

carbon trichloride: *see* hexachloroethane

carbonylation: introduction of a carbonyl group (C=O) into a molecule, e.g., production of Koch acids

carbonyl compounds: compounds that contain –C=O, e.g., aldehydes, ketones

carbonylchloro-*bis*(triphenylphosphine)rhodium (I): $RhCl(CO)(P[C_6H_5]_3)_2$; homogeneous catalyst; *see* carbonyls

carbonylhydro-*tris*(triphenylphosphine)rhodium (I): $RhH(CO)(P[C_6H_5]_3)_3$; homogeneous catalyst; *see* carbonyls

carbonyls: derivatives of CO that is an important π-acceptor ligand; most transition metals form metal carbonyls, e.g., $Fe(CO)_5$, $Ni(CO)_4$, and $W(CO)_6$; they are generally produced from metal salts by treatment with CO under high pressure in the presence of reducing agents; use: homogeneous catalysts for reactions such as olefin isomerization, coupling and cyclization, and the water-gas shift reaction; transition metal complexes involving carbonyls and other ligands such as phosphines are widely used; for example $RhCl(CO)(P[C_6H_5]_3)_2$, $(Rh(CO)_2Cl)_2$ and $RhH(CO)(P[C_6H_5]_3)_3$ catalyze hydroformylations and hydrosilylations of olefins; carbonylations ($H_2C=CHCH_2NH_2$ to the cyclic five-membered amide, nitroalkanes to isocyanates); Fischer–Tropsch synthesis; the water–gas shift reaction; homologation of methanol to ethanol in the presence of CO

and hydrogen; oxidation of 1,2-olefin to 2-ketone; polymerization of the methacrylate to its polymer

carboxin: systemic fungicide, DCMO

CARBOXIN

carboxymethylated cotton: CM; produced by treating cotton fiber first with monochloroacetic acid followed by strong aqueous NaOH; material is stiff and crease resistant

carboxymethylcellulose: CMC; produced by treating cellulose with chloroacetic acid; use: surfactant additive to prevent redeposition of dirt after washing in the machine, protective coating for paints, temporary binder for glazes

carboxymethyl oxysuccinate: CMOS; used by Lever Brothers Co. as builder in detergent formulation

carburization: *see* carburizing

carburizing: carburization; dissolution of carbon in molten metal

carcinogens: cancer-causing agents; one distinguishes physical (e.g., radiation, asbestos fibers), viral and chemical carcinogens; cancers arise from aberrations in cellular DNA; chemical carcinogenesis involves an electrophilic attack on DNA usually resulting in the alkylation of one of the bases; all chemical carcinogens are electrophiles (primary carcinogens) or are converted to electrophiles in the body (secondary carcinogens); cancers arise from the action of initiators or promoters and are evidence of inefficiency in DNA repair mechanisms

card: in textile manufacture a combing machine for opening and orienting staple fibers; *see* staple fiber processing

carded adhesive bonded nonwoven: *see* nonwoven fabrics

cardiovascular: pertaining to the blood circulating system that encompasses the heart, arteries, and veins

carotenes: carotenoids; naturally occurring conjugated hydrocarbons (polymethine or polyene dyes) and their derivatives; their carbon skeleton is built up from between 6 and 10 isoprene units $[CH_2=C(CH_3)CH=CH_2]$; representative members are β-carotene, vitamin A, and lycopene, the red coloring matter of tomatoes; use: food coloring, nutritional factors, potential anticancer agents

β-CAROTENE

carriers: in dye manufacture, swelling agents that improve dyeing rate of hydrophobic fibers, for example, poly(ethyleneterephthalate)

caryophilline: component of essential oil that has an aroma somewhere between cloves and turpentine; used as standard for flavor testing

cascade: in process design, a sequential system of operating units and/or stages; use: in purifications, i.e., distillations and separation by diffusion; in biochemistry events emanating from one reaction such as the arachidonic acid cascade that leads to prostaglandins and leucotrienes

casein: CS; milk protein; produced from milk by precipitation with acid or the enzyme rennet; use: nutrient

casein plastics: soft plastic materials made by the action of formaldehyde on casein; used for textiles and ornamentals

casing head gasoline: *see* naphtha

castile: mild soap made from vegetable oils

casting: in polymer technology the production of films or sheets generally from concentrated polymer solutions through evaporation of solvent, from melts or by in situ polymerization

cast iron: iron alloys containing carbon and silicon; cast irons contain 2–4% carbon

castor oil: viscous oil derived from the castor plant; consists of 87% ricinoleic, 7% oleic, 3% linolenic, 2% palmitic, and 1% stearic acid; use: as raw material for paints, coatings, lubricants, medicinals and cosmetics

catalase: oxidizing enzyme that decomposes hydrogen peroxide; use: in milk production to remove excess peroxide if it had been used as sterilizing agent

catalysis: *see* catalysts

catalyst deactivation: catalysts decay during the course of a process through poisoning, fouling, heat or loss of material; poisoning results from strong chemisorption of reactants, products or impurities on the active catalyst surface sites; fouling results from the physical deposition of materials from the liquid phase onto the catalyst surface and in its pores; thermal degradation generally is due to sintering; loss of catalyst usually occurs through emission of vapors; catalyst deactivation can be prevented or slowed with use of pure material and additives that absorb the poison and by reduction of temperature

catalyst poisons: substances that destroy catalyst activity by occupying active catalyst sites; *see* catalyst deactivation

catalysts: agents that permit reactions to proceed under milder conditions and that are recycled in the process; catalysts may be homogeneous, i.e., dissolved in the reaction medium, or heterogeneous, i.e., insoluble in the medium; catalysts may be attached to (supported on) an insoluble surface such as a polymer, mineral, film or biological cell (immobilized catalysts); catalytic activity depends on the catalyst as well as the catalyst support; heterogeneous and supported catalysts are more easily removed from the reaction medium for recycling than homogeneous catalysts; phase-transfer catalysts permit catalytic activity of inorganic ions in organic media through use of, for example, lipid-soluble quaternary ammonium salts and crown ethers that facilitate the transport of the inorganic ions into the organic phase; *see* catalysts, industrial; enzymes

catalysts, industrial: most heterogeneous industrial catalysts are metals or metal oxides, most homogeneous catalysts are transition metal complexes or carbonyls; major uses are in petroleum refining, chemical manufacturing and emission control (automotive and to a lesser extent industrial); refinery catalysts are used (in order of decreasing dollar value) for catalytic cracking, alkylation, hydrotreating, hydrocracking and reforming; consumption for industrial processes (in order of decreasing dollar value) is for polymerization, organic synthesis, oxidation, syngas, hydrogenation and dehydrogenation; for emission control catalysts, *see* catalytic converter

catalytic converter: in the automotive industry, device that reduces the emission of carbon monoxide CO, nitrogen oxides (NO_x), and hydrocarbons (HC) as well as the small amounts of SO_2, lead, and phosphorus that appear in the exhaust; the ceramic alumina containing catalyst support consists of pellets or a thin-walled honeycomb monolith that provide large surfaces; noble metal catalysts used are Pt/Pd/Rh; addition of cerium oxide to the catalyst improves catalyst activity at lower temperature; addition of Ni serves as scavenger for sulfur, which is a catalyst poison

catalytic cracking: cat cracking; fluid catalytic cracking, FCC; a petroleum refinery process that uses acid catalysts to convert higher boiling distillation fractions into saturated branched aliphatics, cyclohexane and aromatics; catalysts include Cr_2O_3, MnO and crystalline silicon aluminates (zeolites); process temperatures are 450–500°C; used primarily for production of gasoline

catalytic reforming: process to improve the octane number of the naphtha fraction of petroleum by converting paraffins and naphthenes to aromatics; the reactions occur at high temperature; to prevent cracking and coke formation the reactions are run in the presence of bimetallic platinum containing catalysts and hydrogen-rich gas

catalytic supports: solids onto which catalysts are attached; supports include minerals, polymers, biological cell membranes; the supports as well as the means of binding the catalyst to the support affect catalytic activity; an advantage of supported catalysts is their easy recovery

cat cracking: *see* catalytic cracking

catechol: 1,2-dihydroxybenzene; derived from plants or produced from coal tar; used as tanning agent

cathartics: *see* laxatives

cathode: the electrode in a battery or electrolytic cell at which reduction occurs, e.g., during electrolysis of brine hydrogen is liberated at the cathode

cathodic protection: *see* corrosion

catholyte: electrolyte near or at the cathode of an electrolytic cell; the catholyte is reduced during the reaction

cationic dyes: *see* basic dyes

cationic surfactants: specialty surfactants consisting of a hydrophobic group and a positively charged nitrogen moiety; most cationic surfactants are quaternary ammonium ions such as alkyldimethylbenzyl ammonium chloride; use: fabric softeners, disinfecting and sanitizing cleaners, antistatic agents; not used as detergents per se caustic: *see* caustic soda

caustic ash: mixture of 70% NaOH (caustic soda) and 30% Na_2CO_3 (soda ash)

caustic lime: quick lime; CaO; produced by heating limestone ($CaCO_3$) to 1000°F to drive off the CO_2

caustic magnesite: impure MgO that contains a smalll amount of $MgCO_3$; produced by calcining $MgCO_3$ at 700–1200°F

caustic potash: KOH; use: scouring bath for metals, for making soft soaps, for bleaching textiles

caustic soda: NaOH; by-product in the chlorine manufacture using the chlor-alkali process; use: during conversion of bauxite into aluminum, for neutralization of waste streams in chemical manufacture, scouring baths for metals, in chemical manufacturing, petroleum processing, pulp and paper industry, soap and detergent manufacture

cefazolin: cephalosporin antibiotic used in the hospital

cefoxitin: cephalosporin antibiotic used in the hospital

ceiling temperature: in polymer technology, the temperature above which polymerization does not occur and where the polymer decomposes

cell cycle: in biotechnology, the period between one cell division and the next; during that period the chromosomes double, separate, and finally enter the daughter cells produced during cell division; misfunction of the process may result in cell death or in the uninhibited growth typical of cancer cells

cellophane: regenerated cellulose film; used for packaging; cellulose-coated fabrics are used for household articles

cellulases: enzymes that hydrolyze cellulose; produced by numerous molds (fungi); the most promising for commercial utilization are *Aspergillus* and *Thermoderma reesei*

celluloid: *see* nitrocellulose

cellulose: $(C_6H_{10}O_5)_n$; a long-chained natural polymer consisting of glucose monomer; the structural constituent of plants; produced from wood by removal of lignin during pulping processes; indigestible by humans but digestible by animals that harbor cellobiase-producing bacteria in their gut; use: in foods to provide body and gel stability, the textile and chemical industry to produce regenerated cellulose and cellulose derivatives such as cellulose acetate and cellulose nitrate

cellulose acetate: CA; esters of cellulose having various degrees of esterification; produced by treating wood pulp or cotton linters with acetic acid and acetic anhydride in the presence of sulfuric acid; the heterogeneous reaction leads initially to the triester that is then hydrolyzed to the more soluble acetate (2 to 2.5 acetyl group substitution); mixed acetate–butyrate and mixed acetate-propionate are made by adding the appropriate acids to the acetylating mixture; use: thermoplastic resins, fibers and films; binder for pigments and coatings; cellulose acetate fibers are dry spun from acetone/water solutions; these silk-like fibers are used in apparel fabrics (need dry cleaning) and cigarette filters; transparent acetate film is made by casting from CA solutions; use: photographic film, laminates, pressure-sensitive tape, packaging; in cellulose triacetate fibers (CTA) ≥92% of the OH groups are acetylated; synthesis needs perchloric or sulfuric acid as catalyst; fibers are dry-spun from methylene chloride–methanol solution to give crystallizable yarns for apparel fabrics; these can be heat-set and are water washable with "ease of care" properties

cellulose methyl ether: *see* methyl cellulose

cellulose nitrate: CN; *see* nitrocellulose

cellulose nitrate process: obsolete process for producing rayon fibers; process was invented by de Chardonnet

cellulose xanthate: cellulose–O–C(SNa)=S; produced by first boiling wood pulp with caustic soda and after allowing it to age for 2–3 days mixing the soda cellulose with carbon disulfide in an air-tight xanthating churn; cellulose xanthate is dissolved in dilute NaOH to viscose solution and then converted to regenerated cellulose (rayon) by extruding this solution into an acidic bath to form viscose rayon yarns

cellulosics: plastics and fibers based on cellulose

cement: any powdered material that becomes plastic when wetted with water; hydraulic material that can be molded or poured and sets into a hard mass; in the building industry a building material consisting of CaO-SiO$_2$-Al$_2$O$_3$ in varying proportions, e.g., Portland cement consists of "Portland clinkers" and CaSO$_4$ (gypsum or anhydrite); used as binder in concrete; in rubber technology, a solution of rubber in hydrocarbon

centipoise: measure of viscosity; *see* poise

centralite I: [C$_2$H$_5$(C$_6$H$_5$)N]$_2$CO; diethyldiphenylurea; stabilizer for gun powder, especially in nitroglycerin powders; also centralite II (dimethyldiphenylurea) and centralite III (ethylmethyldiphenylurea)

centrifugation: separation process for materials with different densities or removal of moisture from solids using centrifugal force; industrial centrifuges are designed to provide accelerations to ~20,000 times gravity; laboratory centrifuges reach ~300,000 g and ultracentrifuges that are used for biological materials go as high as 500,000 g

cephalosporins: group of antibacterial antibiotics; cephalosporin C that has modest antibiotic activity was first isolated from seawater near a sewer discharge in Sardinia; im-

proved fermentation methods yield cephalosporin C with higher antibiotic titer; the semisynthetic cyclosporins cefoxitin and cefazolin were the most frequently used hospital drugs in 1987; cephalosporins are resistant to the penicillin-hydrolyzing enzyme penicillinase and are useful for treating patients allergic to penicillin

CEPHALOSPORIN C

ceramics: nonmetallic inorganic solids; properties depend on the chemical composition (silicates, oxides, nonoxides), on the homogeneity (coarse: pore or crystallite size >0.2 mm; fine: pore or crystallite size <0.2 mm), and the porosity; produced by physical processing of raw materials (grinding, mixing) followed by molding, drying, firing, and finishing (mechanical processing, glazing); silicate ceramics consist of mixtures of clays, quartz, and feldspar; they are used for building materials, fine earthenware and porcelain, consumer goods, insulators, bricks, fireclay, and refractories; most oxide ceramics consist of a single crystalline phase of a metal oxide; pure oxides must be manufactured by chemical processes and the ceramic is then made by molding, drying and sintering; oxide ceramics must be plasticized with binders and plasticizers before they can be cast; they are solidified by sintering 200–300°C below their melting points; most oxide ceramics have high temperature and chemical resistance, as well as good electrical resistivity that determines their general utility; specific uses depend on their chemical composition (e.g., zirconia is used as neutron reflector in nuclear reactors, BaTiO$_3$ has ferroelectric and piezoelectric properties); the most important nonoxide ceramics are silicon carbide (SiC), silicon nitride (Si_3N_4), and boron nitride (BN); carbides are produced by reaction of the oxides with carbon at high temperatures (*see* Acheson process), the nitrides by reaction of the oxide with ammonia; nonoxide ceramics are stable at high temperature but unstable in the presence of oxygen; use: in composites and engineering ceramics; *see* superconductors

ceramics, engineering: *see* engineering ceramics

ceramics, superconducting: *see* superconductors

ceresan L: pesticide containing mercury

cermets: composites of ceramics and metals having properties intermediate between the two; they are hard, stable at high temperature and malleable

chafer strips: in tire manufacture, protection of wire reinforcement where the tire touches the rim

chain branching: *see* branched polymer

chain polymerization: *see* addition polymerization

chain reaction: reaction that proceeds by creating new reactive species; examples are neutrons during nuclear fission or free radicals during chain polymerization

chain termination: interruption of a chain reaction; in polymerization termination occurs through chain transfer, the interactions of two radicals or use of an inhibitor

chain transfer: in polymerization reactions, a process in which a polymer chain stops growing by abstracting a group or atom from the solvent, an additive, or another polymer; if a hydrogen atom is abstracted from another polymer chain it leads to branching of that chain; if the hydrogen is abstracted from an additive the resulting radical may dimerize (e.g., mercaptans) or not be active enough to initiate further polymerization (e.g., diphenols)

chain transfer agents: regulators; additives used to control the molecular weight of polymers or to prevent polymerization; *see* chain transfer

chamber acid: H_2SO_4 made formerly by the lead chamber process; process is obsolete

channel black: *see* carbon blacks

char: charcoal; charred material resulting during destructive distillation and during coal gasification

charcoal: amorphous form of coal; made by the destructive distillation of wood; use: fuel, decolorizer, deodorant and purifier

Chardonnet silk: nitrocellulose rayon fiber; term is no longer used; process is obsolete

charge: amount of material used in each fabrication cycle, e.g., polymer, metal or clay

Charpy impact test: notched bar lying on a split anvil is broken by a pendulum–hammer; the rise of the pendulum after breaking the bar is used to measure the impact strength of the material

charred peat: partially heat decomposed peat; *see* peat; soil conditioners

chelates: complex ions that consist of a metal ion bonded to ligands (sequestering agents, electron-pair donors, Lewis bases); chelation is used for removal or solubilizing of metal ions (e.g., use of cyanides in gold mining operations); many complex ions serve as catalysts (e.g., carbonyls, phosphines); chelatable dyes use metal ions (e.g., Ni^{+2}) that have been incorporated into the fiber as dye sites

chelating agents: sequestering agents; *see* chelates

chemical fibers: synthetics such as nylons, polyesters, acrylics

chemical oxygen demand: COD; measure of water purity in waste treatment; defined as the amount of oxygen needed to oxidize inorganic and organic materials contained in water

chemical pulping: *see* pulping

chemical spinning: *see* reaction spinning

chemical vapor deposition: CVD; metal vapor deposition; gas plating; pyrolytic plating; production of a metallic coating by chemical reaction of vaporized metallic salts on the surface of the object to be coated; the gases consist of vaporized metallic salts and a carrier gas, the object to be coated must be preheated and the process is carried out in a vacuum

chemical wet spinning: *see* reaction spinning

chemiluminescence: emission of light at ambient temperature due to chemical reactions; use: in emergency lighting, for monitoring pollutants

chemimechanical pulping: *see* pulping

chemisorption: formation of strong bonds between a gas or liquid and a solid surface; unlike in adsorption the bond strength is comparable to covalent bonding

Chemithon process: in the detergent industry, manufacture of alkylaryl sulfonates using a continuous air–SO_3 sulfonator

chemmod cotton: cotton that has been chemically modified, e.g., by acetylation, cyanoethylation, or esterification; treatments lead to improved properties such as resistance to heat, chemicals, or fire

chemotherapeutics: pharmaceuticals; drugs

CHEMTREC: chemical emergency service in the United States provided by the American Chemical Manufacturers Association telephone numbers (1) information: (800) 282-8200 (2) emergency (800) 424-9300

chemurgy: branch of chemistry that covers the industrial utilization of raw materials

chicle: chewing gum derived from latex of the Central and South American trees *Achras zapota* and *Achras chicle china*

Chile saltpeter: *see* sodium nitrate

China clay: *see* kaolin

China wood oil: *see* tung oil

Chinese blue: *see* ferrocyanate pigments

Chinese white alumina: *see* kaolin

chirality: property of molecules that have an asymmetric configuration so that they can exist as structural isomers; chiral molecules have the charateristic ability to rotate plane polarized light clockwise or counterclockwise; *see* isomers

chitin: polymer isolated from exoskeletons of shellfish and insects; use: in wound-healing formulations; deacetylated chitin (chitosan) is used in photographic emulsions, removal of heavy metals from water, to improve dyeability of fabrics

CHITIN

chitosan: *see* chitin

chloral: CCl_3CHO; produced by chlorination of acetaldehyde or by treating ethanol with NaOCl; use: organic synthesis (notably for DDT), a hypnotic

chlor-alkali: chlorine (Cl_2) and caustic soda (NaOH)

chlor-alkali process: production of chlorine and NaOH by electrolysis of NaCl brine or of NaCl derived from seawater by solar evaporation; in a diaphragm or a membrane cell the anode and cathode compartments are separated; in a mercury cell the cathode is a thin

Hg film that is not separated from the anode; the mercury process gives pure chlorine, whereas chlorine from the other two processes contains oxygen; the diaphragm process has the least stringent requirements for pure feed; the membrane process has the lowest energy requirements

chloramine: *see* hydrazine

chloramphenicol: chloromycetin, CAP; antibiotic used against urinary and *Salmonella* infections

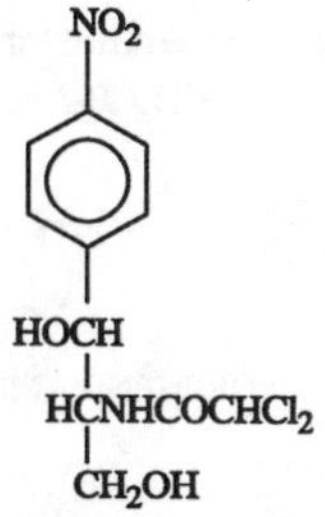

CHLORAMPHENICOL

chlordane: insecticide, herbicide; use restricted in the United States; hazardous chemical waste (EPA)

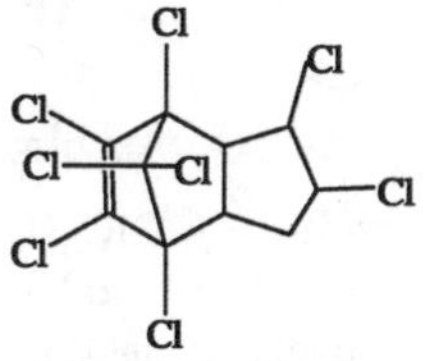

CHLORDANE

chlordiazoepoxide: antianxiety drug, tranquilizer

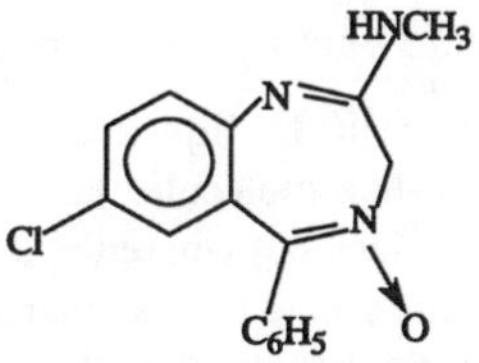

CHLORDIAZOEPOXIDE

chlorella: green algae; use: energy research, food source

chlorendic anhydride: produced from hexachlorocyclopentadiene and maleic anhydride by Diels–Alder condensation; use: fire retardant for polyesters

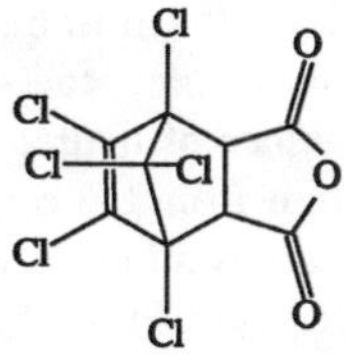

CHLORENDIC ANHYDRIDE

chlorinated camphene: *see* toxaphene

chlorinated hydrocarbons: chlorinated paraffins are manufactured by chlorination with Cl_2; chlorinated and brominated paraffins are used as flame retardants for polymer products

chlorinated isocyanurates: use: dry chlorine bleaches (usually as Na and K salts), present in dishwasher detergent formulations

chlorinated trisodium phosphate: chlorinated TSP; a dry chlorine bleach

chlorinator: in waste treatment, a device for adding chlorine to waste water

chlorine: Cl_2; gaseous element; manufactured by electrolysis of NaCl (*see* chlor-alkali process), also by catalytic (Cu) oxidation of HCl; strong oxidizing agent; use: chemical manufacture, paper and pulp industry, production of $TiCl_4$, water treatment, bleach, disinfectant

chlorine bleach: *see* bleach

chlorine-contact chamber: in waste water treatment, the container in which the effluent is exposed to chlorine before the water is released

chlorine dioxide: Cl_2O; bleaching agent used in paper manufacture

chlorodicarbonylrhodium(I) dimer: (Rh$(CO)_2Cl)_2$; *see* carbonyls

chlorofluorocarbons: CFC; $C_xCl_yF_z$; re-

99

chloroformates

frigerants; formerly used as propellants; use discontinued because CFCs are believed to be responsible for the destruction of stratospheric ozone; nomenclature: $C_xCl_yF_z$ (the final digit gives the number of fluorine atoms in the molecule; the second to last digit is the number of hydrogen atoms plus 1; the third digit from the right is the number of carbon atoms minus 1, thus CCl_2F_2 is CFC-12, C_2ClF_5 is CFC-115); for halons, related compounds that also contain bromine the numbering system simply counts the atoms: CF_2ClBr is halon-1211, $C_2F_4Br_2$ is halon-2402 indicating the absence of chlorine

chloroformates: ClCOOR; produced by reaction of phosgene with alcohols; use: chemical reagent, for the synthesis of urethanes

chlorohydrin process: manufacture of propylene oxide

chloromethane: *see* methyl chloride

chlorophenols: *see* flame retardants

chloropicrin: CCl_3NO_2; soil insecticide, soil fumigant, oxidizing reactant for synthesis of triaryl methine dyes, e.g., methyl violet

chloroprene: $CH_2=C(Cl)-CH=CH_2$; manufactured by the catalytic conversion of acetylene to monovinylacetylene, $CH\equiv C-CH=CH_2$, which is then reacted with HCl in the presence of cuprous chloride solution; use: as monomer

chlorostyrene: *see* flame retardants

chlorosucrose: nonnutritive sweetener

chlorothalonil: systemic fungicide

chlorothiazide: diuretic agent; used to reduce body fluid volume by increasing urine excretion; lowers blood pressure; the 3,4-dihydro derivative, hydrochlorothiazide, that is produced by reduction of chlorothiazide with alkaline formaldehyde, was the most frequently prescribed drug in 1987

CHLOROTHIAZIDE

chlorotrifluoroethylene: monomer; *see* fluorocarbon polymers

chlorotris(triphenylphosphine)rhodium(I): $RhCl[P(C_6H_5)_3]_3$; homogeneous catalyst; *see* catalysts, industrial

chlorpheniramine: antihistamine, H_1 blocker

CHLORPHENIRAMINE

chlorpromazine: antipsychotic agent; the first phenothiazine tranquilizer; its introduction in the 1950s resulted in an approximately 65% drop of patients in mental institutions

CHLORPROMAZINE

chlorpropamide: $p\text{-}ClC_6H_4SO_2NHCON$ $HCH_2CH_2CH_3$; oral blood sugar-reducing (hypoglycemic) agent; antidiabetes drug

chlorosulfonic acid: $ClSO_3H$; produced by treating SO_3 with HCl; use: sulfation agent for the manufacture of detergents

chlorsulfuron: sulfonylurea herbicide

cholesteric crystal: liquid crystal phase, where the rod-like molecules have a nematic structure in a plane but where the direction of the rods changes in a regular manner due to the asymmetric structure of the molecules; *see* liquid crystals

cholesterol: naturally occurring steroid alcohol; essential constituent of cell membranes; implicated in artherosclerosis and heart disease

CHOLESTEROL

chromated zinc chloride: fire retardant consisting of 20% CrO_3 and 80% ZnO

chromatin: structure within the cell nucleus that forms during cell division and consists of DNA and tightly bound small proteins; this material can be dyed and observed under the light microscope; *see* histones

chromatography: separation, detection, and preparative techniques based on the selective adsorption of the components of complex mixtures; used in the laboratory and in industrial production; during the process a mobile phase (liquid or gas) passes over a stationary phase (solid or liquid dispersed over a solid); differences in the continuous adsorption and desorption lead to the separation of components in the mobile phase; the sorbant bed consisting of finely divided material may be packed in a column (column chromatography) or spread on a surface (thin layer chromatography, TLC); paper may serve as sorbant (paper chromatography); process chromatography is completely automated and under computer control; *see* column chromatographic methods; for additional details; *see* affinity chromatography, annular liquid chromatography; gas–liquid chromatography; gas–solid chromatography

chrome dyes: mordant acid dyes that contain chromium complexes; characterized by improved fastness; *see* dyes

chrome greens: inorganic pigments: blends of phthalocyanine blue and chrome yellow

chrome tanning agents: basic chromium (III) salts; produced by reduction of chromates with molasses or with SO_2; used to cross-link collagen fibers in hides

chrome yellow: $PbCrO_4$ or $Pb(Cr,S)O_4$; the mixed sulfate is produced by mixed phase formation of $PbCrO_4$ and $\leq 50\%$ $PbSO_4$; *see* ocher

chromium oxide: Cr_2O_3; produced by reduction of sodium dichromate using carbon or sulfur; use: green pigment, abrasive, oxidizing agent (SO_2 to SO_3, CO to CO_2, methanol to formaldehyde); widely used as industrial heterogeneous catalyst (e.g., for alkylation of phenol with methanol to give *o*- and di-*o* methyl products); for the aromatization of 1,3,5 hexatriene to aromatic products; for the decomposition of hydrazine; for dehydrogenation of ethylbenzene to styrene, formation of diolefins from olefins; for Fischer–Tropsch synthesis of aromatics; for oxidations, e.g., of CO, SO_2; for reforming of butyl alcohol to thiophene in the presence of carbon disulfide; for the carbon monoxide shift reaction

chromocene: chromium containing sandwich compound; *see* ferrocene

chromogen: *see* chromophore

chromophore: chromogen; the structural unit responsible for a molecule's color; chromophores absorb light in the visible spectrum and are always unsaturated; *see* dyes

chrysolite asbestos: commercial asbestos fiber; *see* asbestos

-cide: suffix denoting to kill, e.g., pesticide

cimetidine: antiulcer agent; antihistamine that blocks the secretion of HCl in the stomach; H_2 blocker

$CH_3NHCNHCH_2CH_2SCH_2$
‖
NCN

CIMETIDINE

circulating generator: trickling generator; used for vinegar manufacture; the generator consists of a wooden tank that contains birchwood shavings in its upper level; air is pumped through an ethanol–water–vinegar mixture containing *Acetobacter*; the mixture is agitated and vinegar trickles through the wood shavings to a receiver from which it is drawn off

cisplatin: $Pt(Cl)_2(NH_3)_2$; the square planar *cis* isomer is an anticancer drug that is believed to act by intercalation in the DNA of the cancer cell

citric acid: $HOOCCH_2C(OH)(COOH)CH_2COOH$; industrial production from sucrose using the mold *Aspergillus niger*; used in the food, beverage, cosmetic, polymer, and chemical industries; antioxidant, blood anticoagulant, emulsifier, buffer, detergent additive, plasticizer, and for secondary oil recovery

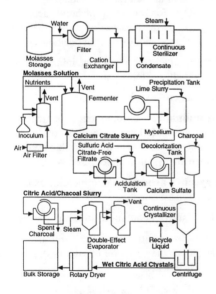

CITRIC ACID PROCESS FLOWSHEET.
[Copyright *Riegel's Handbook of Industrial Chemistry,* 8th ed., Van Nostrand-Reinhold Co., Inc.]

clarification: removal of turbidity and suspended materials by settling or centrifugation

class A explosives: detonating explosives, maximum hazard explosives such as dynamite, nitroglycerin, and lead azide; use: for detonating primers and blasting caps

class B explosives: flammable explosives; use: as propellants, photographic flash powders, fireworks

class C explosives: explosives that contain limited amounts of class A and class B explosives

Claude principle: expansion engine cycle for the liquefaction of gases; expansion occurs in the engine that does useful work and is more efficient than Joule-Thompson expansion; liquid is not produced during main expansion phase but in a series of condensers and expansion valves

Claus process: conversion of hydrogen sulfide to elemental sulfur by oxidation with a stoichiometric amount of air/oxygen (~30% oxygen); use: emission control of refinery and coking oven gases, sulfur recovery

clay: crystalline or amorphous silicate minerals composed mainly of hydrated alumina and silica, $Al_2O_3 \cdot mSiO_2 \cdot nH_2O$; finely divided material; many clays contain Mg instead of Al; use: for ceramic and brick manufacture; industrial adsorbents, ion exchangers, catalysts, fillers and decolorizing agents

clay ceramics: *see* silicate ceramics

clinker: partially vitrified clay or brick

clofibrate: $p\text{-}ClC_6H_4OC(CH_3)_2COOC_2H_5$; a hypolipemic, i.e., a drug that reduces blood cholesterol and blood lipids

clonidine: antihypertensive drug that reduces blood pressure by action on the central nervous system

CLONIDINE

Clostridium thermoaceticum: microorganism that produces acetic acid from glucose

cloud point: temperature at which a solution separates into two phases; characterized by the sudden appearance of turbidity on cooling as the cloud-point temperature is reached; in polymer technology the temperature at which a polymer begins to precipitate from solution upon cooling

coacervation: coazervation; coexistence of two liquid phases—usually of macromolecules, colloids, or highly concentrated solutions; the phases may contain different solutes or different concentrations of the same solute; a coacervate (coazervate) is a liquid aggregate of colloidal droplets that forms before flocculation

coagulum: product of coagulation, as in rubber latex

coal: an organic rock composed of amorphous carbon, hydrocarbons, organic compounds containing oxygen, nitrogen, sulfur, other heteroatoms and inorganic minerals; it is classified by rank in the following decreasing calorific value: anthracite, bituminous coal, subbituminous coal and lignite; higher rank coals are glossier due to the presence of vitrinite, which was formed from the woody parts of ancient plants; methods for cleaning coal to reduce the organic and inorganic sulfur content are being developed; methods include production of soluble sulfates in aqueous medium using complexing agents, exposure of pulverized coal to a high-gradient magnetic field to separate paramagnetic pyrites and linear accelerator technology to remove ash by centrifugal force; coal reserves in the United States identified from geological evidence are estimated at 1.7 trillion tons, proved recoverable reserves are 237–300 billion tons; use: fuel and chemical feedstock

coal beneficiation: coal treatment; cleansing large coal particles from heavier mineral particles by jigging, washing with high-density liquids, floatation; cleansing fine coal particles by use of cyclones or froth flotation; *see* beneficiation

coal gas: gaseous mixture produced by the destructive distillation of bituminous coal

coal gasification: processes that convert coal into fuel gas (high Btu and/or low Btu gas); the reaction of coal with steam and air yields CO, H_2, CO_2, CH_4, tars, liquids, char and ash; the next step is removal of nongaseous products and H_2S (Claus process); shift conversion then adjusts the CO/H_2 ratio; CO_2 and more sulfur are removed to yield purified syngas (synthesis gas, $CO + H_2$); catalytic methanation (carbon monoxide shift, $CO + H_2 = CH_4 + H_2O$) yields CH_4 and other hydrocarbons; final clean-up gives high Btu gas; individual processes use various reactors, process conditions, and catalysts leading to different primary and secondary products; thus the Fischer–Tropsch process produces syngas that is then converted to liquid fuel; the Lurgi, Synthane, Bigas, and Hygas processes give high Btu gas; Lurgi also produces tar, oils, char, naphtha, and coal fines; Hygas produces naphtha; the CO_2 Acceptor process makes low and high Btu gas; CE process makes low Btu gas and electric power, and the Westinghouse process makes electric power and medium Btu gas. *see* FLOW DIAGRAM FOR COAL GASIFICATION PROCESSES

coal gasifiers: these fall into three main categories (1) fixed bed gasifiers in which the coal moves downward through vertical reactors while the product gases move upward countercurrent (Lurgi); (2) fluidized bed gasifiers in which coal particles are suspended in the upward flow of gases that are introduced at the bottom of the reactor (Winkler); and (3) entrained flow gasifiers in which finely divided coal is entrained in a horizontal flow of reactant gases and the flow of gases and coal is concurrent (Koppers–Kotzek); *see* COAL-GASSIFICATION SYSTEMS

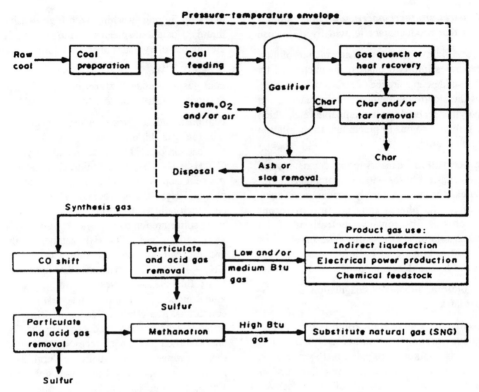

FLOW DIAGRAM FOR COAL GASIFICATION PROCESSES
[Copyright M.B. Bever. *Encyclopedia of Material Science and Engineering,* The MIT Press]

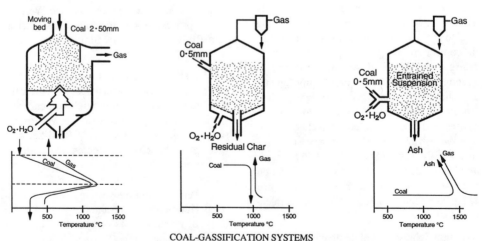

COAL-GASSIFICATION SYSTEMS
[Copyright *Perry's Engineer's Handbook,* 6th ed. (1984), McGraw-Hill, Inc.]

coal liquefaction: direct liquefaction of coal is achieved by pyrolysis at 900–1200°F and ≤1000 psig in the presence of a carrier gas; products differ depending on operating conditions, e.g., the H-Coal process uses a Co-Mo catalyst and yields fuel oil as well as naphtha and gas; the Exxon Donor Solvent (EDS) and the Solvent Refined Coal (SCR I) processes use molecular hydrogen and yield liquid boiler fuel as well as gas, naphtha, and liquid-pressurized gas (LPG); the Solvent Refined Coal II (SCR II) gives solid boiler fuel; and the H-Coal process yields a crude oil mixture (syncrude); indirect liquefaction uses gasified coal as raw material: in the original Fischer–Tropsch process syngas was reacted over an iron or cobalt catalyst at 330°C and 350 psi to produce liquid hydrocarbons and waxes; the process has been modernized particularly by use of different catalysts (e.g., rhodium complexes, transition metal carbonyls, and oxides of Al, Cu, Mo, V, and Zn); the SASOL process, which produces gasoline, exemplifies a modern Fischer–Tropsch synthesis

coal rank: coal is classified by rank based on fixed carbon, volatile matter, and heating value. Ranks in order of decreasing calorific value are: anthracitic, bituminous, subbituminous, and lignitic

coal tar: *see* coal

coal tar creosote: a wood preservative consisting of a mixture of aromatic hydrocarbons (fluorene, anthracene, phenanthrene, and naphthalene) containing ≤5% tar acids (phenols, xylenols, cresols, and naphthols) and ≤5% tar bases (pyridines, quinolines, and acridines); creosote derives its usefulness because it is highly toxic to insects, penetrates wood well, and is water insoluble

coarse filament yarn: yarn weighing 800–1200 denier; use: for reinforcement in tires and conveyor belts

coatings: protective and/or decorative layers; produced by metering a liquid (solution) onto a moving surface, by spraying or by vapor deposition; coatings serve to protect against abrasion, bacteria, corrosion, fire, fouling, gas permeability, static, or weathering (light, mildew, moisture and water, oxidation); they confer wash-and-wear characteristics to fabrics, dissipate heat for reentry space vehicles, reduce surface friction, absorb sound; coating formulations may contain resins in oil-modified alkyds (e.g., diisocyanates, epoxies, nitrocellulose, polystyrene, silicone, urethanes, vinyls), thermosetting resins may be used without a vehicle; anticorrosive marine coatings contain biocidal organotin compounds to protect against fouling; decorative coatings provide color and gloss; widely used in industries including building, industrial equipment, metal, textile, and wood

coazervation: *see* coacervation

cobalt: Co; metallic element; the metal is widely used in alloys and as catalyst, e.g., for the aromatization of hexane; for cracking of methanol to CO and hydrogen; for dehydrosulfurization of naphtha and thiophene; for Fischer–Tropsch synthesis; for hydroformylation of olefins; for hydrogenation of ketones and nitriles; for oxidation of $ArCH_3$ to $ArCOOH$ (*o*-xylene to phthalic anhydride); for hydrogenation of coal liquids

cochineal: red dye stuff (carmine) obtained from dried bodies of insects

cocoa butter: fat derived from the cocoa bean; the major ingredients are the glycerides of oleic, palmitic, and stearic acids; use: confectionary coating

coconut oil: important vegetable oil; its high lauric acid ($CH_3(CH_2)_{10}COOH$) content makes it a good raw material for soap manufacture; coconut oil soap is firm and is resistant to oxidation

codeine: methylmorphine; opiate; use: analgesic, cough suppressant

codon: the sequence of three bases on messenger RNA (mRNA); 61 of the 64 possible triplets encode the message to synthesize one of the amino acids; *see* genetic code, anticodon

COED/COGAS process: fluidized-bed coal gasification technology that uses air at 10 psi in four pyrolysis steps

coenzyme: cofactor; nonprotein moiety essential for catalytic activity; most water-soluble vitamins are components of coenzymes, e.g., members of the vitamin B complex

coenzyme A: CoA; biological cofactor essential for acetylation, e.g., in the metabolism of fatty acids and the biosynthesis of cholesterol

coextruded film: film produced by the simultaneous extrusion of two or more polymers

cofactor: *see* coenzyme

cogeneration: energy conservation method in which process steam of one process is used in a second process; for example, a utility can send steam from its turbines to a nearby plant that then needs not build its own steam generator; or a manufacturing plant can use its process steam to produce electricity for its own use or for sale to a utility; cogeneration is most effective when the two plants are near each other and when the second plant uses large amounts of low-pressure steam

cohesive energy density: CED; term is used in polymer technology; CED equals the square of the solubility parameter; its value is available from thermodynamic data; however, because polymers do not vaporize without decomposition CED is measured by comparative swelling experiments; the solvent with a CED closest to that of the polymer will swell it most or dissolve it completely; *see* solubility parameter

coke: solid residue remaining after devolatization of bituminous coal; produced in fluid coking (Exxon) and flexicoking processes; in fluid coking, the coke particles are sus-

pended in a circulating fluid, in flexicoking part of the coke is gasified; coke has a heating value of 25 million Btu/ton; use: fuel, reducing agent (especially for smelting iron ore)

coking: severe thermal cracking of heavy oil fractions and residuals in which lighter liquids are produced; coke is tolerated as a by-product, is removed in coke drums, and used as fuel

cold drawing: *see* fiber drawing

cold-process soap manufacture: batch process in which melted fat is hydrolyzed by stirring with caustic soda; the resulting solution is run into frames where saponification is completed over several days after which the solidified soap is cut into shapes

cold rubber: copolymer of butadiene and styrene polymerized at 41°F

cold set ink: inks containing plasticized waxes melting at 150–200°F; the inks are solid at room temperature

coliform index: measure of water purity based on the count of bacteria (e.g., *Escherichia coli*) found in the intestinal tract of animals and humans

collagen: polypeptide that occurs in the skin, connective tissue, and bone of mammals; produced by dissolving the inorganic material from bones; use: in photographic emulsions, food casing, gels; *see* tanning

collodion: a solution of pyroxylin in a mixture of alcohol and ether; use: protective film for wounds or burns, agent for differentiating phenol and creosote; *see* pyroxylin

colloids: stable dispersions, having particles in the size range between 1 nm and 1 μm, roughly the size visible under high magnification

collophanite: phosphate rock used for fertilizer production

colorant: materials that absorb visible light;

the resulting color is described by color chroma that identifies the intensity of color, i.e., dull or bright, by color, hue, or tone that describes the individual color, i.e, red, blue, yellow, and by value that describes lightness or darkness

color chroma: *see* colorant

color hue: *see* colorant

color value: *see* colorant

Colour Index: C.I.; classification of dyes used by the Society of Dyers and Colourists (England) and the American Association of Textile Chemists and Colorists; each dye or pigment is given two reference numbers: a generic reference number based on color and a constitution number based on chemical structure; thus C.I. vat dye blue 4 is the generic name of indanthrone; the C.I. constitution number for indanthrone is 69800; *see* dyes

color value: *see* colorant

Column chromatographic methods: when the mobile phase is a liquid (liquid column chromatography [LCC]), separation is due to partitioning between an immiscible mobile and stationary liquid phase (liquid–liquid [or partition] chromatography [LLC]) or due to retentive power of the solid phase (liquid–solid [or adsorption] chromatography [LSC]). When the mobile phase is a gas (gas chromatography [GC]) separation is due to the retentive power of the solid (gas–solid chromatography [GSC]) or liquid stationary phase (gas–liquid chromatography [GLC]). When the stationary phase is a swollen gel, separation is due to the exclusion of molecules because of size and/or geometry (exclusion chromatography [EC], called gel permeation chromatography [GPC] by polymer chemists and gel filtration by biochemists). In ion-exchange chromatography (IEC) separation is due to selective exchange of ions in the mobile phase with counterions in the stationary phase; in affinity chromatography the stationary phase contains a selective adsorbant (e.g., an antibody, polymeric inhibitor, enzyme) for a specific component in the mobile phase; in ligand exchange chromatography (LEC) ligands form complexes with the stationary phase and are selectively displaced by the mobile phase; *see* gas chromatography; gas-liquid chromatography; gel-permeation chromatography

commercial dyes: *see* direct dyes

commercial moisture regain: *see* standard moisture regain

commercial polymer range: polymers with molecular weight high enough to permit good physical properties but low enough to permit economic processing

comminutor: device to grind up solids

commodity chemicals: chemicals produced in high volume; the highest volume inorganics produced in more than a billion pounds/year are sulfuric acid, nitrogen, oxygen, lime, ammonia, NaOH, chlorine, phosphoric acid, sodium carbonate, nitric acid, ammonium nitrate, carbon dioxide, HCl, ammonium sulfate, carbon black, potash, aluminum sulfate, titanium oxide, sodium silicate, calcium chloride, sodium sulfate and sodium tripolyphosphate; inorganics sold at between 100 million lbs and a billion pounds are hydrogen, phosphorus, and sodium chlorate; all organic commodity chemicals are derived from the primary products of coal or petroleum refining, namely: ethylene, propylene, benzene, methanol, toluene, xylenes (*p*-xylene), butadiene, and urea; the secondary and tertiary organic compounds sold in volumes of ≥1 billion pounds are ethylene dichloride, ethylbenzene, vinyl chloride, styrene, terephthalic acid, ethylene oxide, formaldehyde, ethylene glycol, cumene, acetic acid, phenol, propylene oxide, acrylonitrile, vinyl acetate, methyl *tert*-butyl ether, cyclohexane, acetone, adipic acid, isopropyl alcohol, and caprolactam; organic commodities with sales between 100 million lb and one billion lb are aniline, bisphenol A, 1-butanol, carbon tetrachloride, chloroform, dioctylphthalate, dodecylbenzene, ethanol, ethanolamines, ethyl chloride, 2-ethylhexanol, maleic anhydride, methyl chloride, methyl ethyl ketone, methylmethacrylate, methyl chloroform, methylene chloride, perchloroethylene, phthalic anhydride, propylene glycol, toluene, and *o*-xylene;

common salt: NaCl; sodium chloride; produced from brine and by mining

compaction: consolidation of solid particles by applying pressure, e.g., between rollers, with tamps, pistons, or screws

compatibility: ability of two or more substances to form a homogeneous mixture such as a solution; *see* compatibilizing agent

compatibilizing agent: in polymer technology, a block or graft polymer that permits two or more polymers to form one phase; the compatibilizing agent acts as an organic surfactant between two incompatible polymer phases; *see* ionomers

complex ions: *see* chelates

composite explosive: mixtures in which fuel and oxidizer (nitrates and perchlorates) are mixed with a plastic binder (polysulfides, polyurethane, polybutadiene-acrylic acid, polybutadiene-acrylonitrile) to form an explosive product; solid rocket fuels

composites: physical mixture of mutually insoluble solids; usually materials (e.g., resins) must be added to permit the formation of strong interfaces between the matrix and filler and/or reinforcement; composites usually are stronger than the individual constituents; in particular, many advanced engineering materials are composites consisting, for example, of fibers (e.g., graphite, plastic, cellulose) embedded in metals, glass or plastics; in addition to greater strength composites can be

designed to have other properties, e.g., color, ferromagnetism, opacity; ceramic–metal composites are a class of new construction materials; in fiber technology, composite fibers (or fiber blends) are mixtures of staple fibers

composition cap: in the explosives industry, an electric detonator cap

compound: in chemical science, a homogeneous entity consisting of atoms covalently bound in definite proportions, e.g., a molecule of CO consists of carbon and oxygen in an atom ratio of 1:1; in rubber, plastics, or pharmaceutical technology a compound is the formulated product

compounding: formulating; the process of mixing several ingredients to produce a material having a composition to serve a desired function, e.g., cosmetics, pharmaceuticals, rubbers, polymers

compreg: compressed resin treated wood; *see* wood–plastic composites

compressed fiberboard: hardboard; used for structural and insulation materials; *see* fiberboard

compressed wood: *see* densified wood

compression molding: process involves applying heat and pressure on a thermoplastic resin within a closed mold so that the softened plastic fills the mold; use: molding thermosets into large parts like car panels; not suitable for complicated parts or where very low tolerances are required

compressive strength: measure of resistance to crushing forces

condensation polymerization: step polymerization; process to make polymers by condensation of two or more functional organic monomers (e.g., glycols and dibasic acids, hydroxyacids, lactones, amino acids, diamines, diisocyanates, phenol or urea and formaldehyde, siloxanes, sulfides, sulfones); the reaction proceeds by elimination of water, alcohol or ammonia or by ring opening; condensation polymers include urethanes, nylons, polyesters, phenol-formaldehyde and urea-aldehyde resins, polyacetals, polyamines, imides, imidazoles; condensation of three or higher functional monomers leads to three-dimensional, cross-linked, insoluble structures

conditioning: in sugar manufacture the control of moisture, temperature, and grain size; conditioning optimizes handling, packaging, and storage

congo: Congo copal; a natural resin

coniferous wood: soft wood

conjugate spun fibers: *see* bicomponent fibers

contact process: manufacturing process for sulfuric acid; SO_2 is converted catalytically (V_2O_5) to SO_3 gas in a fixed bed at 410–440°C; the gas is then absorbed in water

containment: in nuclear technology the shielding of the reactor vessels

continuous filament yarn: *see* filament yarn

continuous processing: chemical processing in which reactants and products are continuously added and removed; reactants, catalysts and solvents are often recycled; opposed to batch processing

continuous splitting: fatty acid manufacture by high-temperature, high-pressure fat hydrolysis; fats enter a vertical splitter at the bottom and water from the top; the fats and fatty acid products rise countercurrent to the descending water and product glycerol; fatty acids are removed from the top of the splitter

controlled release materials: agricultural and pharmaceutical materials encased in polymeric membranes that permit slow diffusion

copal: natural resin

COPE process: Claus oxygen-based process expansion; modification of the Claus process used for removal of hydrogen sulfide from waste gases; COPE process uses 80–100% oxygen and can purify sulfur-rich oils

copolymers: materials produced by polymerization of two or more different monomers

copolymerization: polymerization of two or more monomers to make a single polymer; composition of copolymers produced by free radical and ionic copolymerizations depend on whether a monomer preferentially reacts with itself or with the other monomer; this tendency is described by the reactivity ratios r; if r ≥ 1 the monomer radical prefers its own monomer, if r $\leq$ the other monomer is preferred; in ideal systems copolymers have the composition of the monomer feed; to obtain uniform copolymer compositions the rate and ratio of monomer feeds must be controlled

copper: Cu; metallic element; produced by mining ore or native copper; widely used in electric wiring, as catalyst, in alloys (brass, bronze), coatings, instrumentation, cooking utensils; compounds of copper are used as pesticides, for copper plating, as catalysts (*see* copper chromite), as pigments; a notable presence of copper is in high-temperature ceramic superconductors (e.g., $YBa_2Cu_3O_x$, $YBa_2Cu_4O_8$)

copper chromite: CuO-Cr_2O_3; black solid; catalyst; has been used for alkylation of phenols; for aromatization of cyclohexenes; for Fischer–Tropsch synthesis of aromatics; for hydrogenation of unsaturated fatty acids to saturated fatty acids; for ester hydrolysis; for isomerization of cycloalkenols to cycloalkanones; for burning of rocket propellants and fireworks

copper oxide: CuO; *see* copper chromite for catalytic activity

cordite: explosive mixture of nitroglycerin and nitrocellulose; prepared by evaporation of acetone from a solution

core spinning: in textile technology, a spinning process in which a filament is covered with a sheath of yarn; used for making stretch yarns by keeping an elastic core (rubber or Spandex) under tension during manufacture; when tension is released the material contracts but retains elasticity of the core and the surface properties of the yarn sheath

core yarn: *see* core spinning

corn oil: by-product of corn starch production; use: lubricant, raw material for margarine production, salad oil

corona wind effect: *see* electrohydrodynamic enhancement

corrosion: electrochemical oxidation of metals; corrosion of iron (rusting) is destructive, formation of the transparent oxide on aluminum or chromium is protective, air oxidation of copper on buildings is considered decorative; corrosion inhibition of metals is achieved by coating or corrosion inhibitors; another method is cathodic protection in which a more readily oxidized metal is plated onto the substance to be protected

corrosion inhibitors: materials that interact with the metal surface to prevent corrosion; organic inhibitors (e.g., amines, sulfur compounds, carbonyls) act by adsorption in the area of the corrosion and decrease either the cathodic or anodic reaction, or both, and thus decrease a loss of metal; inorganic inhibitors (e.g., phosphate and bicarbonate) act by causing precipitates, such as phosphates or carbonates of divalent ions, to form on the metal surface; other inhibitors are redox reagents that can shift the potential to create an oxidized surface; organic corrosion inhibitors are added to liquids such as detergents and fuel oil

cortisone: steroidal antiinflammatory agent

CORTISONE

corundum: hexagonal Al_2O_3; hard mineral (hardness is 9 on the Mohs scale) is used as abrasive; crystalline corundums when colored with oxides are used as gem stones, i.e, ruby, sapphire, and oriental topaz

coruscatives: in explosives technology, pairs of materials that react with each other without gas formation, e.g., Ti-Sb-Pb, Mg-Sn, Mg-P

Cosorb process: selective absorption of CO from synthesis gas using a solution of CuCl and $AlCl_3$ in toluene at ~25°C and 20 bar; CO is released at 100°C and 1–4 bar

cotton: cellulose fiber derived from the plant species *Gossypium*; the staple fiber must be spun into yarn for textile use: *see* yarn spinning

cottonseed oil: obtained as by-product from the manufacture of cotton fiber; use: raw material for the manufacture of hydrogenated fat; cooking oil

Cottrell precipitators: *see* electrostatic precipitators

cough suppressing agents: drugs that raise the threshold for sensory cough impulse; most commonly of these drugs are narcotics such as codeine that act on the central nervous system; other common cough medicines are expectorants that may increase the fluid in the respiratory tract or reduce the viscosity of the mucus

coumarin: 1,2-benzopyrone; used in perfume and soap manufacture; *see* brightening agents

COUMARIN

coumarone: benzofuran; *see* fiberboard

countercurrent extraction: separation of two or more solutes by a series of partitions between liquid phases, e.g., beet sugar is extracted by countercurrent water diffusion

cracking: petroleum refinery processes that convert high boiling fractions to lower boiling ones by breaking C–C bonds of hydrocarbon molecules; thermal cracking occurs at temperatures >650°F, catalytic cracking occurs at lower temperatures, and hydrocracking combines hydrotreating with cracking

crazing: failure of a solid material, such as a ceramic or glassy polymer, due to stress or a combination of stress and solvent; cracks and crazes form perpendicular to the stress direction; in polymers, crazes with numerous fibrils and voids parallel to the stress direction are precursors to a crack and may cause failure by fracture

crazy glue: *see* cyanoacrylate adhesives

creaming latex: in rubber technology, the addition of a small amount of gum or brine to the latex to produce reversible agglomeration of the rubber particles; latex turns into solid crumb and serum

creep: long-term flow of a solid under mild stress (e.g., below the yield point of the fiber); creep rate measures the deformation under constant load, an important indicator in elastomer and plastics technology, e.g., a polyethylene fiber can have a very high break strength when tested at a fast rate of elongation (100%/min or more), the same fiber will creep to break at 10–20% of that load

creosote: *see* coal tar creosote

cresol: methylphenols; *o*-cresol is produced by catalytic (Al_2O_3) methylation of phenol with methanol at 40–70 bar and 300–360°C; used for manufacture of pesticides; *m*-cresol and *p*-cresol are produced from the corresponding isopropyltoluene by oxidation and cleavage; used for manufacture of plasticizers, as disinfectants

crimp: in textile technology, wavyness in fibers such as wool; synthetic fibers are often crimped to improve wearing comfort of woven materials; *see* false twisting process, texturing

critical chain length: in polymer technology, the minimum chain length required for the physical entanglement of polymer chains

critical diameter: in explosives technology, the minimum diameter of a charge at which detonation takes place

critical micelle concentration: characteristic concentration of an emulsifier in water above which the emulsifier forms micelles

crocking: *see* bleeding

cross dyeing: process in which a specially constructed fabric can be given a colored pattern using a single dye bath that contains several types of dyes; such fabrics contain fibers that are selectively dyed by acid, basic or neutral dyes; examples are fabrics consisting of various synthetic and natural fibers

crossflow filtration: separation by flowing a solution tangentially across the surface a membrane; flow in this direction mitigates membrane fouling by sweeping the surface

cross-links: covalent bonds between two or more polymer chains maintain rubbery properties above the glass transition temperature; for example -S-S- links between elastomers in vulcanized rubber; cellulose (cotton, rayon) can be cross-linked by reaction with dimethylol dihydroxy ethylene urea (DMD-HEU) to provide permanent press materials

crotonic aldehyde: $CH_3CH=CHCHO$; crotonaldehyde; produced by acid catalyzed de-hydrogenation of β-hydroxybutyraldehyde (acetaldol); precursor of *n*-butyraldehyde and *n*-butanol

crown ethers: macrocyclic polyethers that can complex cations within their cavities; the size of the cavity determines what cation can be complexed; nomenclature: "18-crown-6" denotes that there are 18 carbon atoms and 6 oxygen atoms; used to assist transport of inorganic ions into organic solutions; *see* phase transfer catalysis

crude oils: the liquid form of petroleum; often used as a synonym for petroleum; the oils range from clear and yellow to solid and black depending on the region from which they are derived; for example, Arabian Light from the Middle East is brown, has medium sulfur content and yields gasoline, heating oil, and lubricant feedstock; Venezuelan black crude, Boscan, is the world's heaviest crude and is used for high-grade asphalts

crumb: in rubber technology, solid material separating from latex upon stirring with concentrated brine; in rayon manufacture, soda cellulose that is shredded to promote uniform aging

crushers: machines that start the size reduction process from very large mining pieces (about 60 in. or smaller) to a size about 1 in.; there are jaw type, hammer and rotating cone

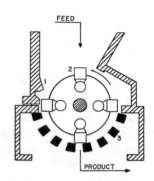

SCHEMATIC REPRESENTATION OF A HAMMER
CRUSHER
(1) BREAKER BAR; (2) HINGED HAMMER; (3) GRATE BARS.
[Courtesy Jeffrey Division, Dresser Industries, Inc.]

crushers; the crushing process depends on the size, hardness, and temperature sensitivity of the raw material and on the size of the end product; crushing is followed by grinding and pulverization; also *see* mills

crutcher: a mixing vessel in soap manufacture where blending with perfume, air, preservatives is done; air is added so that the soap can float; after blending the soap is chilled, extruded and given its final shape

cryo-: prefix denoting cold

cryogenics: technology of handling materials below −200°C

cryogrinding: *see* freeze grinding

cryometry: measurement of molecular weight using freezing point depression

crysolite: asbestos fiber; *see* fibers, inorganic

crystal: homogeneous solid having an orderly arrangement of atoms, ions, or molecules; crystallinity describes the property of having an orderly three dimensional arrangement of ions, atoms or molecules

crystallite: small crystal

crystallization: formation of crystals from solution or melt; used for isolation and pu-

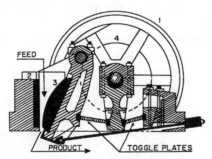

BLAKE-TYPE JAW CRUSHER
(1) flywheel; (2) fixed-jaw plate; (3) swing-jaw plate; (4) pulley. [Courtesy D.M. Considine, P.E.]

rification of products; in fiber technology, used to increase the useful temperature range and stabilize the fiber structure

C-stage: final thermosetting stage of a resin; *see* phenol formaldehyde resin

cumene: $(CH_3)_2CHC_6H_5$; isopropylbenzene; produced by catalyzed (H_2SO_4, $AlCl_3$, or HF) liquid phase (Hüls process) or catalyzed (H_3PO_4/SiO_2) gas phase (UOP process) alkylation of benzene; use: for manufacture of phenol (Hock process), to improve octane number of motor fuel

cuprammonium process: manufacture of rayon by dissolving cellulose in an ammoniacal copper salt solution and spinning the resulting blue cuprammonium cellulose into water or a dilute ammonia water solution

cure: conversion of raw products to finished material; in rubber technology: vulcanization; in textile manufacture heat treatment of fibers with chemicals or resins to impart properties such as water and crease resistance or durable press; in thermosetting resins the final cross-linking procedure; also *see* radiation curing

curing agent: substance that effects the polymerization of a monomer and the cross-linking of a polymer

cyanamide process: reaction of calcium carbide (CaC_2) with atmospheric nitrogen at 1000°C to give calcium cyanamide (CaNCN); use: fertilizer production

cyanoacrylate adhesives: "super glue, crazy glue, miracle glue"; $H–[CH_2C(CN)(COOEt)]_n–R$; produced by aldol condensation of ethylcyanoacetate followed by spontaneous polymerization; polymerization is reversible by heat; the monomer sets as rapidly as in 0.1 sec; use: exceptionally strong glue for machine parts, surgery

cyanocobalamin: vitamin B_{12}; produced by fermentation using *pseudomonas denitrificans* cultures; fermentation broths consist of

beet molasses to which dimethylbenzimidazole and cobaltous nitrate are added

cyanoethylated cotton: CN; produced by treating cotton with acrylonitrile and caustic soda; cyanoethylated paper has improved electrical insulating properties, used in electric condensers; cyanoethylated fibers have improved dyeability and resistance to heat and rot

cyanoguanidine: *see* dicyanodiamide

cyanohydrin: compound that contains –CN and –OH groups

cyanuramide: *see* melamine

cyanuric chloride: *see* 2,4,6-trichloro-1,3,5-triazine

cyclamate: salt of cyclohexylsulfamic acid; artificial sweetener, banned in the United States

cycles per second: cps

cycloaliphatic chlorine: prepared from cyclooctadiene and hexachlorocyclopentadiene; fire retardant

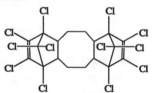

CYCLOALIPHATIC CHLORINE

cyclodextrins: cycloamyloses; naturally occurring crown ethers consisting of rings of 6 to 8 amylose units bonded at 1 and 4 positions by α glucose oxygen links; use: sequestering agents

cyclododecanol: *see* Bashkirov oxidation

cyclododeca-1,5,9-triene: CDT; produced by trimerization of butadiene using a Ziegler catalyst ($TiCl_4$ and $(C_2H_5)_2AlCl$); CDT is the precursor for open and cyclic C12 compounds; can be converted to cyclododecanone which is starting material of lauryl lactam, leading to nylon 12

cyclohexanone: produced by oxidation of cyclohexane or reduction and dehydrogenation of phenol; used for manufacture of caprolactam; *see* caprolactam, nylon 6

cyclohexanone oxime: produced from cyclohexanone; precursor of captrolacatam

cyclone: device for dust collection and for separation of particles according to weight; consists of a cylindrical chamber placed above a cone-shaped receptacle; particles enter the cylinder on top at a rapid speed so that they obtain a free-vortex flow; larger particles move downward along the wall and are discharged through the underflow orifice, smaller particles can be removed from the cylinder through a tangential outlet; for example, a dense-medium cyclone with a 24-in. diameter can clean 75 tons of fine coal per hour; hydrocyclone is a cleaning device that uses only water as the medium

cyclonite: RDX; hexogen; T 4; symmetrical trimethylenetrinitramine; an explosive; prepared by nitration of hexamethylenetetramine with concentrated nitric acid, or by reacting formaldehyde with ammonium nitrate and acetic anhydride (E-process); important high-brisance explosive that is 50% more powerful than TNT; use: for manufacture of boosters, as additive for smokeless powders

cyclophosphamide: antineoplastic (anticancer, chemotherapeutic) agent that acts as alkylating agent

CYCLOPHOSPHAMIDE

cyclosporins: naturally occurring cyclic peptides that depress the immune system (immunosuppressors, immunomodulators); used to suppress rejection of organ transplants; antifungal agents; produced by fermentation from *Tolypocladium inflatum*

cyclotrimethylene-trinitramine: *see* hexamethylenetetramine

cycluron: urea herbicide

CYCLURON

cypermethrin: pyrethroid herbicide

cysteine: $HSCH_2CH(NH_2)COOH$; natural amino acid; constituent of proteins

cystine: $(-SCH_2CH(NH_2)COOH)_2$; natural aminoacid; the constituent of proteins that forms -S-S- bridges and thus influences the molecular shape

cytokinin: *see see* plant hormones

cytosine: one of the four bases present in DNA and RNA

CYTOSINE

Czochralski process: CZ process; method for growing single Si crystals for use in semiconductors by pulling the crystal from the melt; the commercial apparatus, called the puller can hold a melt charge of 60 kg Si to yield a 3-m long crystal with a 10-cm diameter

D

D-: prefix denoting chiral molecules whose spacial configuration is related to D-(+)-glyceraldehyde; this compound rotates plane polarized light clockwise (to the right)

2,4-D: $2,4\text{-}Cl_2C_6H_3OCH_2COOH$; herbicide and growth regulator

114

D-stoff: German name for phosgene

dalapon: 2,2-dichloropropionic acid; poste-mergence herbicide

dalton: unit of mass equal to 1/16 of the mass of an atom of oxygen-16; equivalent to the atomic mass unit (amu)

dammar: damar; naturally occurring resin obtained from trees of the genus *Shorea, Balanocarpus,* and *Hopea;* use: for varnishes, lacquers, adhesives and coatings

damping: mechanism for dissipating excessive energy changes e.g., heat, vibration, or noise

dancer roll: independently moving tension-sensing and tension-maintaining device; use: in wire coating, plastic tape winding, fiber operations, motion control of materials moving through tanks

dandy roll: in paper manufacture, a roller that can be used to impart a water mark; *see* fourdrinier machine

dapsone: 4,4'sulfonyldianiline; DDS; antibacterial and antiprotozoal agent, curing agent for epoxy resins

dash-pot: vibration-damping device

DBCP: *see* dibromochloropropane

DDD: dichlorodiphenyldichloroethane; insecticide; *see* DDT

DDT: dichlorodiphenyltrichloroethane; insecticide; banned in the United States

DDT

deactivation: *see* catalyst deactiavtion

dead fold: in plastic manufacture, a fold that does not spontaneously unfold

dead oil: *see* creosote

debarking of wood: the removal of tree bark using hydraulic barkers or mechanical knives; debarking is needed before wood pulping

decamethrin: decis; pyrethroid insecticide; *see* pyrethrin

decis: *see* decamethrin

deckle: deckle rod; cut-off plate; a device to control the width of films or paper: in an extruder a small rod or plate attached to each end of the die to adjust the length of the die opening; in paper manufacture, a strip placed along the sides of the fourdrinier wire to equalize the flow of the pulp and thus give the paper a straight edge

deckle-box method: wet-felting process used in fiberboard manufacture in which a bottomless frame (deckle box) is placed on a wire screen, sufficient stock to form one sheet is pumped into the box and the water is drained using vacuum below the screen and pressure from above; after sufficient water has been expressed the deckle box is raised and the sheet proceeds to driers

decomposition: chemical breakdown of materials due to bacterial action, radiation, heat, or air

decortication: removal of hulls from grains; preliminary step in seed oil extraction

decrystallized cotton: cotton that has been treated with anhydrous liquid ethylamine; treatment reduces the crystallinity of cellulose, opens up its structure and results in cotton that is more absorbant and easier to dye

deep drawing: formation and shaping of a thermoplastic sheet in a mold involving a high draw ratio

defecation: clarification; in sugar cane purification the removal of soluble and insoluble impurities from the expressed juice using lime and heat

deferoxamine: complexing agent isolated from *Streptomyces pilosus*; used for treatment of iron or aluminum poisoning

deferred cure: treatment of fabrics to produce wash-and-wear materials; fabric fibers are cross-linked after the garment is made and pressed; this treatment locks in the shape of the garment and locks out wrinkles

defibering: in pulping, the mechanical and chemical reduction of fibers before paper manufacture; also removing flax fiber from the plant stalks, noncellulosics from wood

deflagrating explosive: *see* propellant explosive

deflagration: rapid autocombustion of surface particles

deflocculants: substance that prevents particles from agglomerating, e.g., sulfonic acids, clays; *see* dispersants

defoamer: agent that controls foams by causing small air bubbles of a foam to coalesce into larger ones that then rise to the surface; defoamer formulations are highly insoluble hydrophobic particulate mixtures that may contain fatty alcohols, acids and esters, silicones and silica; use: processing aids to improve dewatering, drainage, filtration, and washing of suspensions; used in industries such as pulp and paper, paint and latex, textile, fermentation, adhesive manufacture and chemical processing

degating: in injection molding, removal of waste material left in the area between the injection nozzle and the molded part; in thermoplastic molding this waste is ground and reused

degreaser: typical formulations contain surfactants as well as an alcohol or organic solvent to penetrate and emulsify grease spots; use: for pretreatment, not as general detergents

degree of cure: extent to which curing of a thermosetting resin has progressed: at the A and B stages the polymer still flows, at the C stage the material is being thermoset

degree of orientation: *see* orientation

degree of polymerization: D.P.; average number of monomer units per polymer molecule; a measure of chain length and of molecular weight

degumming: in silk manufacture the process of removing natural gums, sericin, or wax

Degussit: German name for beryllium oxide

delamination: separation of layers of a laminate due to failure of the adhesive bonds

delustrant: agent that removes sheen from synthetic fibers; the most common delustrants are the TiO_2 pigments rutile and anatase (rutile has better covering power and is UV stable, anatase needs an additional coating to make it UV stable)

demetallizing: in hydrotreating of petroleum, the removal of metal ions and metal compounds

demeton: a systemic insecticide consisting of a mixture of $(C_2H_5O)_2P(S)O(CH_2)_2S(C_2H_5)_2$ and $(C_2H_5O)_2P(O)S(CH_2)_2S(C_2H_5)_2$; isosystox; especially active against aphids and cotton mites; acetylcholinesterase inhibitor; highly toxic, (absorbed through skin)

demulcent: material that softens and soothes inflamed parts of the body; typically an oil or mucilage

den: *see* denier

denaturants: materials such as bittering agents, used to denature ethanol, i.e, to prevent its use as beverage; denaturants such as benzene, kerosene, methanol, pine oil and pyridine are poisonous

denier: den; unit used in the textile industry: 9000 m of a 1 denier filament weighs 1 g; single fibers are 1–15 den, yarns consisting of many filaments are hundreds and thousands of denier; the lower the denier per filament the more flexible is the yarn; nylon hosiery is 10–15 den (usually monofilament); mechanical properties such as tensile strength and modulus (stiffness) may be defined in g/den

denitrification: in waste water treatment, the removal of nitrogenous BOD (biological oxygen demand) using continuous fermentation

denitrogenation: in hydrotreating, the removal of nitrogen containing compounds from petroleum by using nickel–molybdenum on alumina

dense ash: soda ash; sodium carbonate; prepared by dehydrating light ash, which is the product from the Solvay process

densified hardboard: superhardboard; fiberboard that weighs >85 lb/cu ft; made by using high pressure during felting; used for instrument panels, templates, jigs, and die stock; *see* fiberboard

densified wood: compressed wood; wood impregnated with resins to give it good dimensional stability and strength

dental amalgam: mercury mixed with finely divided alloys of Ag and Sn

deodorants: agents such as antimicrobials, insecticides, adsorbants, or other chemicals that eliminate unpleasant odors, e.g., deodorant toilet soap contains an antimicrobial agent to retard growth of microbes on the skin and thus prevent body odor; chemicals, such as chlorophyll, react with odiferous impurities; powders contain antiagglomerating agents to maintain high surface areas; activated carbon is used to deodorize kerosine, which is used as solvent for pesticide sprays

depolymerization: reversion of a polymer to its monomers or to a polymer of a lower molecular weight

desalting: in petroleum refining, crude oil is washed with water at the refinery site before distillation; the water–oil mixture is emulsified and washing is carried out at >40 psig and 100–300°F; after washing is completed

the emulsion is passed through a high-voltage electrostatic field that separates the emulsion and permits separation of the liquids

desiccants: drying agents; these materials remove water completely or to constant humidity; in continuous processes desiccants are being regenerated; common desiccants are calcium chloride, phosphorus pentoxide, sulfuric acid, alumina, silica gel, and molecular sieves; $CoCl_2$ is added to desiccants as a humidity indicator as the pink hexahydrate reverts to blue tetrahydrate during desiccant regeneration

desorption ionization: *see* mass spectrometry

destructive distillation: pyrolysis, thermal decomposition; high temperature treatment of organic materials in the absence of air

detergent builder: material that upgrades the cleaning efficiency of the detergent by inactivating water hardness, increasing pH, or helping emulsification action; *see* detergents; builder; built soap

detergents: surface-active agents, surfactants; term usually restricted to cleaning agents other than soap; biodegradable detergents and low-phosphate detergents have been introduced to protect the water supply; detergent formulations may contain alcohols, anionic and nonionic surfactants, antiredeposition agents, bleaches, builders, cellulose derivatives, colorants, corrosion inhibitors, emollients, enzymes, foam inhibitors (suds control agents), fluorescent whitening agents (optical brighteners), fragrances, opacifiers, phosphates, sodium sulfate, water, and zeolites; paper impregnated with detergent has been manufactured for various applications such as washing windows and other surfaces; *see* surfactants

Detol process: *see* benzene

detonating explosive: extremely sensitive explosives that are too sensitive to be used in bulk; *see* high explosive

detonation: rapid self-propagating decomposition of high explosives

developed dyes: *see* diazo dyes

devolatilized coal: coal freed from volatile constituents by heating before contact with reactant gases during fixed-bed coal gasification

dew-cycle weathering test: accelerated weathering achieved by exposing materials alternatively to unfiltered light and dew caused by cold water

dextran: polysaccharide consisting of αD-glucose bonded 1 → 6; produced by *lactobacteria* growing on sucrose; used as blood plasma extender

dextranase: an enzyme found in *penicillium* species; cleaves the glycosidic bond of dextran; used in the sugar industry to reduce the viscosity of dextrans

dextrin: starch gum; hydrolysis products of starch containing very low amounts of reducing sugar; use: in adhesives, thickening agents, additive for pills, and lactose substitute in penicillin manufacture; used in the production of paper, felt, food, explosives, matches, inks

dextroamphetamine sulfate: dexedrine sulfate, *d*-1-phenyl-2-aminopropane sulfate; appetite depressant, relaxes intestinal muscle, stimulates brain activity, and increases blood pressure; may lead to tolerance and physical dependence

dextronic acid: *see* gluconic acid

dextrose: glucose; used for intravenous infusion

dextrose equivalent: D.E.; the reducing sugar content of hydrolyzates or syrups expressed as percent dextrose on a dry weight basis

D-glass: high boron content glass

diacetin: glycerol diacetate; *see* glycerol

diacetone: $CH_3COCH_2C(OH)(CH_3)_2$, diacetone alcohol; prepared from acetone by action of $Ba(OH)_2$ or $Ca(OH)_2$; use: solvent, especially for nitrocellulose, cellulose acetate, gums, resins, lacquers and ink removers; preservative for pharmaceutical preparations; hazard: narcotic in high concentration, causes kidney and liver damage in experimental animals

diacetone alcohol: *see* diacetone

diacetylmorphine: heroin; powerful narcotic, leads to addiction; importation and production prohibited in the United States; *see* morphine

diafoam: foam that contains gas and glass bubbles; *see* syntactic foam

dialdehyde: *see* glyoxal

diallylphthalate: monomer for thermosetting resin

dialysis: process of removing ions or molecules from a solution through a semipermeable membrane; used in industry and in medicine

3,3'-diaminobenzidine: raw material for the manufacture of the high-performance fiber polybenzimidazole (PBI); toxic, carcinogen

4,4'-diaminodiphenylmethane: MDA; produced by condensation of aniline and formaldehyde; raw material for manufacture of 4,4'-diphenylmethane diisocyanate (MDI)

diaminostilbene: *see* fluorophore

2,4-diaminotoluene: MTD; *see* toluene-2,4 diamine

diammonium hydrogen phosphate: DAP; $(NH_4)_2HPO_4$; diammonium phosphate; manufactured by treating phosphoric acid with ammonia; use: fertilizer, for making phosphors, flameproofing papers and pH control of bakery products

diammonium phosphate: *see* diammonium hydrogen phosphate

diamond: cubic carbon crystals with a hardness of 10 Mohs; gem stone; synthetic diamonds are produced from graphite at 800,000–1,800,000 psi and 2200–4400 °F in the presence of molten Fe, Cr, Co, Ni, Mn, their alloys and carbides; diamond coatings are produced by chemical vapor deposition; industrial diamonds (carbonados, ballas, boort, boart) are used for cutting, grinding, abrading, polishing, and as temperature control element for ovens and in chemical processing; precision cutting tools

diaphoretics: drugs that stimulate sweat glands to cause perspiration, e.g., the alkaloid aconite; use: to lower fever (antipyretic)

diaphragm cell: *see* chlor-alkali process

diatomaceous earth: diatomite, DE, fossil flour, kieselguhr, siliceous earth, tripolite; a fine siliceous natural deposit of skeletons of small aquatic animals (diatoms); use: as filler for rubber and plastics, inactive ingredient in dry formulations, filter during waste disposal and air purification

diatomite: *see* diatomaceous earth

diazepam: benzodiazepine tranquilizer; valium; antianxiety agent, a minor tranquilizer; hazard: abuse may lead to addiction

DIAZEPAM

diazinon: insecticide; hazard: a cholinesterase inhibitor

DIAZINON

diazo compound initiators: used to initiate free radical polymerizations, e.g., azobisisobutyronitrile decomposes to isobutyronitrile radicals $(CH_3)_2C \cdot CN$ and thus acts as initiator

diazodinitrophenol: DDNP; high explosive; prepared by diazotization of picric acid amide with $NaNO_2$ in HCl solution; DDNP is more powerful than mercury fulminate but slightly less than lead azide; used as initiating explosive

diazo dyes: developed dyes; *see* azoic dyes

diazo hemicyanines: lightfast bright dyes for acrylics

diazoquinone-novolac: DQN; commonly used resist material for the manufacture of integrated circuits; *see* resists

diazotization: exothermic process used to manufacture azo dyes; batch process takes place in a brick-lined tub; $NaNO_2$ is added to an acid solution of the aromatic amine with stirring and addition of ice; the resulting diazo salt is added to a solution of the intermediate with which it is to couple; the coupling process takes up to 3 days or until coupling is complete; final purification is by filtration and recrystallization

dibasic lead phosphite: white pigment

dibromochloropropane: DBCP; $CH_2BrCHBrCH_2Cl$; soil fumigant, nematocide

dibutyl phthalate: high boiling point (bp 340°C) solvent for plastics; plasticizer for Buna rubber and polyvinyl chloride; hazardous chemical waste (EPA)

dibutyl succinate: insect repellent especially against biting flies, ants, and roaches

dicalcium phosphate: $CaHPO_4$; produced from lime by treatment with phosphoric acid or precipitated from bones; use: fertilizer, animal feed, baking powder

dicamba: 3,6-dichloro-2-methoxybenzoic

acid; made from the hydroxy acid by treatment with $(CH_3)_2SO_4$; herbicide and growth regulator

dichlorobenzene: *see* fumigants

dichlorodifluoromethane: CCl_2F_2; CFC-12; Refrigerant 12; chlorofluorocarbon refrigerant; one of the freons; may form toxic substances on contact with flame or hot surface; use regulated by government because of damaging effects on the atmospheric ozone layer; *see* chlorofluorocarbons

1,1-dichlorethylene: $C_2H_2Cl_2$; solvent for extraction of fats and rubber, insecticidal fumigant (only for use on bare soil); strong irritant for skin, eyes, mucous membrane; hazardous chemical waste (EPA)

dichloroethyl ether: *see* fumigants

dichlorethyl formal: $CH_2(OCH_2CH_2Cl)_2$; water-insoluble solvent for cellulose, fats, oils, and resins

dichloro-*bis*(triphenylphosphine)palladium (II): $PdCl_2(P[C_6H_5]_3)_2$; catalyst; yellow solid, moderately soluble in chloroform amd dichloromethane, insoluble in alcohols; has been used to catalyze alkylations, carbonylations, hydroformylations, hydrogenolyses, for polymerization of norbornene, reductive coupling

dichloromethane: *see* methylene chloride

dichloro-*tris*(triphenylphosphine)ruthenium (II): $RuCl_2(P[C_6H_5]_3)_3$; catalyst; black solid, moderately soluble in chloroform, acetone, and benzene, (solutions are air sensitive), insoluble in water and methanol; has been used to catalyze additions, cyclizations, reductions of ketones, aldehydes, and olefins, hydrogenolyses, for polymerization of 2-norbornene

dichroism: exhibition of different colors; in solids, depending on the direction of viewing; in liquids, depending on concentration; circular dichroism, which is the effect of changing linearly polarized light, into elliptically polarized light is used in the analysis of optically active substances

dicofol: dichlorodiphenyltrichloroethanol; releases DDT on hydrolysis; pesticide (acaricide, miticide)

dicy: *see* dicyanodiamide

dicyanodiamide: "dicy," cyanoguanidine; $H_2NC(=NH)NHCN$; produced by dimerization from cyanamide in water in the presence of base; use: for manufacture of melamine, barbiturates, guanidine derivatives

dicyclopentadiene: produced from cyclopentadiene by dimerization at 100°C and normal pressure; use: chemical intermediate (insecticides, metallocenes, paints, flame retardants), monomer for polymers and copolymers, chemical intermediate

endo-DICYCLOPENTADIENE

dicyclopentadienyl iron: *see* ferrocene

die: device, usually made of steel, to impart a specific shape to a material by stamping, casting, or extrusion

dieldrin: broad spectrum insecticide; readily absorbed through the skin

DIELDRIN

dielectric: nonconductive material that separates the conductive elements of a condenser; material with electrical conductivity $<10^{-6}$ mho/cm; use: insulator

dielectric constant: the ratio of capacitance

of a material to the capacitance of a vacuum or air; a measure of the tendency of a molecule to align itself to an applied electric charge; a measure of molecular polarity

dielectric strength: dielectric voltage; maximum voltage that can be applied to a material before it breaks down

dielectric voltage: *see* dielectric strength

diene polymers: polymers made from olefins having two double bonds, e.g., polybutadiene, polyisoprene (rubber, guttapercha), polychloroprene, copolymers such as ABS

diesel oil: refinery blend produced from light gas oil after distillation of crude oil; used for compression–ignition engines, space heating

diethanolamine: DEA; $HN(CH_2CH_2OH)_2$; produced from ethylene oxide and aqueous ammonia at 60–150°C and 30–150 bar; used in manufacture of amphoteric surfactants, foam boosters, emulsifiers, and dispersants

diethyleneglycol: $(HOCH_2)_2O$; made by heating ethylene oxide and glycol; use: antifreeze, solvent, for softening cotton and wool fibers and as additive to glues, gelatin, and pastes to prevent them from drying out

diethyleneglycol dinitrate: $(O_2NCH_2 CH_2)_2O$; explosive prepared by batch or continuous nitration of diethylene glycol; rocket propellant

diethylenetriamine: $(NH_2CH_2CH_2)_2NH$, DETA; produced by treating 1,2-dichloroethane with excess ammonia; use: crosslinking agent, raw material for urethane polyether polyols

diethylether: $(C_2H_5)_2O$, ethyl ether; produced by catalytic (WO_3) gas phase dehydration of ethanol at 120–130°C and 100 bar; use: solvent, primer for gasoline engines, anesthetic; hazard: explosive (forms peroxide on standing), flammable

diethylphenyl urea: *see* centralite I

diet supplements: *see* agonists

differential scanning calorimetry: DSC; analytical method that measures the rate of heat release or absorption of a specimen during a programmed temperature change; used to study thermal properties, annealing processes, and cure characteristics

differential thermal analysis: DTA; analytical method to compare thermal energy changes of a specimen and a standard control while they are both heated; resulting graph gives information on phase changes and chemical transformations; *see* differential scanning calorimetry

diffuse reflectance infrared Fourier transform spectroscopy: DRIFT; microanalytical technique to obtain molecular chemical information; the beam source and the detected species are IR photons; resolution is of the order of microns and the method is applicable to almost any sample form

diffusers: *see* diffusion

diffusion: spontaneous spreading of a substance throughout a phase due to random molecular motion; spontaneous intermingling of molecules or particles across permeable membranes; industrial operations based on diffusion include countercurrent separation, dialysis, distillation, drying, electrodialysis, extraction, freeze drying, gaseous diffusion, ion exchange, osmosis and reverse osmosis, sorption, sublimation, thermal diffusion, and uranium-235 isotope enrichment; the rate of gaseous diffusion is inversely proportional to the square root of the molecular weight as described in Graham's Law; the mass transfer occurring during diffusion of two nonreacting species is described by Fick's Law; diffusion processes may use batch or continuous diffusers and often use cascade systems

digesters: vessels in which materials are broken down, for example, in wood pulping wood chips, caustic, and sodium sulfate are placed into a digester and cooked under pressure; in waste water treatment a closed tank is used to decrease volume of solids and stabilize raw sludge by bacterial action

digestion: chemical or biochemical decomposition of organic matter

digitoxin: cardiovascular drug, blood vessel dilator; extracted with alcohol from leaves of the purple foxglove plant *Digitalis purpurae*

diglycerides: fatty acid esters of glycerol in which only two of the glycerol hydroxyl groups are esterified

dihydrocodeinone: *see* hydrocodone

diisocyanate: *see* isocyanates

diisodecylphthalate: DIDP; *see* orthophthalate plasticizers

diketene: *see* ketene

dimensional stability: ability of a material to retain its precise shape; in the electrochemical industry, dimensionally stable anodes (DSA) made of titanium with a coating of iridium, platinum, or ruthenium have largely replaced graphite electrodes in the United States chlor-alkali process

dimer: molecule formed from two identical molecules

dimerization: formation of a dimer

dimethoate: $CH_3O_2P(S)SCH_2CONHCH_3$; systemic insecticide; a cholinesterase inhibitor

N,N-dimethylacetamide: DMA, DMAC; $(CH_3)_2NCOCH_3$; produced from methyl amine and acetic anhydride; industrial solvent for cellulose derivatives, linear polyesters, polyacrylonitrile, styrene polymers and vinyl resins; paint remover; hazard: can emit irritating fumes on decomposition

dimethylamine: $(CH_3)_2NH$; used for organic synthesis; raw material for manufacture of dimethylformamide

4-N,N-dimethylamino-4'nitrostilbene: *see* nonlinear optics

3,3-dimethyl-1-butene: *see* neohexene

dimethyldichlorosilane: $(CH_3)_2SiCl_2$; prepared by passing methylchloride over powdered silicon and a copper catalyst; raw material for the manufacture of methylsilicone rubber

dimethylformamide: DMF, DMFA; $HCON(CH_3)_2$; produced by hydroformylation of dimethylamine; use: solvent for organic and inorganic compounds, industrial solvent for acrylic fibers and other polar polymers; reaction medium for ionic and nonionic reactions; hazard: vapor is irritating to skin, eyes, and mucous membranes, liver injury in experimental animals on prolonged inhalation; implicated in testicular cancer

dimethyl glycol: *see* ethylene glycol ethers

dimethylketone: *see* acetone

dimethylphenol: *see* xylenol

dimethylphenyl urea: *see* centralite

2,2-dimethyl-1,3-propanediol: *see* neopentyl glycol

dimethylsulfide: $(CH_3)_2S$; *see* dimethylsulfoxide

dimethylsulfoxide: $(CH_3)_2SO$, DMSO; produced by oxidation of dimethyl sulfide or methyl mercaptan; use: solvent; caution is necessary because of rapid penetration of skin

dimethylterephthalate: DMT; $p–CH_3OOC$ $C_6H_4COOCH_3$; made by a two-step liquid phase oxidation (Co/Mn catalyst) of p-xylene to p-toluic acid followed by esterification with methanol; used as intermediate for polyester fiber production

3,4-dimethyltoluene: *see* pseudocumene

dinitroanthraquinones: produced by nitration of anthraquinone using nitric acid; use: starting materials for dye manufacture

dinitrocellulose: *see* nitrocellulose

N,N'*-dinitroso-*N,N'dimethyl terephthalamide**: NTA; produced by reaction of *p*-CH$_3$NHOCC$_6$H$_4$CONHCH$_3$ with nitrous acid; blowing agent with a low decomposition temperature (80–90°C); the decomposition product is dimethylterephthalate

***N,N'*dinitroso-pentamethylene tetramine**: DNPT; blowing agent; produced from formaldehyde and ammonia; limited use in PVC, polyethylene, epoxy foam manufacture; decomposition product is an amine residue with a "fishy" smell; main use for foamed rubber

dinitrotoluene: produced by nitration of toluene; raw material for manufacture of tolylene diisocyanate

dioctylphthalate: DOP; *see* orthophthalate plasticizers

diol: alcohol containing two hydroxyl groups

dioxin: 2,4,7,8-T; 2,4,7,8-tetrachlorodibenzodioxin; by-product in herbicide manufacture; highly toxic to certain test animals; causes chloracne in humans

DIOXIN

dioxyethylnitramine dinitrate: (O$_2$NOCH$_2$CH$_2$)$_2$NNO$_2$; DINA; explosive prepared by nitration of ethanolamine with nitric acid in the presence of acetic anhydride; use: to gelatinize nitrocellulose; explosive power comparable to cyclonite and PETN

dipentenes: C$_{10}$H$_{16}$ hydrocarbons derived from plants; use: solvents, dispersing and wetting agents, odorants

diphenhydramine: (C$_6$H$_5$)$_2$CHOCH$_2$CH$_2$N(CH$_3$)$_2$; antihistamine

diphenylcarbonate: (C$_6$H$_5$)$_2$CO$_3$; raw material for production of polycarbonate

diphenylhydantoin: *see* phenytoin

diphenylisophthalate: produced from *m*-xylene by oxidation via the acid chloride; raw material for organic synthesis; reacts with 3,3'-diaminobenzidine to give polybenzimidazole

DIPHENYLISOPHTHALATE

4,4'-diphenylmethane diisocyanate: MDI; produced by reaction of phosgene and 4,4'-diaminodiphenylmethane (MDA) (which is made by condensation of aniline and formaldehyde); raw material for manufacture of urethanes

dipole: molecule in which atoms of different electronegativities are separated so that one end of the molecule carries a relatively negative charge, whereas the other end is relatively positive; *see* polar molecules

dipole moment: $\mu \geq$ measure equal to the charge (e) multiplied by the distance (d) between the centers of the charges; e is measured in electrostatic units, d in cm, and μ in Debye units (D); the dipole moment depends on the electronegativities of the atoms that form the bonds as well as on the molecular shape; the dipole moment of molecules that contain polar bonds but have a center of symmetry is zero; *see* polar molecules, dipole

diquat: contact herbicide; minimal absorption from gastrointestinal tract; *see* paraquat

DIQUAT

direct dyes: commercial dyes; substantive dyes; class of dye stuffs that are applied directly to fibers in a neutral or alkaline bath; they have poor wash qualities; used primarily for natural fibers

direct printing: process in which pattern on the fabric is produced using a thickened paste of dye

direct screw transfer: DST; *see* injection molding, thermosets

disc: in textile manufacture, cam wound or zero twist package of yarn, distinct from ring-spinning that puts a slight twist onto a yarn; *see* ring spinning

disc and cone agitators: mixing devices with discs and cones rotating at 1200–3600 rpm; used to prepare pastes and dispersions

discharge printing: process in which pattern on the previously dyed fabric is produced by using an oxidizing or reducing agent to destroy the dye

disinfectant: antimicrobial agent such as chlorine bleach, phenols, quaternary ammonium salts, and pine oils; used directly on surfaces or in formulations such as soaps and detergents

dispersants: dispersing agents; deflocculants; antiredeposition agents; suspension agents; surface active agents that keep particles in suspension by overcoming cohesive forces such as electrostatic or van der Waals forces or surface tension; common dispersants include aluminum compounds, polyphosphates, sterols, and fatty acids; used in colloidal formulations (paints, inks), oil drilling muds; carboxymethylcellulose (CMC) is a common antiredeposition agent in detergents

disperse dyes: nonionic dyes that are slightly soluble in water and can dissolve in some synthetic fibers; used for polyester and other synthetic hydrophobic fibers; *see* dyes

dispersing agents: *see* dispersants

disproportionation: reaction in which one compound can oxidize and reduce itself, i.e, undergo auto-oxidation, e.g., H_2O_2 converts to H_2O and O_2; in the termination step of a free radical polymerization $RCH_2CH_2\cdot$ converts to RCH_2CH_3 and $RCH_2=CH_2$

dissipation factor: in an AC circuit this factor represents the imperfection of a solid capacitor that causes a leakage current; the ratio of parallel reactance to equivalent AC parallel resistance; ASTM D150; also *see* power factor

dissolved oxygen: D.O.; oxygen required in waste water or sewage for the preservation of aquatic life and prevention of offensive odors; insufficient DO is due to excess organic materials and indicates insufficient waste treatment

distillation: a unit operation to separate liquids from solutions or to purify products; the process consists of evaporation of the the lower boiling component; if that component is the desired product it is recovered by condensation; if it is a solvent used in the reaction it is recycled; as evaporation requires heating the process is energy intensive and is at times replaced by membrane processes

disulfide bond: –S–S–; in rubber technology, the bond is formed during vulcanization and serves as cross-link between the polymer chains; in proteins the disulfide bond of the amino acid cystine forms cross-links within or between protein chains

dithiocarbamates: *see* fungicides

dithiocarbamic acid: H_2NCSSH; *see* ultra-accelerator

dithionite: *see* sodium sulfite

ditridecylphthalate: DTDP; *see* orthophthalate plasticizers

diuretics: substances that promote the excretion of urine

diuron: urea herbicide

DNOBP: 2-*sec*-butyl-4,6-dinitrophenol; insecticide, herbicide

DNOC: 4,6-dinitro-*o*-cresol; insecticide, herbicide; hazard: cumulative poison, skin contact may lead to necrosis and dangerous systemic effects

DNOCHP: 2-cyclohexyl-4,6-dinitrophenol; insecticide, herbicide

doctor: to spread a uniformly thick coating on a substrate

1,12-dodecanedioic acid: $HOOC(CH_2)_{10}COOH$; produced by nitric acid oxidation of a cyclododecanol/cyclododecanone mixture; the diester is used in lubricants and as monomer for polyester; *see* cyclododeca-1,5,9-triene

domain: small region within a solid

dope: in the textile industry, solutions of cellulose derivatives (acetates, nitrates); cellulose ester lacquer is used as adhesive or coating to make fabrics taut, less permeable to gases and more waterproof; originally used for airplane and balloon fabrics; now solutions of cellulose acetate in acetone are used for dry spinning of fibers or casting of films. In the semiconductor industry, trace impurities introduced into crystals to produce special properties, e.g., *p*-Si is made by addition of group III elements (B, Ga) to ultrapure silicon to cause electron-deficient "holes"; *n*-Si is made by addition of group V elements (e.g., As), which produce electron rich areas. In the explosives technology; *see* balanced dope

dope-dyeing: *see* solution dyeing

dosimeter: detector worn by workers to measure exposure to pollutants, including radioactivity

double base propellants: explosives containing two main components such as nitroglycerine and nitrocellulose; rocket propellants

double refraction: *see* birefringence

Down's cell: electrolytic cell that has a graphite anode and a steel cathode; used to manufacture Na metal and chlorine gas from electrolysis of molten $NaCl/CaCl_2$

downstream: final operations in processing

doxepin: antidepressant

DOXEPIN

doxorubicin: *see* adriamycin

drag flow: in an extruder the component of material flow caused by the motion of the screw in the cylinder

dram: unit of weight equal to 1/16 ounce

drape: in the textile industry, term denoting the ability of fabrics to form graceful folds

draw down: draw down ratio; in polymer technology, the difference between the velocity of the melt-extruded polymer at the extrusion position and the wind-up velocity of the solid yarn; depending on the draw down ratio, temperature and cooling rate, the finished yarn may have different degrees of orientation, may be used "as spun" or after additional drawing under various conditions; mechanical properties strongly depend on draw down conditions; also *see* spinning; orientation; fiber drawing; poly(ethyleneterephthalate)

draw down ratio: *see* draw down

drawing: in fiber technology, the process of stretching thermoplastic material to align the polymer chains; stretching reduces the cross section and improves the mechanical properties of the material; *see* fiber drawing

draw ratio: in fiber technology, the ratio of the length of the drawn filament to that of the undrawn filament or yarn; for yarns of synthetic polymers the ratio is 4:1 or higher.

driers: *see* desiccants

DRIFT: *see* diffuse reflectance infrared Fourier transform spectroscopy

125

drop-weight test: impact test in which weights are dropped from various heights; the toughness of a polymer specimen is the value where 50% of the test bars break

drugs: physiologically active substances derived from natural sources or made synthetically; major drug classes are: analgesics (pain killers), antacids, antiallergic agents, antianxiety agents, antibacterial agents, antibiotics (antifungal, antiviral), anticancer agents (chemotherapeutics, anticarcinogens), anticonvulsants, antidepressants, antihistamines, antihypertensive agents (blood pressure lowering), antiinflammatory agents, antiulcer agents, cardiovascular agents, central nervous system stimulants, cholesteremic (cholesterol lowering) agents, cough preparations (expectrorants, depressants), diuretics ("water pills"), hypoglycemic (blood sugar lowering) agents, sedatives, steroidal sex hormones, tranquilizers (major and minor), vaccines, vitamins. Drugs are formulated to optimize absorption and utilization in the body; they are dispensed as aerosols, emulsions, ointments, solutions, and tablets as well as through the skin as transdermal delivery systems; tablets may be sugar or film-coated; modern slow-release drug systems allow a steady diffusion of drugs through polymer membranes; slow-release allows the maintainance of a constant dose over long periods

drug resistance: common phenomenon that occurs with antibiotics and antibaterial agents; resistance to the drug is due to mutations in the microorganisms that permit them to destroy the active drug form, e.g., the development of the enzyme β-lactamase that hydrolyzes the lactam ring of penicillin thus making penicillin inactive; resistance is also exhibited against other biocides, e.g., pesticides; resistance can be reduced by avoiding excessive use of a single agent

dry-forming process: air-felting; manufacture of fiberboard using a dry fiber-blend containing wax and occasionally phenolic resin; the blend is compressed between belts or rollers

dry ice: solid CO_2

drying: in general the removal of water; in coating technology, the term is applied to cross-linking of a polymer; see drying oils

drying agents: desiccants; materials used to increase drying speed of paints and varnishes, e.g., naphthenates

drying oils: unsaturated vegetable oils that harden when oxidized by exposure to air; used as binders for coatings, paints

dry spinning: in synthetic fiber manufacture, extrusion of polymer solutions through a spinnerette into a heated tower where the rate of formation of the fibers is controlled; at the bottom of the tower the fibers are collected into a yarn which is lubricated and then wound up; see poly(acrylonitrile); cellulose acetate

dry jet-wet spinning: fiber spinning process in which polymer solutions are extruded from a spinnerette through a small air gap into a precipitating solvent that removes the solvent from the system; used first for extrusion of liquid crystalline aramid solutions in 100% H_2SO_4; a water bath removes the sulfuric acid from the solid, oriented aramid-sulfuric acid structure that had formed in the air gap; this spinning process is much simpler than wet spinning and additionally provides considerably improved properties; see aramids, wet spinning

dry-laid nonwovens: see nonwoven fabrics

Ducasse shredder: Unigrater; a chopper–shredder used for extraction of sugar from sugar cane

Duplex-Beken kneader: see Beken mixer

duprene: original name for polychloroprene

durene: see 1,2,4,5-tetramethyl benzene

dyed printing: process in which pattern is

printed on the fabric using a mordant; the treated material is then immersed into a dye bath and the mordant-containing pattern becomes colored

dyeing processes: reaction conditions (e.g., temperature, pH) and equipment used for dyeing depends on the nature of the dyes and the material to be dyed; during the dyeing process the material may move through the dyebath (e.g., jig dyeing), the dye solution may sweep through the material (e.g., package dyeing) or both material and solution may be in motion (e.g., paddle dyeing); *see* dyes, beck

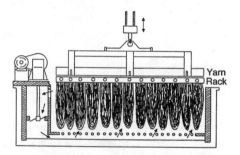

SKEIN DYEING (HUSSONG MACHINE)
A rack from which skeins are hung is lowered into the dye bath. A propeller circulates the dye through the skeins. In some machines, the cover may be closed to permit dyeing under pressure. [Courtesy D. M. Considine, P.E.]

dyes: natural or synthetic coloring agents used for materials such as textiles, food, leather, paper, petroleum, and plastics; dyes are applied by dispersion or by reaction with the substrate; modern dyes are organic molecules containing unsaturated groupings which absorb light of specific wavelengths in the visible spectrum (chromophores); dyes may be classified by their dyeing properties: for example, the hydrophilic acid and basic dyes become attached to polar fibers such as cotton and wool, hydrophobic disperse dyes are used primarily for synthetic fabrics such as polyesters, cellulose acetate, nylons and acrylics, mordant dyes require pretreatment of the fiber with a chelating salt, direct dyes are held to the fiber by hydrogen bonds and require special processing

to become colorfast, vat dyes are applied from alkaline solution in their reduced state and then are oxidized on the fiber; classification by chemical structure distinguishes between molecules such as those containing azo, triphenylmethane, or anthraquinone moieties; also *see* acid dyes, azo dyes, basic dyes, chrome dyes, neutral dyes, disperse dyes; *see* Colour Index

dynamite: explosive material consisting of nitroglycerin (absorbed on diatomaceous earth, wood pulp, flour, starch, or sawdust), oxidizers (ammonium or sodium nitrate), and a small amount of antacid ($CaCO_3$ or ZnO); commercial straight dynamites contain 20–60% nitroglycerin

E

ease-of-care textiles: *see* heat setting

EB cap: electric detonator cap; composition cap

ebonite: a hard rubber; *see* rubber

ebulliometry: in polymer technology, a method to determine the number-average molecular weight by measuring the boiling point elevation of a polymer solution

EDB: *see* ethylene dibromide

EED: electroexplosive device; detonator initiated by electric current

effluent: in chemical processing liquids or gases discharging from an outlet, for example, waste water or stack gases; outflow from liquid column chromatography

egg shell: *see* gloss

eicosapentaenoic acid: derived from fish oil; a C20 fatty acid with *cis*-double bonds at positions 5,8,11,14, and 17; dietary ingredient that reduces lipoprotein content of blood

elastic deformation: *see* elasticity

elasticity: ability of a material to recover its original shape after release of stress; that change in dimensions is called the elastic deformation; the elastic limit is the maximum stress a material can experience and still return to its original dimensions; elastic memory is the tendency of certain materials that have been deformed by stress to return to the original dimensions upon heating; elastic recovery is the fraction of a given deformation that can be recovered upon release of stress

elasticizer: additive that imparts elasticity to a material; also *see* plasticizer

elastic limit: *see* elasticity

elastic memory: *see* elasticity

elastic recoil detection analysis: ERDA; technique to analyze surfaces down to 1 μm; the incoming ions that weigh 10–40 MeV penetrate the surface and a small number collide with target atoms; the recoil of the atoms is proportional to their mass

elastic recovery: *see* elasticity

elastomers: cross-linked (vulcanized) or thermoplastic high molecular weight polymers with elastic properties; examples are natural rubber, synthetics such as polybutadiene, polyisoprene, butyl rubber (BR, from isobutylene with some isoprene for cross-linking), nitrile rubber (NBR), neoprene (polychloroprene), styrene-butadiene copolymers (SBR), polyethers (based on epichlorhydrin and other epoxides), fluorinated elastomers (mostly copolymers of vinylidenefluoride, tetrafluoroethylene, and hexafluoropropene), acrylic esters and copolymers with ethylene; thermoplastic block copolymers: styrene-butadiene, urethane-esters or -ethers, ether-ester; use: tires, belts, foot wear, hoses, O-rings, seals, adhesives, chewing gum, tough molded parts, wire and cable insulation; the specific application depends on the elastomer's resistance to temperature, specific chemicals, oxidation, ozone

elastomeric fibers: *see* Spandex; urethane block copolymers

elastomeric seals: *see* elastomers

Elbs persulfate oxidation: oxidation of phenols to hydroquinones using persulfate in an alkaline medium; useful for oxidation-sensitive compounds such as aldehydes; yields are improved by use of chelating agents such as ethylenediamine tetraacetate (EDTA)

electrets: electrically polarized polymeric disks that have one positively and one negatively charged side; formed by heating and cooling poor conductors such as nylon, poly(methyl methacrylate), polypropylene, and polystyrene in a strong electromagnetic field

electric cells: *see* batteries

electric furnace phosphoric acid: *see* furnace phosphoric acid

electrically conducting polymers: these fall into two classes (1) conducting composites that are produced by incorporating metal or carbon flakes into the polymer matrix; (2) highly conjugated systems to which electron donors or acceptors are added or in which bulk properties have been modified by pyrolysis; examples are polyacetylene, poly(*p*-phenylene), poly(*p*-phenylene vinylene), poly(*p*-phenylene sulfide), polyphosphazenes, polypyrrole, polythiophene, polyfuran; applications include their use in batteries, biomedical membranes, sensors, indicators, and ion-exchange devices

electric detonator cap: EB cap, composition cap; used to initiate secondary explosives; unlike in fuse caps, the primary explosive in the cap is ignited by an electric spark

electroacoustics: process in which ultrasound energy is used to augment electrical energy, e.g., in electroacoustic dewatering, ultrasound (20,000 cycles/second) is passed through a slurry to promote diffusion and migration of the liquid away from the filter; use: in food, waste water, pulp and paper, mineral processing industries

electrocoating: in textile technology electrically charged floc is shot onto an adhesive treated base to produce decorative materials

electrodecantation: continuous process used to concentrate rubber latex by adding the latex to a tank lined with electrodes; when an electric current is applied the latex particles build up and float to the top where they are decanted

electrodeposition: in coating and paint technology the use of an electric charge to deposit solid or dissolved material on a metal surface

electrodes: batteries and electrolysis cells have two electrodes that have different electromotive potentials so that a current can flow through the cell via the electrolyte; *see* batteries

electrodialysis: ED; unit operation that uses semipermeable membranes and a direct electric current to remove solutes from solutions; use: purifying brackish water; *see* inverse electrodialysis

electroexplosive device: *see* EED

electrography: nonimpact printing process in which first a uniform electric charge is deposited onto a photoreceptor in the dark, then an electrostatic latent image is created on the photoreceptor either digitally with a laser or by reflecting an image from the original document; the exposed areas become conductive because the photoreceptor loses surface charge, unexposed areas retain the charge; the latent electrostatic picture is developed with fine dry toner powder; *see* printing

electrohydrodynamic enhancement: corona wind, an effect that occurs when a high voltage discharge ionizes the gases within an oven and forces them to the surface of the products being heated; process improves convective heat transfer without use of fans

electroless plating: deposition of metal on a surface from solution without use of electric current; the material to be plated is first dipped into a reducing agent and then into a metal ion solution, e.g., dipping a plastic into an acidic $SnCl_2$ solution followed by $PdCl_4$ solution causes palladium metal to precipitate onto the surface of the plastic

electrolytes: substances that provide ionic conductivity; *see* batteries

electromagnetic waves: light is propagated through space as electromagnetic waves; the shorter the wavelength the more energy the light contains; spectroscopic methods use this difference in energy: for example, atoms and molecules characteristically absorb the short wave UV light of specific wavelengths to move electrons further away from the nucleus; on the other hand the longer wave infrared light is absorbed by molecules causing chemical bonds to vibrate or rotate; by knowing the relation between the energy absorbed and the molecular structure one can use absorption spectroscopy for analysis; *see* atomic absorption spectroscopy

electromembrane: *see* electrodialysis

electrometathesis: unit operation that combines electrodialysis and metathesis reactions by passing more than two streams through a stack; *see* metathesis

electronic chemicals: materials used in the manufacture of high-tech devices such as integrated circuits and printed wiring boards; they include materials to produce semiconductors (elements, alloys and oxides), resists (e.g., photopolymers), optical storage devices (e.g., conducting polymers and liquid crystals) as well as those needed for packaging, encapsulating, insulating and laminating

electron spectroscopy for chemical analysis: ESCA; also X-ray photoelectron spectroscopy (XPS); surface analysis of a solid by bombardment with monochromatic X-rays; by measuring the energy of the electrons ejected from the inner electron shells one can determine the elemental composition and chemical states; sensitivity range is to a depth of a few atom layers, lateral resolution ranges from a few 100 μm to a 1 mm

electron spin resonance spectroscopy: ESR; method to detect free radicals by measuring the magnetic moment of their unpaired electrons; energy source is microwave radiation rather than radio frequency as in nuclear magnetic resonance

electron volt: eV; unit of energy equal to 1.6×10^{-19} joules; the million electron volt (MeV) equals 1.6×10^{-13} joules; used to describe the energy of particles employed in analytical methods

electroosmosis: the transport of water through a membrane under the influence of an electric current

electrophoresis: migration of colloidal particles in the presence of an electric field; used for protein analysis and purification, coating of surfaces

electroplating: deposition of metals in an electric cell in which the object to be plated serves as cathode

electroporation: in biotechnology a process that facilitates the introduction of DNA molecules into host cells; the cells are briefly exposed to an electric shock of several thousand volts that does not kill the cells but opens temporary holes in the cell wall through which the DNA molecules enter

electrostatic precipitators: Cottrell precipitators; devices to remove particles from gases; the particle-laden air is charged with electricity, the electric charge is transferred to the particles that then cling to the container surface and can be removed upon discharge; use: for cleaning factory stacks, for removing fly ash from combustion gases and for dust removal during manufacturing processes

electrostatic printing: transfer of powdered

ink from an electrically charged stencil to a plastic film

electrostatic spraying: transfer of atomized paint to a conductive object by producing an electrostatic potential between the nozzle and the surface to be painted; useful for painting irregular surfaces; electrodeposition of paint on a conductive surface can also be done by placing the object into a water bath that contains the paint, the surface to be painted becomes the anode when current is applied

elemi: natural resin

elongation: in polymer technology the increase of length of a material at the point of rupture; *see* break elongation

Emersol process: purification and separation of fatty acids by fractional crystallization with methyl alcohol

emery: impure variety of corundum consisting of Al_2O_3 crystals embedded in an iron oxide matrix; hardness 8 Mohs; emery paper is produced by gluing ground emery onto paper or cloth; emery cake consists of mixtures of Al_2O_3 and iron oxide and is designed for buffing (higher percentage of Al_2O_3) or polishing (higher percentage of iron oxide); use: abrasive

emmission spectroscopy: *see* atomic emission spectroscopy

emollient: in the cosmetic and surfactant industry, an ingredient such as a fatty acid or lanolin; use: to make skin soft or supple; *see* detergents

empty-cell process: wood preservation that achieves deep penetration by air; wood is first subjected to air pressure, a hot liquid preservative is injected under pressure; when the pressure is released the air in the cells forces any free liquid out of the cells

empyreuma: acetone soluble portion of carbon black

emulsifiers: emulsifying agents that stabilize emulsions by preventing the suspended droplets from coalescing; *see* dispersants, emulsions

emulsions: dispersions of one liquid phase in another; the droplet size generally exceeds 100 nm; emulsions are intrinsically unstable and can break up in two ways (1) the droplets may coalesce into larger drops that eventually causes a complete phase separation ("breaking") or (2) they may flocculate to form aggregates that retain individual droplets followed by "creaming" in which all droplets are in one single aggregate; emulsifying agents that stabilize emulsions act by adsorption to the dispersed phase; they include surfactants, ions, particulate solids, or polymers; emulsions are commonly described as oil in water (o/w) or water in oil (w/o) in which the first phase mentioned is the dispersed phase and the second one mentioned is the continuous phase; examples of emulsions are commonly found in consumer goods such as foods, pharmaceuticals, paints, and photographic products

emulsion polymerization: polymerization in a low viscosity medium in which monomers are held in suspension by emulsifiers; the process has good temperature control and uses redox initiator systems; high molecular weight is achieved with high polymer concentrations; direct use as emulsion or latex paints

enamel: hard, glossy coating

enantiomer: one of the isomers of a pair of chiral (optically active) isomers; *see* chirality

enantiomeric excess: ee; measure of excess (percent) of one of the enantiomers in a chiral product: use of chiral ligands in catalysts leads to large ee; important in synthesis and in biologically active products; *see* ligands

encapsulation: *see* immobilization

end-group analysis: in polymer technology, the determination of the number of end-groups per mass of polymer; used to find the number-average molecular weight

endocrine glands: *see* hormones

endonuclease: *see* recombinant DNA

endorphins: brain polypeptides that inhibit the transmission of pain impulses

endosulfan: polychlorinated broad spectrum insecticide; hazardous chemical waste (EPA); also *see* chlordane

endothermic process: reaction taking place with the absorption of heat

energy density: in a battery, the ratio of the energy available to the volume (watt-hour/L) or weight (watt-hour/kg) of the battery

energy efficiency: in a battery, the ratio of watt-hours delivered on discharge (output) to the watt-hours needed to restore it to its original condition (input) under specified conditions

enflurane: CHF_2OCF_2CHClF; *see* anesthetics

engineering ceramics: ceramics that provide high-temperature resistance, electrical resistivity, dielectric strength, corrosion and abrasion resistance; use: in electronics (capacitors and resistors, integrated circuit substrates and packages, piezoelectrics, insulators), wear products (tumbling media, wire guides, nozzles and cyclones), aircraft (radomes, turbine coatings), tools and dies, pump and valve components, instrumentation and control (sensors, automotive choke heaters), process equipment (heat exchangers, reactors and vessels, filters), dentistry, and catalyst substrates.

engineering plastics: materials having high mechanical strength, and good chemical, moisture, and thermal resistance; examples: acetals, polycarbonates, polyphenylene sulfide, polysulfone, modified polyphenylene oxide, polyimide, and polyamide-imide; properties are often further improved with additives; *see* plastics

enhanced oil recovery: EOR; processes to

reclaim petroleum after primary (pumping) and secondary recovery (water flooding) processes have removed up to 35% of the oil; enhanced (or tertiary) recovery processes promise to be able to remove about an additional 10%; EOR includes flooding the well with steam, CO_2, or surfactant/polymer mixtures followed by polymer/water injection to drive the oil out of the rock cracks; the use of microorganisms is showing promise

enkephalins: brain polypeptides that inhibit transmission of pain impulses

enthalpy: H; heat content; thermodynamic quantity

entrained-flow gasification: *see* coal gasification

envenomation: process by which the surface of a plastic is deteriorated by close contact or vicinity with another surface (ASTM D 883-65T)

environmental impact statement: document required by the United States National Environment Policy Act; used as a tool for legislative decision making

enzyme multiplied immunoassay technique: EMIT; *see* immunoassays

enzyme-linked immunosorbent assay: ELISA; *see* immunoassays

enzyme immunoassay: EIA; *see* immunoassays

enzymes: proteins that act as biological catalysts; produced by extraction from natural sources, especially from microorganisms; often used as crude preparations, as immobilized whole cells or immobilized on supports (e.g., glutaraldehyde, polyacrylamide); the most important industrial enzymes are: proteases that hydrolyze the peptide bond in proteins (e.g., rennin for cheese manufacture, alkaline protease for detergents), carbohydrases that hydrolyze starch are mainly

used for high-fructose corn syrup (others are α-amylase, glucoamylase and glucoisomerase), lipases that hydrolyze fats, and lactases that hydrolyze milk sugar; products that are manufactured with enzymes are phenylalanine, aspartic acid, novel penicillins, and other amino acids; use: for industrial catalysis, organic synthesis, and process aids in biotechnology, in detergent formulations to remove stains, waste treatment, food processing; highly purified enzymes are used as analytical tools; some industrially important enzymes are amino acylase, catalase, cellulase, dextranase, glucose oxidase, hemicellulase, invertase, lactase, lipase, pectinase, penicillin acylase, pullulanase, rennins; *see* immunoassays

enzyme nomenclature: names identify the reaction the enzyme catalyzes or the substrate on which it acts; the name ends with the suffix -ase, e.g., hydroxylase, amylase; a formal system (E.C.) ascribes four numbers to each enzyme; the first number is based on the following six classes of enzymes (1) oxireductases; (2) transferases; (3) hydrolases; (4) lyases; (5) isomerases; and (6) ligases; three additional numbers indicate the subclass, the sub-subclass and the serial number of the enzyme in the sub-subclass; thus the industrially used α-amylase an enzyme that hydrolyzes starch has the code E. C. 3.2.1.1 denoting that it is a hydrolase, belonging to the subclass of endoamylases; numerous enzymes have retained their former trivial names such as rennin which is used in cheese making

eosin: xanthine dye; use: for dyeing textiles; staining biological specimen

epichlorohydrin: epichlorhydrin; produced by HOCl addition to allyl chloride followed by dehydrochlorination; use: for manufacture of bisphenol-A glycidyl ethers, which are precursors for epoxy resins; epichlorohydrin rubber is oil and ozone resistant and is nonflammable due to the high chlorine content

$$H_2C \overset{\displaystyle{}}{\underset{O}{\diagdown\diagup}} CHCH_2Cl$$

EPICHLOROHYDRIN

epidermis: in plants the outer layer of seeds and stems, in animals the outer layer of skin

epinephrine: adrenaline; isolated by extraction of cattle adrenal glands; neurotransmitter that constricts blood vessels, heart stimulant; use: medication against bleeding, cardiac arrest; an antiallergic agent, antiasthmatic

HOCHCH$_2$NHCH$_3$

EPINEPHRINE

epitope: molecule on the surface of cells that is recognized by an antibody; important factor in immune response

epoxidation: formation of epoxy compounds by oxidation of olefins

epoxy-: prefix indicating presence of an epoxy group in a compound

$$-\overset{\displaystyle{}}{\underset{O}{C\diagdown\diagup C}}-$$

EPOXY–

epoxy: short for epoxy resin

epoxy adhesives: *see* epoxy resins

epoxy binders: *see* binders

epoxy foams: *see* epoxy resins

epoxy paints (1) pigments held by epoxy binders that can be oil-modified to dry by oxidation; (2) epoxy resin that is mixed with amine or polyamide to harden and cure; epoxy paints are highly durable because of their good adhesion, hardness, flexibility, and resistance to abrasion; used for outdoor coatings and as protection to corrosive environments

epoxy plasticizers: unsaturated fatty and vegetable oils epoxidized by oxidation (e.g., H_2O_2) and epoxidized triglycerides of linseed and soy bean oils as well as oleic acid and its esters can be used as primary or secondary plasticizers for polyvinylchloride (PVC); alone or with metal soaps, they can prevent light and heat degradation in PVC; they also act as stabilizers because the epoxy group can neutralize acids (HCl), especially in the presence of metal soaps

epoxy resins: thermosetting resins that contain the epoxy group; they are condensates of bisphenol A (p-HOC$_6$H$_4$)$_2$C(CH$_3$)$_2$ and epichlorohydrin, other diols or chlorine-free di-epoxides; their properties depend on the relative concentrations; large excess of epichlorohydrin leads to a liquid compound of 1 mole of bisphenol A and 2 moles hydrin, higher phenol concentration gives viscous and solid materials of higher molecular weight; liquid epoxies are used for casting, potting, coating, and adhesives, and are cured (cross-linked) with amines, polyamides, anhydrides; solid resins can be modified with epoxidized phenol-formaldehyde resins or unsaturated fatty acids; epoxyresins have excellent wetting and adhesion properties and are resistant to heat, water, corrosive materials; use: as adhesives, as foams, in composites with a wide variety of fillers, in electric printed circuits, in structural units requiring strength at elevated temperatures, in the construction industry as bonding agent for concrete blocks, epoxy pipes, floors

epoxy stabilizers: *see* epoxy plasticizers

eptam: $C_2H_5SCON(C_3H_7)_2$; preemergence carbamate herbicide

ERDA: *see* elastic recoil detection analysis

ergot: *see* alkaloids

erythorbic acid: isoascorbic acid; used as antioxidant

erythritol: $CH_2OHCHOHCHOHCH_2OH$; 4-carbon sugar (tetrose); vasodilator

erythritol tetranitrate: explosive that blows up on percussion; produced by nitration of erythritol with concentrated nitric acid; when mixed with milk sugar it is nonexplosive and is used as blood vessel dilator

erythromycin: macrocyclic antibiotic (macrolide antibiotic); produced by *Streptomyces erythreus*

esparto: raw material for cellulose fiber

essential oils: aromatic plant oils used for perfumes and flavors

esters: compounds formed from an organic acid and an alcohol with the elimination of water: $R'COOH + ROH \rightarrow R'COOR + H_2O$; esters of higher fatty acids and glycerol occur naturally in vegetable and animal oils, fats and waxes; esters provide flowers and fruit with their aroma; use: solvents, flavors, perfumes, fats, lubricants and pharmaceuticals; they serve as raw materials for the manufacture of soaps and chemicals; esterification of dibasic acids (HOOCR'COOH) and diols (HOROH) leads to polyesters -(OCR'CO–)n (ORO-)m, an important class of polymers

esterification: reaction leading to ester formation; most industrial esterifications are catalyzed, usually with an acid catalyst

estradiol: female sex hormone; use: for estrogen hormone therapy; the 17-ethynyl derivative is used as contraceptive and for breast cancer therapy

ESTRADIOL

estrogens: female sex hormones that produce secondary female sexual characteristics such as breasts; antiestrogens are used as drugs for sex-linked cancers in women

esu: abbreviation for electrostatic unit in the cgs (cm-g-sec) system

etching materials: materials used to cut into the surface; for metals: acids, for glass: HF or white acid [$HF + (NH_4)FHF$]

ethanal: *see* acetaldehyde

ethane diacid: *see* oxalic acid

ethane dial: *see* glyoxal

ethanol: *see* ethyl alcohol

ethanolamine: $H_2NCH_2CH_2OH$; produced from ethylene oxide and ammonia; use: chemical raw material particularly for manufacture of amphoteric surfactants, scrubbing agent for removal of acidic compounds from gases

ethene: *see* ethylene

ethenoid plastics: polymers made from monomers containing –C=C–

ethers: compounds having the generic formula ROR'; produced by the catalytic hydration of olefins; diethyl ether [$(C_2H_5)_2O$] and tetrahydrofuran are important solvents; polyethers are used as commercial polymers; crown ethers are large cyclic polyethers that can serve as ligands and chelating agents for metal ions, polyoxyethylenes are used in nonionic detergents, polytetramethylene oxides are the flexible block in spandex polyurethanes

ethirimol: *see* fungicides

ethrel: $ClCH_2CH_2PO_3H_2$; plant growth regulator; releases ethylene gas on decomposition within the plant

ethyl alcohol: ethanol, grain alcohol; C_2H_5OH; 95% (190 proof) ethyl alcohol is produced by; (1) acid-catalyzed vapor-phase addition of water to ethylene (Shell Direct

Hydration Process); (2) indirect hydration of ethylene by hydrolysis of ethylene sulfates; or (3) fermentation of agricultural materials such as molasses, using the yeast *Saccharomyces cerevisiae*; absolute alcohol is produced by distillation with benzene; industrial alcohol is denatured with a bittering agent to prevent consumption; industrial use: solvent, fuel (gasohol), raw material for production of glycol ethers, ethyl chloride, acetaldehyde, amines, ethyl acetate, single cell protein and vinegar; used as beverage

ethyl benzene: $C_6H_5C_2H_5$; produced by reaction of benzene and ethylene in the presence of alumina on silica gel, solid phosphoric acid or synthetic zeolite; use: as feedstock for styrene synthesis; hazardous chemical waste (EPA)

ethyl cellulose: EC, cellulose ether; produced from wood pulp or cotton linters by reaction with strong alkali to alkali cellulose followed by treatment with ethylchloride in a solvent at 110–150°C; degree of substitution 2.2 to 2.5 OH groups out of three per glucose unit; EC is tasteless and nontoxic; use: coatings, compression and injection moldings of high impact strength and heat distortion temperature, tough flexible sheets, food storage containers

ethylene: $CH_2=CH_2$; ethene; produced by thermal (steam) cracking and catalytic cracking of hydrocarbons ranging from ethane to heavy gas oil; use: chemical feedstock to manufacture ethanol (Shell and UCC processes), ethylene glycol, ethylene oxide, styrene, and vinyl chloride; major end-uses are polymers such as polyethylene and polyvinyl chloride, antifreeze, fibers and solvents; used as plant growth stimulator

ethylene cyanohydrin: $HOCH_2CH_2CN$; produced from (1) propylene, ammonia and air; (2) ethylene oxide and HCN; or (3) acetylene and HCN; use: intermediate in the production of acrylates

ethylene diamine: EDA, $NH_2(CH_2)_2NH_2$; prepared from ammonia and ethylene dichloride or ethylene oxide; use: feedstock for manufacture of ethylenediamine tetraacetate (EDTA), carbamate fungicides and other chemicals; stabilizer for rubber latex, antigelation agent, solvent

ethylene diamine tetraacetic acid: (EDTA, $(HOOCCH_2)_2NCH_2CH_2N(CH_2COOH)_2$); manufactured by addition of NaCN and formaldehyde to ethylenediamine; use: as chelating agent in detergents, for metal cleaning, removal of metal ions from solutions (including blood), analytical reagent, processing aid, stabilizer for latex

ethylene dibromide: EDB; $BrCH_2CH_2Br$; manufactured by bromination of ethylene; use: as lead scavenger in leaded gasoline; fumigating pesticide particularly used against nematodes; use controlled by the FDA

ethylene dichloride: EDC; $ClCH_2CH_2Cl$; produced from ethylene via catalyzed ($CuCl_2$) oxychlorination with HCl and O_2 in air; liquid phase variation of the process employs aqueous $CuCl_2$/HCl solution at 170–185°C and 12–18 bar; older method: liquid phase chlorination of ethylene with Cl_2 at 40–70°C and 4–5 bar and using $FeCl_3$, $CuCl_2$, or $SbCl_3$ as catalysts; use: for manufacture of vinylchloride, vinylidene chloride, tri- and tetrachloroethylene, as solvent for asphalt, bitumen, resins and tars, for extracting fats and oils

ethylenedinitramine: *see* haleite

ethylene glycol: EG; CH_2OHCH_2OH; glycol; colorless, sweet, syrupy liquid; prepared by (1) hydrolysis of ethylene oxide; (2) catalytic oxidation of ethylene to the diacetate followed by hydrolysis (Halcon's oxirane process, Celanese, duPont, Kuraray, ICI, Teijin); and from (3) syngas (CO + H_2O) using $Rh(CO)_n$ catalyst at 1300–3400 bar and 125–350°C (UCC process) or using $H_2Ru(CO)_3$·N-methylimidazole (Sumitomo,

Mitsubishi, Mitsui); use: antifreeze, for synthesis of polyethylene terephthalate (PET)

ethylene glycol dinitrate: EGDN; high explosive

ethylene glycol ethers: $R(OCH_2CH_2)_nOR'$, R=alkyl, R'=H,alkyl produced by reacting an alcohol with ethylene oxide in the presence of acid or base; products are isolated by fractionation; use: de-icing of jet fuels (n=1), brake fluid diluent (n=2 or 3), solvents for printing inks and high boiling mixtures (e.g., tetraglyme, where R=R'=CH_3, n=4)

ethylene oxide: oxirane; produced by (1) air oxidation of ethylene; (2) treatment of ethylene chlorohydrin with alkali; used for manufacture of ethylene glycol, nonionic surfactants, and ethanolamines

$$H_2C \overset{}{\underset{O}{\diagup\diagdown}} CH_2$$
ETHYLENE OXIDE

ethylene urea: *see* 2-imidazolidone

ethylene-vinyl acetate: EVA, copolymers; *see* vinyl acetate

2-ethylhexanol: 2EH; iso-octanol; $CH_3(CH_2)_3CH(C_2H_5)CH_2OH$; produced from propylene in the one-step Aldox process; the process consists of hydroformylation of propylene to butyraldehyde followed by an aldol condensation of the butyraldehyde to the unsaturated aldehyde and finally gas phase hydrogenation; the single-step process is possible due to the use of oxo catalyst (Co, cobalt salt, $Co(OH)_2$ or cobalt phosphine complexes) together with cocatalysts (Zn, Sn, Ti, Al, or Cu compounds, or KOH); use: for manufacture of esters of adipic, phosphoric, sebacic, and phthalic acids

ethylene-propylene rubbers: EPR; these saturated elastomers can only be cross-linked with peroxides or by radiation; to permit cross-linking with sulfur nonconjugated dienes must be grafted onto the polymer; *see* ethylenpropylene terpolymer

ethylene-propylene terpolymer: EPDM, EPT; ethylene-propylene elastomers containing 2–4% of nonconjugated diene such as dicyclopentadiene or 1,5-hexadiene to permit vulcanization with sulfur as well as with peroxide; they have good low-temperature properties and excellent resistance to oxygen and ozone and are used for tire manufacture

ethylmethyldiphenyl urea: *see* centralite

ethyltetryl: 2,4,6-trinitrophenylethylnitramine; explosive; similar to tetryl but it has a lower melting point so that it can be used in energy-rich pourable mixtures

17-α-ethynylestradiol: contraceptive; used in breast cancer therapy

eugenol: 2-methoxy-4-allylphenol; isolated from cloves; use: raw material for synthesis of vanillin, dental analgesic, used in perfumery

eutrophication: process by which a lake becomes enriched in nutritive compounds resulting in excess growth of algae and plants and, over time, in the conversion of the lake to a marsh and finally to dry land; eutrophication is sometimes accelerated by human activity

evaporation: change of a liquid into a gas; used for separation and purification processes

excipient: binder used in pharmaceutical industry for making tablets

exclusion chromatography: *see* gel permeation chromatography

exinite: dull bands in coal consisting of finely divided material that is derived from leaves, pollen, spores, seeds, and resins

expansion engine: gas refrigeration unit in which the engine does useful work; liquefaction occurs in condenser units; *see* Joule-Thompson expansion cycle

exothermic process: reaction that proceeds

with the evolution of heat; scale-up presents special problems of heat dissipation

expectorant: *see* guaifenesin

explosion process: in wood pulping, process in which lignin is softened and partially hydrolized; wood chips are steamed at 285°C and 600 psi for 30 sec, the pressure is raised to 1000 psi for 5 sec before the softened chips are ejected through a quick-opening valve; the quick expansion causes the chips to explode into a fluffy mass of fibers

explosives: metastable solids, gels or liquids that react with violent expansion upon application of shock, heat, friction, or impact; the ease with which the chemical reaction can be initiated is known as sensitivity; propagation rate from initiation site may be slow (deflagration) or rapid (detonation); important explosives include nitro compounds (TNT), aromatic nitramines (tetryl), aliphatic nitramines (cyclonite), nitric acid esters (nitroglycerin), initiators (lead azide); also *see* black powder, detonating explosive, dynamite, gelatin, high explosive, mercury fulminate, trinitrotoluene, tritonal, class A, B, and C explosives; use: for military and industrial purposes as explosives, blasting agents, pyrotechnics, propellants, igniters, and detonators;

explosophore: chemical group or radical whose decomposition or rearrangement generates large amounts of heat, e.g., azide ($-N_3$), acetylide ($-C_2$) and fulminate ($-ONC$)

expression: in processing the removal of liquid from a solid–liquid mixture by application of pressure; *see* filtration; in biotechnology, the cellular process of producing compounds from information stored in genes (DNA). *see* PLATE PRESS, CONTINUOUS SCREW PRESS

extended X-ray adsorption fine structure spectroscopy: EXAFS; microanalytical technique to determine elemental composition and chemical structures by X-ray bombardment and measuring the energies of the

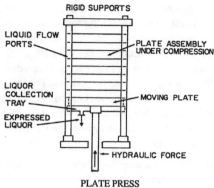

PLATE PRESS
[Courtesy D. M. Considine, P.E.]

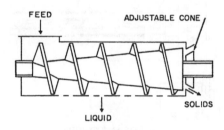

CONTINUOUS SCREW PRESS
SCREW PRESS DECREASES IN THE DIRECTION OF FLOW.
[Courtesy D. M. Considine, P.E.]

emitted X-rays; technique is applicable for in situ analyses and gives a lateral resolution in the millimeter range

extender plasticizer: *see* secondary plasticizer

extenders: materials that modify inks, paints, plastics, and rubber by controlling viscosity, texture, suspension, and gloss; diluents often used to decrease cost

externally heated-oven process: distillation of hardwood; batch process using retorts that were heated with fossil fuel; no longer used in the United States

extraction: unit process to remove products with solvents; when a heavy solvent is used it is introduced at the top of the column, the feed is introduced at the bottom, and the extract is removed at the bottom; conversely a light solvent is introduced at the bottom, the feed enters on top and the extract is removed from the top; extraction columns may be packed and/or have mechanical agitation and/or be heated; extraction with supercritical fluids takes advantage of their superior solvent action

extruder: machine for forcing a heat-softened polymer through a die by means of a ram or a screw, use: to extrude sheets, rods, bars, tubes, etc.

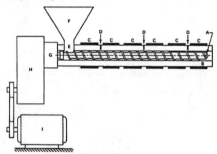

PARTS OF AN EXTRUDER
(A) screw; (B) barrel; (C) heater; (D) thermocouple; (E) feed throat; (F) hopper; (G) thrust bearing; (H) gear reducer; (I) motor.
[Courtesy of Society of Plastic Engineers. Copyright Kirk-Othmer. *Encyclopedia of Chemical Technology,* 3d ed., John Wiley & Sons, Ltd.]

extrusion: process for converting polymers or their solutions into products: fibers, films, sheets, rods, tubes; for fiber production, the solution or melt is pushed through a spinneret, a plate with a number of fine specially designed holes (one to thousands depending on the spinning method); sheets and films are extruded through adjustable slits, e.g., a cellulose solution is pushed through a slit into a regenerating bath to give cellophane film; many polymers are melt extruded into sheets that may be biaxially stretched, oriented, crystallized, annealed, and cut into tapes (*see* polyethylene terephthalate); most thin films (polyethylene, polypropylene) are melt

extruded into a tube that while warm is air blown into a wide balloon, then is flattened and wound up

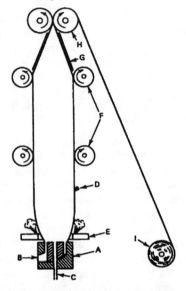

EXTRUSION OF BLOWN FILM
(A) blown-film die; (B) die inlet; (C) air hole and valve; (D) plastic tube (bubble); (E) air ring for cooling; (F) guide rolls; (G) collapsing frame; (H) pull rolls; (I) windup roll. [Courtesy of Society of Plastic Engineers. Copyright Kirk-Othmer. *Encyclopedia of Chemical Technology,* 3d ed., John Wiley & Sons, Ltd.]

Exxon Donor Solvent process: *see* coal liquefaction

F

fabric softener: laundry additive that reduces static electricity and wrinkling; usually a quaternary ammonium salt

FACS: fluorescence-activated cell sorter; *see* flow cytometry

factice: polymerized product of unsaturated vegetable oils and sulfur; it is resistant to ozone, improves extrusion and calendering, gives smooth finishes to textiles; use: softener, electrical insulator, eraser

fadometer: apparatus to test lightfastness of plastics, rubbers and dyed fabrics by exposing the materials to electric or xenon arc having a spectrum similar to that of the sun

falling dart impact test: *see* hardness

false twisting process: method to manufacture textured synthetic yarns; yarn is twisted by a rotating spindle, then heat set in the twisted state, and finally untwisted to give voluminous, crimped yarn

fast breeders: *see* nuclear reactors

fats: lipids; high melting esters of fatty acids derived from animals; they consist mostly of triglycerides with lesser amounts of mono- and diglycerides, and esters of other alcohols such as sterols (e.g., cholesterol), free fatty acids and phospholipids; low melting lipids are known as oils; fats are distinguished from waxes that contain no glycerides; use: nutrient in food and feed, source of fatty acids, for the manufacture of soap; *see* vegetable oils

fat splitting: process to obtain free fatty acids (FFA) and glycerol from fats; fat splitting is achieved by three methods (1) the Twitchell process is a batch hydrolysis of fats in the presence of benzenestearosulfonic acid, (Twitchell's reagent) or sulfonated petroleum products; a mixture of fat, water, acid and the above reagent are boiled for 1–2 days, after which the fatty acids are drawn off and purified; (2) batch autoclave hydrolysis of a charge consisting of fats, water and sometimes a catalyst at 75–450 psi and 300–450°F; the reaction is completed in 2–10 hours depending on reaction conditions; (3) continuous, countercurrent splitting processes are completed in less than two hours at 450–500°F and 600–700 psi; purification: the solid stearic/palmitic acid fraction can be separated from the liquid fatty acid fraction by passing the cooled solid–liquid mixture through a mixer containing a surfactant and separating the suspended solids by centrifugation (Lurgi)

fatigue cracking: when rubber or elastomers are subjected to cyclic stress, cracks appear in the direction normal to the stress; where stress is greatest fatigue failure results in the rupture of the material

fatigue life: number of cyclic stresses before a material fails

fatigue limit: stress below which a material can be stressed cyclically without failure

fatigue ratio: fatigue strength/tensile strength

fatigue strength: maximum cyclic stress a material can withstand for a certain number of cycles without failure

fatty acids: long-chained saturated and unsaturated aliphatic carboxylic acids; obtained from fats by hydrolysis; use: food, detergent, and chemical industries; *see* fat splitting

fatwood: *see* lightwood

FCC: fluid catalytic cracking; *see* catalytic cracking

feedstock: raw material used for chemical production

feed zone: *see* extrusion

feel: fabric characteristic; *see* hand; drape

feldspar: $KAlSi_3O_8$; also a general name for alkali aluminum silicates; used in pottery, as filler for plastics it enables reduced use of pigment and provides clarity and heat resistance; other use: as abrasive and fertilizer

felt: nonwoven material in which fibers are interlocked and held together by physical or chemical forces

fenoprofen: $C_6H_5OC_6H_4CH(CH_3)COOH$; diphenylether propionic acid; antiinflammatory drug

Fenton's reagent: ferrous sulfate-hydrogen peroxide; use: an oxidizing agent

fenvalerate: synthetic pyrethroid insecticide; *see* pyrethrin

ferbam: ([CH$_3$]$_2$NC(S)=S)$_3$Fe; thiocarbamate fungicide

fermentable sugars: hexoses, such as glucose, mannose, and galactose, can be fermented by yeasts; pentoses are not fermentable

fermentation: process to convert raw materials to products with the help of microorganisms; ancient process has been used for manufacture of bread, yogurt, cheese, and wine; modern fermentation is used to manufacture commodity and specialty chemicals and biologicals; fermentation processes include chemical reactions such as oxidations, reductions, polymerizations, and hydrolyses, as well as biosyntheses and the formation of cells; some processes may need the presence of air (aerobic), others the absence of air (anaerobic); rate of fermentation depends on the concentration of microorganisms, cells, cellular components (i.e, macromolecules) and enzymes, as well as on temperature and pH; product recovery always involves the concentration of the dilute solution; products may be excreted into the broth or they may remain within the cells so that the cells must be broken for isolation of the products; the cells must be separated from the broth whether or not products are excreted by the microorganisms

fermentors: fermentation is usually carried out in tanks equipped for aeration, stirring, control of temperature and pH; continuous processes are primarily used for feed or food production and for waste treatment; some continuous process use recycling or step-feeding, i.e, feeding to all reactor stages; batch processes are preferred for production of pharmaceuticals because of frequent mutation of microorganisms; in some processes concentrated substrate is continuously added during the cycle to avoid substrate inhibition (batch-fed fermentation); product inhibition is prevented by repeated draw-off of the product; to avoid biological process contamination, the fermentation medium, air, and equipment are usually sterilized; the shape and size of stirrer impellers are chosen by taking cell fragility into account; modern fermentors are controlled by computer coupling

ferric ferrocyanide: *see* ferrocyanate pigments

ferrimagnetism: a solid state property due to the preferential directional alignment of the spins of the orbital electrons; ferrimagnetic materials are relatively strongly magnetic

ferrocene: (C$_5$H$_5$)$_2$Fe; dicyclopentadienyl iron; a double ring compound with the iron atom between the two planar rings; produced from FeCl$_2$ and cyclopentadiene sodium; use: in fuels as combustion control additive and antiknock agent, in plastics and lubricants for heat stabilization and radiation resistance, in missiles and satellites as coating, high-temperature lubricant

ferrochrome: alloy of iron and chromium (60–70% Cr); use: raw material for manufacture of chromium; also used to add chromium to steel

ferrocyanate pigments: these pigments have the general structure: M$^+$[Fe^{+2}Fe^{+3}(CN)$_6$]·xH$_2$O with M = Na$^+$, K$^+$, NH$_4^+$; examples: Prussian blue, Chinese blue, Milori blue, Turnbull's blue; the pigments are produced by precipitation of ferrous salts with hexacyanoferrate(II) ([Fe(CN)$_6$]$^{4-}$) followed by oxidation with chlorates or dichromates; use: car paint, printing ink, colored paper, coloring fungicides for outdoor application

ferromagnetism: a solid state property due to the one-directional alignment of the electronic spins of the orbital electrons; ferromagnetic materials are highly magnetic

ferrorefining: in refining of lubricating oil, a process to improve color and oxidation stability by mild hydrotreatment; this low-pressure process does not use clay and proceeds with low hydrogen consumption

ferrosilicon: alloy of iron and silicon; *see* killed steel

fertilizers: soil additives to increase plant fertility; nutrients supplied by fertilizers are mainly phosphorus, nitrogen, and potassium with micronutrients added in special formulations; fertilizer grades are characterized by the weight percentage of N, P_2O_5, and K_2O in the formulation, e.g., NPK 10-20-10; fertilizers may be liquid or granular; the most important nitrogen source for fertilizers is ammonia which is derived from coal, oil, and gas reforming; ammonia, may be present as free ammonia, ammonium salts, or urea; other nitrogen products in fertilizers include calcium cyanamide, sodium nitrate, and guano; phosphorus is mined from phosphate rock and is used directly as phosphate rock or as ammonium phosphate (MAP), diammonium phosphate (DAP), or superphosphate (TSP); potassium is derived from mineral deposits or recovered from potassium-containing brines; finished NPK fertilizers are marketed as granulated, bulk blended, or fluid product; granulated fertilizer is produced by ammoniating a mixture of ordinary superphosphate (OSP) and potash in a granulator, and adding sulfuric acid to promote granulation before drying the product; bulk blending consists of mixing granulated fertilizer to desired grades, e.g., urea, ammonium nitrate, diammonium phosphate (18-46-0), and KCl; fluid fertilizers are solutions, suspensions, or slurries and have the advantage of easy application

festooning oven: oven used for drying, curing, or fusing plastic-coated fabrics at constant heat; material is held on rotating shafts with loops ("festoons") between the shafts

Feuerland machine H 50: device used to determine fatigue properties between –20 and 70°C

fibers: materials that have small cross section with lengths that range from <1 in. (staples) to >1000 yards ("endless" fibers); they present a wide range of mechanical properties (strength, stiffness, extensibility) with high lateral flexibility in yarns and fabrics

fibers, inorganic: natural fibers consist of a group of minerals that are collectively called asbestos; of these crysotile, a hydrated magnesium silicate, is the most important for textile uses; asbestos is no longer used for insulation because it is implicated as a cause of cancer; the major inorganic fiber melt spun into continuous or staple fiber is glass, which is used for industrial filtration, insulation, composite reinforcing material and for consumer goods; metallic fibers are meltspun and drawn similar to the organic polymers and are used for wires, tire cords, and fabrics; high-temperature-resistant fibers made of silicon carbide, silicon nitride, and boron nitride, consist of very short staples, whereas the length of refractory alumina-silicon fibers ranges from 1 in. to endless; carbon, graphite, graphite whiskers, or boron are used for high strength and/or high temperature reinforcements in composites; synthetic inorganic fibers such as phosphazenes that have a $(-P=N)_n$ backbone and polyphosphates, $(MPO_3)_n$ are not yet in commercial use: *see* fiber optics, fiber reinforced materials

fibers, organic: natural fibers based on cellulose (cotton, ramie, hemp, flax, jute) and protein (wool) are short (staple) filaments, therefore they must be twisted to produce yarns for fabric manufacture; silk is the only "endless" natural fiber and is a few 1000 yards long; synthetic fibers are extruded from melts or solutions as endless yarns, are oriented to give desirable mechanical properties in fabrics or industrial materials; blending cotton or wool with synthetic fibers requires cutting endless yarns into staple; *see* fiber drawing; spinning

fiber axis: length direction of a fiber

fiberboard: rigid or semirigid sheet materials made from wood or other vegetable fibers; to manufacture fiberboard wood pulp is refined, screened, mixed with additives, formed into sheets and dried; sizing for insulation board includes rosin, cumarone resin, and asphalt; sizing for compressed board is mainly paraffin wax and tall oil; fiberboard may be formed by wet-felting that is similar to paper manufacture or by air-felting in which dry fiber is mixed with wax for sizing and sheets are formed by compression between rollers; compressed fiberboard and hard board are used for structural materials and furniture; noncompressed fiberboard is used for heat and sound insulation

fiber drawing: in synthetic fiber manufacture, the application of tension along the fiber axis either during or after spinning; the process is needed to orient the filaments to obtain useful properties; the extent of drawing is adjusted to the end use (i.e, textile, industrial yarns and fabrics, composites)

fiberfil molding: injection molding of short bundles of fiber surrounded by resin

fiberglass: *see* glass fibers; inorganic fibers

fiber number: linear density of a fiber, given in denier, tex, or decitex; tex = weight of 1000 m of filament or yarn, decitex = weight of 10,000 m; *see* denier

fiber optics: transmission of light by coated glass or plastic fibers that are gathered into bundles and surrounded by a flexible plastic such as polyethylene; use: for transmitting light along curved paths and into areas that are difficult to access, as probes and sensors (e.g., for monitoring oil pipe lines for frost heave and corrosion, mapping stress in structures), in electronics, medicine, and displays; *see* optical fibers

fiber reinforced materials: FRM; incorporation of fibers into plastics, glass, or metal leads to increased strength of the materials; fibers used include boron/B_4C_3, carbon/graphite, silicon carbide, ceramic materials, glass, and metals; *see* composites; MMC

fiber reinforced plastics: FRP; *see* fiber reinforced materials

fibrid: short fibers made from synthetic polymers; use: manufacture of strong, tough papers

fibrillation: in textile technology, splitting of filaments in yarns or fabrics by mechanical action or deformation

fibroin: chief protein constituent of raw silk that remains after processing during which the other major constituent, sericin is removed

Fick's Law: *see* diffusion

field separation: in petroleum refining, the pretreatment of crude oil in which gases, water, and dirt are allowed to settle out in large vessels

filament: long fiber; silk is a naturally occuring filament fiber; synthetic filaments are the individual extrudates as a molten or dissolved polymer is forced through the die of a spinneret; filaments that are thousands of yards long are called "endless"; filament yarns consist of many filaments that can be used with little twist; they transmit tensile properties along the length of the filaments in contrast to short staple fibers that must be twisted into yarns

filament yarn: *see* filament

filler: *see* additives; plastics

filling yarn: in a woven fabric, fibers running perpendicular to the warp

film: in plastic technology, thin sheets or tapes made of thermoplastic polymers or regenerated cellulose; films are cast from solution or melt extruded; films or sheets requiring improved mechanical properties are subsequently biaxially oriented; very thin films can be obtained by melt extrusion of a continuous tubing that is then blown by air pressure to a large thin-walled cylinder and slit to produce flat sheets

filter: device for separation of solids from liquids; *see* expression

filter cake: dried solids recovered by filtration

filtration: separation of solids from the liquid phase; industrial filters include variable-volume filters that use pressure to dry the filter cake after filtration, belt filters that combine gravity drainage with mechanical squeezing between two belts, continuous press filters designed for handling large amounts of mineral materials, cartridge and bag filters used for clarification, and electrofiltration for colloids; microparticles, such as bacteria, can be separated with microfilters having pore diameters of 0.1–10 μm; molecules weighing 500–300,000 daltons can be separated by ultrafiltration. *see* HORIZONTAL-TRAVELING BELT FILTER

fine chemicals: high value-added chemicals produced in small quantity; specialty chemicals

fine filament yarn: yarn weighing 40–100 denier; used for lightweight apparel

fines: powdered material; the size of fines depends on a given technology: in coal purification fines are 1/16–1/8 in. diameter

fire-damp: *see* methane

fire-extinguishing agents: the agents used depend on the type of fire; for wood, paper, fabrics, or plastics one uses water, foams, water-surfactant mixtures, salts, and antifreeze compounds; for flammable liquids and greases water is often unsuitable; foams cannot be used with gases or liquefied gases; electric equipment fires require carbon dioxide, halogenated hydrocarbons, or dry chemicals (e.g., sodium and potassium bicarbonates, phosphates, chlorides); for fires involving combustible metals (Mg, Ti, Zr, Na, K), NaCl and graphite compounds or Met-L-X (a mixture of NaCl and vinylidene copolymer) serve to seal off oxygen from the surface

firefoam extinguisher: system that produces a foam blanket by interaction of a sodium bicarbonate solution and an alum solution in the presence of a foam stabilizer

fire retardant: *see* flame retardant

Fischer–Tropsch process: liquefaction of synthesis gas; the original process used an iron or cobalt catalyst at 350 psi and 330°C; modern catalysts have led to more moderate conditions

fish eye: in plastics technology, small globular fault appearing in plastic sheets or films due to incomplete mixing

fission: in nuclear technology, the reaction that splits an atomic nucleus into smaller fragments with the release of energy; in commercial power plants fission of uranium-235 is initiated by neutrons and the resulting heat is used to generate steam; the probability that a neutron of given energy will cause fission of the nucleus of a given element is

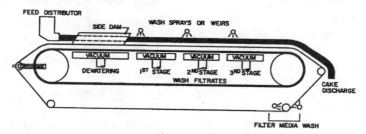

HORIZONTAL-TRAVELING BELT FILTER
[Courtesy D. M. Considine, P.E.]

obtained by its "cross-section" (expressed in barns, 1 barn = 10^{-24} cm^2); the fission products are radioactive and problems of disposing this nuclear waste have not yet been solved

fitting operation: in batch soap manufacture, the separation of the batch into neat soap and nigre

fixed-bed gasification: Lurgi gasifiers; *see* coal gasification

fixed-bed gasifiers: *see* coal gasifiers

fixed bed reactors: industrial operations in which the reaction bed or column (adsorbent, catalyst) is penetrated by a mobile phase (liquid or gas) permitting the coexistence of solid and fluid for a desired time; the design of fixed-bed reactors depends on the equilibrium and kinetic behavior of the reaction; use: for processes such as catalytic reforming, hydrocracking, benzene production, hydrotreating, ion-exchange, vinyl chloride production

flame emission spectroscopy: FES; analytical method to detect the presence of atoms or atomic groups; solution samples are nebulated and then vaporized in acetylene- or hydrogen-oxidant flames. During vaporization some atoms, molecules, or parts of molecules are elevated to excited electronic levels; when returning to the ground state, they emit characteristic radiation, which is analyzed in a spectrometer

flame-over: *see* flashover

flame retardants: fire retardants; materials that reduce the tendency of flammable substances to burn; fire retardants may be applied to the surface, impregnated into materials such as wood, or incorporated during polymerization of plastics and rubbers; fire retardants in ancient times included impregnation of materials with alum and coating of wood with clay and gypsum;

inorganic flame retardants include antimony trioxide, ammonium sulfate, and borax; alumina trihydrate filler dehydrates under flame conditions and injects large amounts of steam (but no HCl) into the flame vicinity; organic flame retardants are highly halogenated compounds (e.g., chlorendic anhydride, tetrabromophthalic anhydride, tetrabromobisphenol A, chlorostyrene) and unsaturated phosphonated chlorophenols; compatible halogenated polymers (e.g., brominated polystyrene) have been used with engineering plastics; however, halogenated organics often release HCl or HBr during fires; combinations of organic halides and antimony oxide act synergistically; intumescent flame retardants such as ammonium sulfonates, swell under fire conditions and block access of oxygen; because carbon is less flammable than plastics, catalysts that promote the pyrolysis of the polymer backbone to carbonaceous char are added to plastics formulations; highly aromatic polymers (aramids, aromatic polyesters, imidazoles, ladder polymers) or compounds such as polytetrafluoroethylene, are intrinsically fire resistant

flammability: ease with which a material will ignite; measured by the oxygen index test (OI) in which a test film is ignited and exposed to increasing concentrations of oxygen; OI equals the percent oxygen at which the material does not self extinguish; an OI of 25–30 is required for most fire retardant systems; the OI of carbon is 65, therefore conversion of a polymer backbone to char during pyrolysis leads to fire retardance

Flanders stone: *see* graphite

flash: in the rubber and plastics industry the overflow when pressure is applied during molding

flashing: distillation process in which the total vapor removed approaches phase equilibrium with the residual liquid; used in flash evaporators

flashover: flame-over; in the plastic industry: ignition of plastic surfaces in buildings that are exposed to hot gases; in the electrical industry: the electric discharge around the surface or edges of electrical insulation

flash point: the lowest temperature at which a substance emits flammable vapor that ignites spontaneously

flat: *see* gloss

flatting agent: colorless additive (e.g., silica) that reduces the gloss of coatings

flavanol: *see* tannins

flavor: sensation derived from a combination of taste, aroma, and texture, as well as appearance and color; quantitative flavor profiles are determined by panels who judge each product by a set of descriptors and standards; for example, the standard "woody note" for black pepper is the aroma of caryophilline, for vanilla it is wood of a freshly sharpened pencil; to compare flavors of natural and synthetic products panels construct profiles by putting the various descriptors on a circular graph

flax: annual plant, *Linum usitatissimum*; fibers are obtained from the plant stalks by retting with bacteria and moisture; the resulting fine fibers are the starting material for linen fabrics; oldest textile material known

flexicoking: process that integrates coke gasification with fluid coking; products are coke gas and hydrogen-rich synthesis gas, which can be used to manufacture methanol or ammonia; *see* coke

flexicracking: catalytic conversion of petroleum oils to lower molecular weight products such as olefins, high octane gasoline, and middle distillates

flexographic inks: *see* flexography

flexography: printing method using raised type, rubber plates, and fluid ink; flexographic fluid inks contain few solids in a solvent that does not affect rubber; inks using polyamide resin binders are widely used for printing on plastics despite their lack of heat resistance because of their high gloss and good adhesion; inks made with acrylic resins dissolved in an alcohol/ester mixture are oil and water resistant, have good heat stability and good adhesion and abrasion resistance; use: for printing on packaging materials such as plastic film, paper, aluminum foils, corrugated boxes

flexural modulus: the ratio of applied flexural stress/highest bending strain within elastic limit

flexural strength: measure of a material's resistance to bending; the maximum stress in the outer fiber at the moment of rupture

float glass process: *see* glass

floc: aggregates forming when a colloidal solution coagulates; in water purification technology, clumps of solids formed by biological or chemical action, i.e, by flocculation

flocculation: first stage of coagulation of a colloidal solution; a finely divided, stable dispersion of solid particles in a liquid is a colloidal solution, the particles show Brownian motion; flocculation causes cluster formation (due to addition of a salt, other solvent, or change of surface characteristics of the particles); flocculated clusters can be redispersed by stirring but will cluster again when stirring stops; greater changes of medium cause particles to adhere irreversibly

flock: in textile technology, short fibers used as reinforcement

flocking: surface coating of flock onto plastic materials to simulate surfaces such as suede

flotation: separation of particles based on

the difference in wetting by the suspending medium; *see* froth flotation

flow cytometry: optical technique to study biological cells; cells are labeled with a fluorescent dye and allowed to flow past a laser beam; scattering of the laser light depends on cell size and shape, DNA and RNA content, as well as interaction between drugs and cells; the technique can be used to identify and separate antibody-cell complexes from a mixture by labeling the antibody with a fluorescent dye; in the fluorescence-activated cell sorter (FACS) fluorescing antibody–cells complexes are given a negative charge and separated from the stream when they flow past a laser beam

flue gases: term usually applied to stack gases (e.g., CO_2, NO_X, SO_2) produced during the combustion of carbon containing fuels; the nitrogen and sulfur oxides are sources of air pollution and their emission is restricted by law; reduction of NO_X emission can be achieved by various methods: flue gas recirculation, selective catalytic reduction with NH_3, water injection, low excess air firing or two-stage combustion; SO_2 can be removed by physical absorption, chemisorption, or partial catalytic reduction (Claus process)

fluff process: method of preparing detergents with a density of 0.25–0.9 g/mL; sodium trimetaphosphate in soap formulation undergoes an exothermic conversion to sodium tripolyphosphate hexahydrate when the soap slurry is heated to 60–70°C; the heat released causes the slurry to expand and to form fluffy granules (Monsanto process); expansion can also be achieved in formulations that release CO_2 on heating (Colgate-Palmolive)

fluid catalytic cracking: FCC; *see* catalytic cracking

fluid coking: *see* coking

fluid fertilizers: *see* fertilizers

fluidity: property defined as 1/viscosity

fluidization: *see* fluidized bed reactor

fluidized bed combustor: FBC; *see* fluidized bed reactors

fluidized bed gasification: *see* coal gasification

fluidized bed reactors: reactors in which combustion gas agitates a bed of granular solids (e.g., fuel, limestone, and ash) so as to behave like a fluid; gas velocity is adjusted so as to keep the particles in motion but not to blow them out of the bed; the temperature can be kept constant and suspended particles remain in close contact with the gas medium; a grate supports the fuel bed and distributes the underfire air that is blown into the bed through nozzles; high turbulence causes the fluidization; particles must be prevented from agglomeration; used in processes such as calcining, combustion, cracking, roasting, coking, steam generation, and chemical syntheses such as the production of phthalic anhydride from *o*-xylene or naphthalene; for example in coal combustion crushed coal is mixed with lime and suspended on gas jets while it burns; SO_2 reacts with the lime and is removed as ash; because combustion occurs below 1800°F, little or no NO_X is formed; in a second generation multiple-bed variation combustion occurs in a lower bed, desulfurization with lime occurs in the upper bed; advantage is said to be that the two beds can be kept at different temperatures taking advantage of the fact that coal burns most efficiently at 1750°F and SO_2 reacts most efficiently with lime at 1500–1600°F

fluorapatite: $Ca_5(PO_4)_3F$; phosphate ore

fluorescein: acid yellow 73; xanthene dye; fluorescent compounds are characterized by stiff structures that do not permit the loss of excited states through torsion; in fluorescein the quinone is reduced to a phenol ring; use: for diagnosis of eye injury

FLUORESCEIN

fluorescence: some materials when illuminated radiate light usually of a longer wavelength; the radiation ceases when illumination ends; fluorescence is caused by the emission of part of the energy absorbed during illumination; *see* phosphorescence

fluorescence-activated cell sorter: FACS; *see* flow cytometry

fluorescent whitening agent: FWA; dyes that brighten by reflecting light beyond the visible spectrum into the UV; *see* optical brighteners

fluoridation: addition of 1 ppm of fluoride to the drinking water to prevent tooth decay; solids added to water are sodium fluoride or sodium silicofluoride; fluosilicic acid in liquid form is prefered as it is easier to meter automatically into the water supply

fluorine: F; element; produced by electrolysis from fused KF-2HF using steel-monel cathodes, carbon anodes, and monel containers and screens; electrolysis proceeds at 8–12 V, a current of 6–15 kA and, a current density of 0.1–0.15 A/cm^2, efficiency is 90–95%; because of the high voltage, the reaction is exothermic (35 MJ/kg F)and requires cooling; the anode is protected with metal against corrosion by fluorine; the product F$_2$ contains ~10% HF, which is largely removed by cooling (< −100°C); the electrolysis product can be used directly for production of UF$_6$ or SF$_6$ or liquefied for storage; use: chemical synthesis, oxidizer for rocket fuel

fluorocarbon polymers: the monomers of industrially useful fluoro polymers are tetra-

fluoroethylene, hexafluoropropylene, chlorotrifluoroethylene, vinylfluoride, vinylidenefluoride; tetrafluoroethylene is polymerized by free radical mechanism in water dispersion or emulsion, the polymer [poly(tetrafluoroethylene), PTFE, Teflon] is crystalline, it has a density of about 2.2 g/cm^3 and a very high molecular weight; it is resistant against corrosion and solvents, thermally stable at 250°C, and feels slippery and waxy; it has a low coefficient of friction and good electrical properties; molding and forming involves compressing the granules or powder at room temperature followed by sintering above the melting point; use: electrical insulation, gaskets, packings, bearings, tray linings, dry lubricants, any application where chemical inertness ≤250°C is essential; the copolymer with hexafluoropropylene has similar inertness and can be processed by conventional extrusion methods; polyvinylfluoride is crystalline, has good chemical resistance, is tough and strong between −180 and 150°C, has low permeability to gases; use: protective coatings; polyvinylidenefluoride is crystalline, has good weatherability, chemical and solvent resistance, has piezoelectric properties; copolymers with hexafluoropropylene are elastomers with good thermal properties to 200°C, resistance to lubricants, fuels, hydraulic fluids; the chlorotrifluoroethylene polymer (Kel-F) is less crystalline, forms noncrystalline transparent sheets, is tough over a wide temperature range, can be molded by normal methods; copolymers with vinylfluoride range from thermoplastics to elastomers, have excellent resistance to strong oxidizing agents such as 90% hydrogen peroxide, fuming nitric acid

fluorocarbons: organic compounds in which hydrogen has been replaced by fluorine; in general they are chemically more inert than hydrocarbons although they react violently with metals such as Ba, K, and Na; use: refrigerants, solvents, blowing agents, coatings, momomers; *see* fluorocarbon polymers

fluoroimmunoassay: FIA; *see* immunoassays

fluorophore: molecular grouping that causes dyes to fluoresce; the most important fluorophore is diaminostilbene

5-fluorouracil: antimetabolite that acts by competing with uracil, a constituent of RNA; use: chemotherapeutic

5-FLUOROURACIL

fluosilicic acid: *see* fluoridation

flurazepam: benzodiazepine tranquilizer; *see* diazepam

fluroxene: $CF_3CH_2OCH=CH_2$; the first fluorinated inhalation anesthetic; potentially explosive; *see* anesthetics

flux: additive such as borax or fluorspar that promotes fusion of metals, minerals, and glass; in the polymer industry, an additive to improve flow properties of plastics

fly ash: solids carried in a gas stream

foam materials: foams are dispersions of gas in a liquid or solid that are characterized by low density and increased compressibility; many foams are also poor conductors of heat, sound, and electricity; solids used for foam materials include ceramics (refractories), glass, metals, plastics, and rubber; formulations of the matrix includes blowing agents that release a gas during foam manufacture and surfactants that control foaming and stabilize the foams; depending on the blowing agent and method of curing foams have open (e.g., sponges) or closed cells; use: important industrial and consumer materials that provide buoyancy, insulation, and cushioning; end products include: padding, roofing, packaging, seals, insulation against heat, sound, electric, vibration, and pressure; *see* blowing agents; surfactants; urethane block copolymer

foam molding: *see* injection molding; thermoplastics; injection molding; thermosets

foam rubber: *see* foam materials

foamed plastics: *see* foam materials, polyurethane, polystyrene

foaming agents: *see* blowing agents

fodder molasses: waste fraction from sugar recovery of beet molasses

fold strength: in the paper and textile industry, a measure of the material's stability upon folding

folic acid: water-soluble vitamin; deficiency causes anemia

folpet: thiourea herbicide with low toxicity for mammals and low persistence; used on food crops and in plastics formulations

food additives: materials that serve to improve the lifetime of foods (e.g., preservatives and antioxidants), texture (e.g., thickeners, stabilizers), nutritional value (e.g., vitamins, proteins), aesthetics (e.g., coloring agents, flavor enhancers), and processing (e.g., anticaking agents, lubricants); The Food and Drug Administration (FDA) has issued a list of additives that are generally regarded as safe (GRAS) or with acceptable daily intakes (ADI)

foots: the bottoms separated from the edible portion of vegetable oils; used for soap manufacture

forensic chemistry: application of chemical, physical, and biological methods of analysis to identify physical evidence observed on a victim or at a location of a crime; microanalytical methods are at present available to determine causes of fire, hair and other body samples, intoxication beverages, drugs, explosives, fingerprints, and faked documents

formaldehyde: HCHO; produced by catalytic oxidation of methanol; the commercial form of formaldehyde consists of a 37–50% aqueous solution (formalin); use: chemical feedstock (notably as raw material for urea, phenol, melamine, alkyd and acetal resins, pentaerythritol, ethylene glycol, urea-formaldehyde fertilizers, hexamethylene tetramine and tetrahydrofuran), preservative, embalming agent; toxic and carcinogenic

formamide: $HCONH_2$; produced (1) from methyl formate and ammonia, and (2) by reaction of CO and NH_3; use: solvent, extraction agent

formic acid: HCOOH; strongest of the aliphatic carboxylic acids; produced (1) as by-product of liquid phase oxidation of hydrocarbons; (2) by reaction of CO with water or alcohol; and (3) from synthesis gas via formamide; use: silage manufacture, food conservation, organic syntheses, formamide, methyl- and dimethyl-formamide

formonitrile: *see* hydrocyanic acid

formulation: specific mixture of chemicals and additives required for optimum properties of a product

fossil flour: *see* diatomaceous earth

fossil fuel: plant remnants including coal, peat, tar sands, shale oil, petroleum, and natural gas

foundry resins: thermosetting polymers used to bind sand in foundry operations

fourdrinier machine: device used for paper manufacture; a screened slurry of pulp is mixed with additives before entering the headbox at the wet end of the machine; the slurry is moved onto a series of wire screens upon which the fibers are concentrated and through which the water is drained; the wet paper passes through presses and driers where a dandy roll smoothes the paper and enters any water marks; sizing is added after the first drying step and the paper is then

dried again over felt blankets, finally passes through calender rolls to be wound into large rolls

fractional distillation: separation of components in a solution based on differences in boiling points

fractionation: separation of a mixture of product components by fractional distillation, e.g., separation of the xylene isomers by the Parex process; in polymer technology the separation of polydisperse polymer mixtures into fractions of similar molecular weight by fractional precipitation of a solution using small amounts of a precipitant solvent, precipitating first the fraction with the highest molecular weight; *see* gel permeation chromatography

fracture toughness: resistance to crack propagation in a solid

franklinite: zinc and manganese ore; use: raw material for ZnO pigment; residue from ZnO manufacture is smelted to form an iron-manganese alloy (spiegeleisen) that is used for steel manufacture

Frasch process: extraction of elemental sulfur by piping superheated water (165°C) into sulfur-bearing strata through the outside ring of three concentric pipes; the sulfur is melted, compressed air is fed down through the inner pipe and the molten sulfur is forced upward through the second pipe. *see* MINING SALT-DOME SULFUR BY THE FRASCH METHOD

free fatty acids: FFA; *see* fat splitting

free radical: molecule or atom with an unpaired electron; formed by UV irradiation or thermal decomposition of compounds such as peroxides, hydroperoxides, azo or diazo compounds

free radical polymerization: reaction that is started when a free radical initiator acts on a monomer; polymerization proceeds via further reaction of the free radical chain and more monomer until the reaction is terminat-

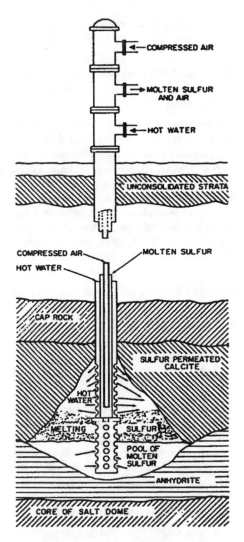

MINING SALT-DOME SULFUR
BY THE FRASCH METHOD

The process entails placing three consecutive pipes inside
a casing that lines a shaft drilled to the base of the sulfur
deposit. Typical diameters for the pipes are 8, 4, and 1 in.
Water heated to a temperature well above the boiling point
is pumped down the space between the 4- and 8-in. pipes
and also, in the initial stage, down the 4-in. pipe. It flows
through the holes in the casing and into the rock, where it
melts the sulfur, which forms a pool at the bottom of the
well and, under the pressure of the water, rises part way up
the 4-in. pipe. Compressed air forced down the 1-in. pipe
lightens the sulfur so that it rises to the surface.
[Courtesy D. M. Considine, P.E.]

ed by combination of two radicals or disproportionation with another free radical

freeze-drying: lyophilization; method of
food preservation allowing future food recon-
stitution and for saving biological materials
without chemical, physical and enzymatic
changes; the process involves freezing, sublimation of ice under vacuum and finally removing bound (nonice) water

freeze grinding: cryogrinding; material is
cooled with liquid nitrogen to a temperature
at which it is brittle enough for grinding;
used for polymers such as rubbers that cannot
be ground at room temperature

French mold: two-piece mold used to make
irregularly shaped objects

freon: dichlorodifluoromethane; *see* chlorofluorocarbons

friction calendering: frictioning; impregnation of a fabric using a calender; the heated impregnating compound enters between
the upper and central rollers of a three-roll
calender, the fabric enters between the lower
and central rollers; the central roller moves
faster than the two outer rollers so that the
compound is forced into the interstices of the
fabric

Friedel–Crafts processes: acid-catalyzed
additions, cracking, eliminations, isomerizations, polymerizations, or substitutions; the
original reaction consisted of the alkylation
of aromatics using $AlCl_3$ as catalyst; modern
Friedel–Crafts processes are used for aromatic as well as aliphatic condensations, conditions can be heterogeneous or homogeneous;
generally the solvents are the hydrocarbons
that undergo substitution; catalysts usually
are Lewis-type acids

Frings acetator: in vinegar manufacture, a
batch process in which room air is pulled
from the bottom by a high-velocity self-aspi-
rating rotor into the fermentor containing the
submerged culture; high production rate is

150

off-set by the need for filtering the final product

FRM: in Japan, fiber reinforced metal; *see* metal material composite (MMC); FRP

froth flotation: separation of small particles (usually less than 1 mm) with the help of chemical agents that attach the particles to air bubbles; the chemical agents include (1) collectors (e.g., xanthates, fatty acids, or amine salts, sulfonates) that coat particles to be floated with a water-repelling surface; (2) modifying agents (e.g., phosphates, aluminum sulfate-tartrate complexes) which can depress flotation of unwanted material while permitting flotation of the desired substance; (3) activators (e.g., ammonium sulfide); (4) pH regulators (acids, bases) to achieve selectivity between species to be separated; and (5) frothers (e.g., amyl and butyl alcohols, methylisobutyl carbinol [MIBC], pine oil, and cresol); use: concentration of minerals (beneficiation, separation of oil from industrial waste, removal of ash and sulfur minerals from coal, separation of bacteria and colloidal particles from water, recovery of post consumer plastics

fructose: $C_6H_{12}O_6$; levulose; fruit sugar; produced from hydrolized sucrose or extracted from inulin in countries where inulin-producing plants are common; the sweetest of all sugars; use: sweetener; *see* invert sugar; high fructose syrup; inulin

β-fructosidase: *see* invertase

fruit sugar: *see* fructose

frying oils: refined oils that have been hydrogenated to resist oxidation at high temperatures

fuel alcohol: ethanol produced from biomass and used as fuel or fuel additive

fuel cells: electrochemical batteries that convert chemical energy directly to electricity and heat; hydrogen or hydrogen-rich gas is fed to the anode, air is fed to the cathode; water is the by-product of the reaction; individual cells produce less than 1 V so that they must be assembled into stacks; hydrogen released from chloralkali electrolysis can be utilized by coupling the electrolysis cell with fuel cells; large-scale fuel cells using phosphoric acid as electrolyte (PAFC) have passed the demonstration stage; cells with solid oxides (SOFC), alkali (AFC), solid polymer electrolyte (SPEFC), and molten carbonate (MCFC) as electrolyte are under development

fuel oil: petroleum fraction, grades No.1–6

fuels: materials that can be used to produce heat and energy by burning solids, coals, wood, liquids: gasoline, oil, petroleum, or gases (natural gas, syngas); waste material can be burned to make use of the heat developed; major considerations are the cost of combustion to useful forms of energy such as heat or electricity; problems to be solved are the control of air pollution by CO_2, NO_X, SO_2, and other reaction products

fuel value: recoverable heat energy

full-boiled kettle process: batch manufacture of soap by pumping melted fat or oil into a 12.6–14.4% solution of caustic (18–20°Bé); initial saponification of the boiling mixture takes ~4 hr; solid soap is separated by addition of NaCl; total work up including additional saponifications, removal of trapped glycerol by boiling and separation of NaCl, requires 4 days after which the neat soap is pumped into dryers, crutchers, or storing frames; soap obtained by this process is known as kettle soap

full-cell process: in wood technology, impregnation of wood with preservatives by subjecting the wood to a vacuum before filling it with a hot liquid preservative at 125–200 psi; this method leads to high retention of preservative but is not as penetrating as the empty-cell process; *see* empty cell process

fullerene: *see* buckminsterfullerene; allotropy

Fuller's earth: porous clay; *see* adsorbants

fulvic acid: *see* humus

fumaric acid: *trans*-HOOCCH=CHCOOH; produced in conjunction with maleic anhydride; by isomerization of maleic acid and by fermentation of glucose using *Rhizopus delemar*; use: food acidulant and flavoring agent, for production of unsaturated polyester resins, paper-size resins, surface coatings, plasticizers and for upgrading natural drying oils

fumed silica: pure SiO_2 made by reacting $SiCl_4$ in an oxy-hydrogen flame

fumigants: chemical gases such as hydrogen cyanide, methylbromide, ethylene oxide used in high enough concentration to kill insects; for slower release ethylenedibromide, β, β'-dichlorethylether are useful as are naphthalene and dichlorobenzene, which have high sublimation pressure

fuming sulfuric acid: *see* oleum

functionality: reacting group in a molecule, e.g., ethylene glycol, $HOCH_2CH_2OH$, which has two OH groups is a bifunctional molecule

fungi: soil organisms that are sources of antibiotics such as streptomycin

fungicides: agents that kill fungi; action may be due to surface protection of plants or due to penetration of plant tissue (systemic agents); nonpenetrative fungicides are most effective when applied before infection occurs; penetrative agents attack internal fungal tissue; systemic fungicides translocate within plants from point of application (e.g., root, stems or leaves); nonpenetrant fungicides include phthalimides (e.g., captafol), phthalonitriles (e.g., chlorothalonil), dithiocarbamates (e.g., maneb), mercurials (e.g., phenyl mercuric acetate), mineral oils, and sulfur (flower of sulfur); site-specific penetrant fungicides include benzimidazoles (e.g., benomyl), triazoles (e.g., flutriafol), imidazoles (e.g., prochloraz), morpholines (e.g., tridemorph), hydroxy pyrimidines (e.g., ethirimol), carboxyamides (e.g., carboxin), dicarboxyamides (e.g., iprodione), guanidines (e.g., guazatine), and organophosphates (e.g., pyrazophos)

furan: furfuran; tetrol; produced by dry distillation of furfural; use: organic raw material

FURAN

furanose: cyclic sugar whose structure resembles furan

furanoside: glycoside whose sugar component is a furanose

furfural: produced by steam-acid digestion of plant waste materials; use: solvent (e.g., for extraction and refining of gas oils), wetting agent, analytical agent, raw material for the manufacture of tetrahydrofuran, furfuryl alcohol, adipates, phenolic and furan polymers, and lysine

FURFURAL

furfuran: *see* furan

furfuryl alcohol: produced by vapor phase hydrogenation of furfural; used as wetting agent, sealant, penetrant and for the manufacture of urea-formaldehyde and furan polymers

FURFURYL ALCOHOL

furnace black: black pigment produced by the partial combustion of furnace gas; recovered in cyclones and electrical precipitators

furnace phosphoric acid: high grade phosphoric acid made by burning phosphorus in air and quenching the resultant P_2O_5 with water

furosemide: diuretic; useful in case of fluid retention in the lungs (pulmonary edema); *see* loop diuretic

fuse caps: detonation devices that contain only mercury fulminate $[Hg(ONC)_2]$ and are triggered by fuses

fused calcium magnesium phosphate: produced by fusing a magnesium ore with phosphate rock in an electric furnace; the pulverized product is used as fertilizer

fusinite: constituent of coal derived from carbonized wood or charcoal

G

G acid: 6,8-di-SO_3H-β-naphthol; dyestuff precursor

gage factor: ratio of the relative change of resistance to the relative change in length of piezoresistive strain

gage–pressure transducer: sensor that measures pressure in relation to the ambient (or atmospheric) pressure

galactose: $C_6H_{12}O_6$; sugar derived from lactose by hydrolysis; use: organic synthesis; *see* saccharides

D-galacturonic acid: hydrolysis product of pectin

galena: PbS; lead sulfide; lead ore; yields lead upon roasting to lead oxide, followed by smelting

gall: gall nuts; nutgalls; excretions of oak and sumac caused by insect punctures; source of tannins for the leather industry

gall nuts: *see* gall

gallic acid: *see* tannins

gallium arsenide: GaAs; used as semicon-ductor in transistors that need faster switching than silicon; can be used in rectifiers >300°C; use: for lasers and modulators

gamma acid: 3-SO_3H-7-NH_2-α-naphthol; dyestuff precursor

gamma-cellulose: *see* hemicellulose; saccharides

gamma-globulin: immunoglobulin G, IgG; antibody

gamma transition: *see* glass transition

gangue: *see* beneficiation

Gardner–Holt viscosity: measured with an air-bubble viscometer, in which one compares the time it takes for a standard bubble of air to rise in a sample tube to that in a set of standard liquids of known viscosities

gas chromatography: GC; gas–liquid chromatography, GLC; gas-solid chromatography, GSC; a separation technique in which the mobile phase consisting of a carrier gas and a complex mixture is driven through a column packed with an adsorbant solid phase; separation occurs by selective adsorption of constituents in the mobile phase on the stationary phase; the carrier gas is usually helium, hydrogen, or argon, but may be any gas that does not react with the system; use (1) as analytical tool; (2) for preparative

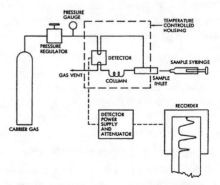

BASIC ELEMENTS OF A GAS CHROMATOGRAPH
[Courtesy D. M. Considine, P.E.]

153

purposes; and (3) to follow the progress of industrial processes; the technique has been completely automated for use in process technology where it is under computer control; *see* chromatography

gas-cooled power reactor: GCR; nuclear power generator built in England and France; gas-cooled fast breeders (GCFBR) are being developed

gaseous diffusion: *see* diffusion

gasification: *see* coal gasification

gas–liquid chromatography: GLC; *see* gas chromatography; chromatographic methods

gasohol: gasoline-ethyl alcohol (90%/10%) mixture used for automotive fuel

gasoline: fuel fraction of petroleum that has been refined for use in the internal combustion engine

gas-phase polymerization: polymerization of a gaseous monomer; used to make high-density polyethylene (HDPE) by feeding gaseous ethylene and a powdered chromium catalyst into a fluidized-bed reactor or by adding small amounts of butene-1 or hexene-1 to ethylene to form linear low density polyethylene (LLDPE); the resulting resins are formed in powder form

gas reburning: method to control NO_x formation by firing coal in the lower part of the boiler and burning natural gas in the upper, cooler, oxygen-deficient region

gassing agents: in explosives technology, materials used to provide enough void space to sensitize explosive for initiation; typical gassing agents, (e.g., $NaNO_2$, H_2O_2, Na_2CO_3, and Unicel [N,N'-dinitropentamethylene tetramine]) release gas bubbles by reaction in acids in the presence of catalysts or heat

gas–solid chromatography: GSC; *see* gas chromatography; column chromatographic methods

gas turbine: device in which hot combustion gases expand to drive a generator or compressor; used in aircraft engines

Gay-Lussac tower: tower used for sulfuric acid chamber process

gel: semirigid solid; colloidal structure consisting of a dispersion of a connected network of particles or large molecules; the dispersion medium is usually a low molecular weight substance (e.g., water); *see* sol; fish eye

gelatin: mixture of high molecular proteins from hydrolysis of collagen from skins, bones, and hides; before hydrolysis the raw materials are cleaned and carefully treated to give pure products; warm water soluble, has amphoteric character, reversible sol-gel formation ~30°C; use: as dispersing agent, food stabilizer, film coating, for pharmaceuticals, culture medium; gelatin comes in edible, photographic, and technical grades; *see* gelatin dynamite

gelatin dynamite: gelatin; high explosive mixtures of gelatinized nitroglycerin and explosively active absorbants (balanced dope); ammonia gelatin dynamites contain NH_4NO_3, straight gelatins contain $NaNO_3$, semigelatin contains $NaNO_3$, NH_4NO_3, and dinitrotoluene oil; gelatins are made by drying the carbonaceous material and dope together and then mixing it with the nitrates in a blender; all equipment must be spark-free and materials are handled in small lots; gelatins have the highest explosive power

gelling agent: thickener

gel permeation chromatography: GPC; exclusion chromatography (EC), gel filtration; a separation technique that uses the structure of a swollen gel to separate substances based on molecular shape and size; *see* chromatography

gel point: the stage in a polymerization reaction at which due to cross-linking the weight average molecular weight has become infinite; at this point the polymer stops flowing

gel spinning: a process for converting a very high molecular weight linear polyethylene, HDPE ($M_w = 10^6$, $M_n = 2 \times 10^5$) to an extremely strong, stiff fiber (modulus: 70–90 GPa, tensile strength: ≥ 3 GPa); a 2–5% solution, prepared at 160°C in decalin is spun into water to form a gel, the solvent is extracted or evaporated during subsequent drawing at 120°C (draw ratio: ≥ 30), the gel system allows orienting chains without entanglements or chain breaks; the method yields tensile properties in flexible polymers similar or better than those of aramids

generator: machine that converts mechanical energy into electrical energy

generic names: trivial names as opposed to trade names; generic names (and generic chemicals such as drugs, materials, pesticides, salts) are based on chemical structure and are not protected by copyright

genetic code: the universal code by which the bases of the nucleic acids are translated into amino acids during the synthesis of proteins in all living organisms; messenger ribonucleic acid (mRNA) carries the information from deoxyribonucleic acid (DNA) in a three-letter code spelled out by four bases, in RNA the bases are: adenine (A), guanine (G), uracil (U), and cytosine (C); in DNA U is replaced by thymine (T); thus, for example the triplet UUU spells phenylalanine (Phe); *see* GENETIC CODE (table)

genetic engineering: branch of biotechnology that is concerned with altering the structure of the cells' genetic material by removing and inserting fragments in the DNA chain

gibberillic acid: plant growth regulator; produced by fermentation; used to enlarge the size of seedless grapes, stimulate the amylase content of barley and initiate flowering

GIBBERILLIC ACID

glass: supercooled multicomponent inorganic liquid; produced by melting a mixture of fine particulate sand (SiO_2) and added cations in the form of oxides or salts; other additives include clarifying agents (Na_2SO_4, arsenic oxides plus Na or K nitrate) and decoloring agents (a mixture of KNO_3, CeO_2, MnO_2 [glassmaker's soap]); industrially the most important glasses are soda lime silicate glass, lead glass and borosilicate glass; soda lime silicate is used for windows, plate glass, bottles, laminated safety glass; most manufacturing is fully automated; molded container glass is made by blowing or pressing; most flat glass is made by Pilkington's float-glass process in which glass is floated under a protective gas over a tank filled with molten tin; lead glass contains a high percentage of lead, has a high refractive index, is used for quality art work and table ware; borosilicate glass has a low thermal expansion, is easy to blow, and is used in industrial equipment and laboratory glassware, for cookware; specialty glasses are solder glass (for joining different types of glass or making metal–glass joints, i.e, electric light bulbs), silica glass which is pure silica, has very high softening temperature and low thermal expansion for special laboratoryware; colored glasses contain transition metal cations (e.g., Cr^{+3}, Co^{+3}, Ni^{+2}) or colloidal metals (e.g., Au, Ag, Cu, Pt); *see* sol-gel process

glass–ceramics: high-performance materials prepared by introducing nuclei of Angstrom size to crystallize molten glass; glass–ceramics are machinable, have low dielectric constants, high strength, and nonlinear optical properties

155

GENETIC CODE

1st position 5' end	2nd position U	C	A	G	3rd position 3' end
U	Phe	Ser	Tyr	Cys	U
U	Phe	Ser	Tyr	Cys	C
U	Leu	Ser	STOP	STOP	A
U	Leu	Ser	STOP	Trp	G
C	Leu	Pro	His	Arg	U
C	Leu	Pro	His	Arg	C
C	Leu	Pro	Gln	Arg	A
C	Leu	Pro	Gln	Arg	G
A	Ile	Thr	Asn	Ser	U
A	Ile	Thr	Asn	Ser	C
A	Ile	Thr	Lys	Arg	A
A	Met	Thr	Lys	Arg	G
G	Val	Ala	Asp	Gly	U
G	Val	Ala	Asp	Gly	C
G	Val	Ala	Glu	Gly	A
G	Val	Ala	Glu	Gly	G

The abbreviations for the amino acids are: Ala (alanine), Arg (arginine), Asn (asparginine), Asp (aspartic acid), Cys (cysteine), Gln (glutamine), Glu (glutamic acid), Gly glycine), His (histidine), Ile (isoleucine), Leu (leucine), Lys (lysine), Met (methionine), Phe (phenylalanine), Pro (proline), Ser (serine), Thr (threonine), Trp (tryptophan), Tyr (tyrosine), Val (valine); STOP means "end the chain"

glass fibers: fiberglass; produced by spinning from molten glass; glass fibers are stiff so that they must be made finer than organic fibers; they have tensile strength in excess of 200,000 psi; they must be dyed before extrusion and must be protected against abrasion by coating with a lubricant; use: for fireproof and acid-resistant textiles, in fiber reinforced plastics (FRP), filters, electrical insulating tape; ultrapure glass fibers are used for telephone communication; *see* inorganic fibers

glassine: thin transparent paper treated with ureaformaldehyde resin

glass maker's soap: decolorizing mixture; *see* glass

glass transition: the reversible change of a viscous, elastic polymer to a brittle, glass like substance

glass transition temperature: T_g; the middle of the temperature range during which glass transition occurs; above the glass transition the glassy amorphous regions of polymers become flexible due to local segmental motion

glassy state: region in which a solid material is hard, brittle and noncrystalline or partially crystalline

Glauber's salt: $Na_2SO_4 \cdot (H_2O)_{10}$; use: for energy storage, air conditioning, as aid in direct dye baths

gloss: specular reflection; the degree to which a surface reflects light, i.e, acts as a perfect mirror; gloss of any surface increases with the angle of incidence that is determined with a goniometer; paints are classified by gloss (angle of incidence): high gloss (70°), semigloss (30–70°), eggshell (6–30°), flat to eggshell (2–6°), and flat (2°); uneven (dull or matte) surfaces cause diffuse reflection because the reflected light is dispersed through many different angles; gloss decreases with the amount of extender or pigment in the paint

Glover tower: tower used in sulfuric acid chamber process

glucoamylase: *see* amyloglucosidase

gluconic acid: $CH_2OH(CHOH)_4COOH$; dextronic acid; produced by glucose oxidation (chemical, fermentation using *Aspergillus niger* or *Gluconobacter suboxydans*); use: in food industry as acidulant, as protein coagulator in tofu (bean curd) manufacture, as metal ion sequestrant in detergent formulations and intravenous therapy and for metal pickling

Gluconobacter suboxydans: microorganism used for manufacture of tartaric, 5-ketogluconic, and gluconic acids from fermentation of glucose

glucose: dextrose; six-carbon sugar; a hexose; the monomeric unit of cellulose and starch; produced by hydrolysis of starch or wood; use: in food manufacture, inks, tanning, as a reducing agent; *see* invert sugar

glucose oxidase: notatin; enzyme produced by *Aspergillus niger* and *Penicillium notatum*; use: as oxygen scavenger in the food industry; combined with catalase for removal of oxygen or glucose; combined with peroxidase for quantitative glucose determination

glucose isomerase: GI; enzyme that converts glucose to fructose; produced by microorganisms such as *Actinoplanes missouriensis*, *Arthrobacter sp.*, *Bacillus coagulans*, *Streptomyces welmorensis*, *S. olivaceus*, and *S olicochromogenes*; used for manufacture of high fructose corn syrup (HFCS)

glucuronic acid: $HOOC(CHOH)_4CHO$; constituent of arabic gum; *see* saccharides

glue: colloidal suspension of polypeptides that forms reversible gels in water; animal glue is a hydrolysis product of collagen and is produced by boiling bones, hides, tendons; contrary to gelatin production, no prior cleaning of the starting materials is required; a vegetable glue is made from soybean protein; use: gummed tape, paints, plasters, liquid glue; *see* adhesives

glutamic acid: $HOOC(CH_2)_2CH(NH_2)COOH$; natural amino acid; produced by fermentation of acetic acid; use: as flavor enhancer in the form of monosodium glutamate

glycerides: esters of glycerol; *see* glycerol

glycerin: *see* glycerol

glycerol: $HOCH_2CH(OH)CH_2OH$; glycerin; present as constituent of fats; produced (1) as by-product of soap manufacture; (2) from epichlorhydrin by alkaline hydrolysis at ~100°C; (3) from allyl alcohol by catalytic (titanium oxide or titanates) oxidation with H_2O_2 (Shell or Degussa process) or via the monochlorhydrin using HOCl (Olin Mathieson process); and (4) from propylene using peracetic acid (FMC, Daicel processes); used as antifreeze, manufacture of organic chemicals, in printing inks, as intermediate of polyurethanes and alkyd resins, in food production

glyceryl monoacetate: *see* acetin

glyceryl trinitrate: *see* nitroglycerin

glyceryl tristearate: *see* stearin

glycidol: 2,3-epoxy-1-propanol; intermediate in glycerol manufacture; use: as chemical intermediate

glycogen: polysaccharide produced by animals; "animal starch"

glycol: *see* ethylene glycol

glycols: alcohols with two hydroxyl groups; used as raw material for manufacture of polyesters, e.g., 1,4-butylenediol or ethylene glycol react with terephthalic acid to give PBT or PET respectively, polymeric glycols are used in nonionic surfactants

glycolysis: in biological systems, the enzymatic conversion of glucose to pyruvate with the concomitant production of ATP; an energy releasing process

glycoside: compound that on hydrolysis gives sugars and other organic compounds (aglycones); widely distributed biologically active substances

glyme: dimethyl glycol; *see* ethylene glycol dimethyl ether

glyoxal: OHCCHO; dialdehyde, ethanedial; produced by (1) catalytic (Ag or Cu plus small amounts of halides) air oxidation of ethylene glycol; or (2) oxidation of acetaldehyde with nitric acid; use: as substitute for formaldehyde, particularly in condensation and cross-linking reactions; important in the paper and textile industries

glyoxalic acid: *see* allantoin

glyphosate: $(HO)_2P(O)CH_2NHCH_2COOH$; postemergence herbicide

godet: idling roller in yarn processing

gold sodium thiomalate: $AuSCH(COONa)CH_2COONa$; an antiinflammatory drug

goniometer: instrument that measures the intensity of radiation as a function of position or angle; *see* gloss; X-ray diffraction

Good Laboratory Practice: GLP; regulations promulgated by the United States Food and Drug Aadministration (FDA) concerning preclinical studies of drugs

Good Manufacturing Practice: GMP; regulations promulgated by the United States Food and Drug Administration (FDA) concerning quality control of drug manufacture

Gordon plasticator: large extruder used to plasticize rubber between a spiral screw and the walls of the machine; process is carried out at $\geq 150°C$; an internal mixer

gossypol: toxic polyphenol found in cottonseed; must be removed during refining of corn oil for edible use: use: stabilizer of vinyl polymers, antioxidant

graft polymer: macromolecule in which the monomer of the backbone differs from those in attached groups

Graham's Law: *see* diffusion

grain alcohol: *see* ethyl alcohol

Grainer process: production of NaCl from brine heated by steam pipes in flat pans open to the atmosphere; *see* Alberger process

grains per gallon: gpg; measure of water hardness; one gpg = 17.1 ppm

grains per liter: gpl; measure of water hardness

grams per denier: in the textile industry, a measure of the tensile strength of fibers and yarns

Gram stain: in biotechnology, bacteria are identified as gram positive or gram negative depending on whether they can be stained with the Gram stain; gram-negative bacteria have a thicker cell wall than gram-positive bacteria and are thus more resistant to antibiotics than gram-positive bacteria

granular polymerization: polymerization of suspended monomer resulting in polymer granules; *see* suspension polymerization

graphite: plumbago, black lead, Flanders stone; naturally occurring and synthetic form of carbon; in contrast to diamond, graphite is soft (0.5–1 Mohs), has a high thermal and electrical conductivity; it occurs mostly in a hexagonal crystal structure; produced by mining or manufactured from petroleum, pitch, furnace coke, anthracite, and carbon black; use: lubricant, pigment, as electrodes for the production of Al, electro-steel, alloys and carbides, as smooth and corrosion resistant lining for pipes or containers; recrystallized molded graphite serves as rocket casings; pyrolytic graphite is used as guiding tubes for fuel rods in nuclear reactors and when mixed with boron for atomic radiation shielding

graphite fibers: *see* carbon fibers

gravure printing: *see* printing

Gray-King test: ranking of coal according to coking properties

green compound: unvulcanized rubber compound

greenhouse effect: heating of the earth's climate due to the accumulation in the atmosphere of gases that absorb infra-red radiation ("greenhouse gases"); these gases permit solar energy to penetrate to the earth's surface but do not allow heat energy to escape; CO_2 and CH_4, which are produced by human activity, are of particular concern

greenhouse gases: *see* greenhouse effect

green liquor: after alkaline wood pulping, a solution produced by dissolving the solids left on the bottom of the furnace

ground water: supply of fresh water under the earth's surface; aquifers

groundwood: in paper and fiberboard manufacture, the mechanical wood pulp produced from debarked wood

GR-S rubber: copolymer of butadiene and styrene

GT/T-A process: pyrolysis of forest residues to produce low Btu gas and oil; process has been developed by the Georgia Gas Institute and Tech-Air Corporation

guaifenesin: guaiphenesin; a noncodeine cough medicine that acts by helping the expulsion of mucus; an expectorant

$OCH_2CH(OH)CH_2OH$

OCH_3

GUAIFENESIN

guanethidine: blood pressure lowering (hypertensive) agent that acts by blocking the release of the neurotransmitter norepinephrine

$N-CH_2CH_2NHC$
NH
NH_2

GUANETHIDINE

guanidines: *see* guanidinium salts

guanidinium salts: guadinine salts; produced from guanidine nitrate by ion exchange; guanidine nitrate ($HN=C(NH_2)_2 \cdot HNO_3$) is produced from urea and ammonium nitrate, or by fusing dicyanodiamide ($NH=C(NH_2)NHCN$) and ammonium nitrate; the nitrate is a strong oxidant, is sensitive to shock and heat and is used for manufacture of explosives; the chloride and sulfate salts are stable and are organic intermediates, e.g., for medicinals (sulfa drugs), disinfectants, photographic chemicals

guanine: one of the four bases in DNA and RNA; a purine base; source: guano, sugar beets, yeast

159

GUANINE

guayule: shrub that grows in northern Mexico and southern California and is a source of rubber; guayule rubber is produced by uprooting and pulverizing the plant every 3–5 years and extracting the rubber by flotation; guayule rubber is softer and tackier than hevea rubber and is used as a coating adhesive

gums: natural gums are sticky exudates of plants that dissolve or swell in water; gum naval stores, which are derived by tapping trees, consist of complex carbohydrates; use: thickeners, adhesives, sizing

gun cotton: explosive; produced by nitrating raw cotton with nitric and sulfuric acid

gutta-percha: natural *trans*-polyisoprene

gypsum: $CaSO_4.2H_2O$; use: extender in paints and plastics

H

HA: heptaldehyde-aniline condensate

H acid: 3,6-di-SO_3H-8-NH_2-α-naphthol; dyestuff raw material

H_1 blockers: antiallergy drugs that block histamine activity; antihistamines

H_2 blockers: pharmaceutical agents that block gastric (stomach) secretion

Haber–Bosch process: *see* ammonia

Haber process: Haber–Bosch process; production of ammonia from nitrogen and hydrogen gas; *see* ammonia

habitat: the environmental conditions of a specific place occupied by a living organism

hafnium: Hf; element; high melting metal; used in nuclear reactors as control material because of its ability to capture neutrons above thermal energies; hafnium oxide (HfO_2), hafnium carbide (HfC), and hafnium titanate $(TiO_2)_n \cdot (HfO_2)_y$ are excellent, though costly refractories

hairlock process: process used to bind loose fibers such as sisal or horsehair with latex resulting in lightweight material that keeps its shape

haleite: $O_2NNHCH_2CH_2NHNO_2$; ethylenedinitramine (EDNA); high explosive

half-life: measure of decomposition kinetics: the time taken for half of a material to decompose, e.g., the half-life of a radioactive isotope or of a drug in the body

Hall process: electrolytic recovery of metallic aluminum from bauxite; *see* aluminum

halocarbon plastics: polymers containing carbon and halogens but no hydrogens, e.g., fluorocarbon resins

haloperidol: butyrophenone tranquilizer; antipsychotic drug

HALOPERIDOL

halothane: $CF_3CHClBr$; *see* anesthetics

hand: in the textile industry the quality of a fabric as judged by touch

hand mold: compression mold requiring removal from the press after each shot; used for short runs or experimental molding

haploids: plants that have only one set of chromosomes; therefore variations arise only through mutations; the group includes all industrially important bacteria and fungi; to prevent mutations, cells or spores of such organisms that are used for fermentation are stored at liquid nitrogen temperature and thawed immediately before use

hardboard: compressed fiberboard; paraffin wax or phenol-formaldehyde resin are added as sizing to the coarse fibers and shheets are formed in a process similar to that used for paper manufacture; in the final step the sheets are dried and pressed; use: heat and sound insulation, panelling, parts for doors, cabinets, furniture, flooring

hard butters: hydrogenated oils suitable for confectionary applications

hard coal: anthracite; *see* coal

"hard elastic" materials: crystalline polymers that have immediate elastic recovery of 80–95% from extensions of 50–100%; repeated cycling approaches work recoveries of ~60%; the elastic characteristics are based on a well-oriented lamellar crystalline morphology; produced by extrusion orientation and subsequent annealing at 30–50°C below the melting point (T_M); a high rate of crystallization under stress is essential; polymers forming this structure include polyethylene, polypropylene, poly-4-methyl pentene, poly-3-methyl butene, polyoxymethylene, polypivalolactone, polyisobutylether; the name "hard elastic" or "springy" is due to their the elastic modulus that is too high for normal elastic applications (i.e, rubber, spandex); the elastic mechanism is based on reversible void formation; use: precursor for microporous films, fibers, hollow porous fibers, semipermeable membranes

hardener: curing agent

hard fiber: natural, stiff fibers from leaves and stems of plants used for mats and ropes

hard glass: Bohemian glass; potash–lime glass with high silica content

hardness: (mechanical) relative resistance of a material to compression, indentation or scratching; tests for hardness include the *Barcol* impressor and the *Shore* instrument for rigid plastics that measure resistance to penetration by a sharp spring loaded steel point; the *Brinell* and *Vickers* tests that measure the area of indentation caused by impression of a steel ball or diamond shape; the *Knoop microhardness* test that measures the length of indentation; the *Mohs* scale that is based on scratch tests; the *Pfund* test that measures hardness of organic coatings such as paint; the *Rockwell* (ASTM D 785) hardness that is based on the depth of indentation; the *Scleroscope* hardness test that measures the rebound after applied pressure; and the *Sward Rocker* that determines the softening of an organic coating caused by contact with another polymer; numerical values of these tests are based on individual standardized scales, e.g., diamond has the value of 10 on the original Mohs scale, 15 on the modified Mohs scale, and 6000–6500 on the Knoop scale

hard rubber: ebonite, vulcanite; dark, hard vulcanizate with high sulfur content; it is viscoelastic >80°C; use: corrosion-resistant coverings, electrical insulation, and mechanical goods, combs

hard water: water containing a high concentration of Ca^{2+} and Mg^{2+}; hard water causes $CaSO_4$ and $MgSO_4$ deposits in kettles and pipes and forms insoluble scum with soap

hardwood: wood usually derived from leaf-bearing (deciduous) trees, e.g., ebony, boxwood, lilac, whitehorn, persimmon, elder, ash, elm

Hargreaves process: production of sodium sulfate as by-product during manufacture of HCl from NaCl by the action of SO_2 and air

H-coal process: liquefaction of coal by hydrogenation over cobalt-molybdenum catalyst in a fluid-bed reactor

HDPE: *see* high-density polyethylene

head: in a blow molding apparatus the end

section where the melt is transformed into a hollow tube

headbox: in the fourdrinier machine for paper manufacture the container from which the slurry is released onto the rollers

head-to-head polymers: they have alternatively reversed monomers (i.e., –CH$_2$CHX –CHXCH$_2$–CH$_2$CHX–); CH$_2$CHF has <30% of this structure, the rest is head to tail.

head-to-tail polymers: polymers with regular repeat of monomers (i.e., –CH$_2$CHX –CH$_2$CHX–CH$_2$CHX–); most polymers have this structure because it forms most easily

heat-distortion point: in the polymer industry a measure for the characterization of thermoplastics: the temperature at which a standard test bar deflects 0.010 in. under a load of 66 or 264 psi

heat distortion temperature: *see* heat-distortion point

heater bands: electrical heating units for heating plastics in extruder barrels and injection molding cylinders

heat exchanger: vessel in which a heat-transfer agent (hot liquid or gas) moving in one direction transfers much of its energy to a heat-transfer agent moving in the opposite direction; heat-transfer agents are fluids used to transport heat between the heat source and the sink or to dissipate heat by radiation; heat-transfer agents should have a wide liquid range, be noncorrosive, nontoxic, and have a low vapor pressure; air, steam, and water are most commonly used in industry because of economics and their wide availability; other typical agents are anisole, C$_6$H$_5$OCH$_3$ (liquid range of –37 to 154°C), hexachlorobutadiene (boils at 210°C), and brines (NaCl or CaCl$_2$); sodium or potassium salts are liquid between 600 and 1400°C, but they are corrosive to metals; tetraryl silicate (–40 to 375°C) and gallium (30–2200°C) are expensive, and mercury is toxic

heating oil: *see* fuel oil

heat of combustion: ΔH_c; usually expressed as the molar heat of combustion: the heat evolved by combustion of 1 mole of a substance

heat resistance: heat stability; the ability to withstand permanent changes in physical properties such as color due to exposure to heat; heat resistance can be enhanced by stabilizers

heat sealing: process of joining two or more thermoplastic materials by heating areas that contact each other

heat-set inks: inks consisting of pigments in resins that boil between 400 and 500°F; use: for large-quantity, high-quality publications that are printed at high speeds

heat setting: in the textile industry a technique to impart or stabilize a definite shape to fibers and fabrics with dry heat or steam; heat setting of synthetic, natural, and blends of synthetic and natural fibers leads to easy-to-care fabrics; heat setting is used to improve dye fastness, reduce shrinkage, and maintain shape

heat sink: device for accepting excess heat from a critical area of a product, apparatus, or plant

heat stability: *see* heat resistance

heat-transfer agents: *see* heat exchanger

heat-transfer printing: in the textile industry the transfer of disperse dyes from a paper pattern onto the material at 400°F

heavy naphtha: petroleum fraction boiling at 150–200°C; use: source of jet fuel, chemical feedstock; *see* naphtha

heavy oil: high boiling petroleum fraction; also the high boiling fraction distilled from coal tar

heavy water: D$_2$O; produced by fractionation of water; use: in nuclear reactors, scientific research

Hehner number: value expressing the percentage of water-insoluble fatty acids in an oil or fat

helium: He; element; noble gas; inert gas; in the United States helium is extracted from natural gas fields located in Texas, Kansas, Oklahoma, Wyoming, and New Mexico; the gas mixture is first scrubbed and cooled to liquefy and remove the natural gas; the resulting crude helium contains up to 88% He, 0.1% H_2, 0.1% CH_4, and the rest is nitrogen; final purification is achieved by oxidation of hydrogen and removal of the resulting water; fractionation of nitrogen and removal of trace impurities with activated charcoal; the 99.995% pure helium is compressed, precooled with liquid nitrogen, and further cooled with recycle He gas; use: as inert gas in gas chromatography and electronics manufacture, as coolant for superconductors and for lighter-than-air balloons

helix: a three-dimensional molecular structure typical of linear polymers such as amino acids, nucleic acids, isotactic polypropylene, and tetrafluoroethylene; a particularly important form is the α-helix of proteins: these polyamides have a right-handed twist along the helix axis, which is 0.54 nm long per turn and has 3.6 amino acid residues per turn

hemicellulase: enzymes capable of breaking down hemicelluloses; used in food processing

hemicellulose (1) a low molecular weight polysaccharide fraction consisting of polymers of pentoses and hexoses, mainly galactose, glucose, mannose, and xylose; (2) a mixture of β- and γ-cellulose; hemicellulose is soluble in dilute acid and base

hemodialysis: membrane separation of solutes from the blood; *see* dialysis

hemp: long, coarse fiber obtained from *Cannabis sativa*; contains about 78% α-cellulose and is a source of cannabis (marijuana); use: cordage, packing, plastic filler and pulp; hempseed oil is obtained by pressing the hemp plant and is used for paints and varnishes; *see* fibers, organic

heparin: polysaccharide; extracted from animal sources; use: anticoagulant

heptachlor: insecticide used for control of cotton boweevil; toxic via skin, breathing, mouth; stimulates nervous system; hazardous chemical waste (EPA)

HEPTACHLOR

heptyl aldehyde: $CH_3(CH_2)_5CHO$; obtained by pyrolysis of ricinoleic acid methyl ester (castor oil) together with undecenoic acid

heptyl aldehyde-aniline: HA; $C_6H_5N=CH$ C_6H_{11}; used for gum stocks, air hose, white wall tires

herbicides: agents that selectively destroy unwanted vegetation; preemergence herbicides generally attack the root system, postemergence herbicides interfere with photosynthesis; the chemical structures of modern herbicides fall into several groups: acetanilides or amides (e.g., alachlor, propanil), aliphatics (e.g., dalapon, glyphosate), benzoics (e.g.dicamba), bipyridiliums (e.g., diquat, paraquat), carbamates (e.g., propham), highly chlorinated organics (e.g., 2,4,5-T), cyclohexadiones (e.g., sethoxydim), dinitro anilines (e.g., benefin, trifluralin), diphenyl ethers (e.g., lactofen, oxyfliorofen), imidazolinones (e.g., imazapyr), nitriles (e.g., bromoxynil), organic arsenicals (e.g., cacodylic acid), phenoxys (e.g., 2,4-D), phenylureas (e.g., monuron), pyrones (e.g., coumarin), sulfonyureas (e.g., chlorsulfuron), thiocarbamates (e.g., butylate, vernolate), triazines (e.g., atrazine, prometrone), uracils (e.g., bromacil), ureas (e.g., diuron, linuron); herbicides are subject to plant resistance with excessive use

hertz: Hz; SI unit of frequency; periodic occurrence of once per second has a frequency of 1 Hz

hetero-: prefix denoting different, e.g., a solid catalyst used in a solution is termed a heterogenous catalyst

heteropolymer: copolymer consisting of two or more different polymers

hevea: plant family from which natural rubber is derived

hexa: *see* hexamethylenetetramine

hexabromocyclododecane: flame retardant additive for polymers, particularly for polyolefins

hexachlorobenzene: HCB, perchlorobenzene; produced by chlorination of benzene at ~600°C and 200 bar; use: fungicide, for organic synthesis; ingestion can cause increased blood pigment production (porphyria); do not confuse with benzenehexachloride

hexachlorobutadiene: C_4Cl_6; use: solvent, heat transfer agent, hydraulic fluid; hazardous chemical waste (EPA)

hexachlorocyclopentadiene: produced by multistep chlorination of cyclopentadiene; herbicide; intermediate for production of herbicides such as chlordane and mirex, and of fire retardants, e.g., chlorendic anhydride; hazardous chemical waste (EPA)

HEXACHLOROCYCLOPENTADIENE

hexachloroethane: C_2Cl_6, carbon trichloride, perchloroethane; produced by chlorination of

ethane; used as rubber accelerator, retarding agent in fermentation, organic synthesis

hexachlorophene: produced by treating 2,4,5-trichlorophenol with formaldehyde; preservative for rubber and latex, bacteriostat incorporated into thermoplastics; use in germicidal soaps and personal deodorants is regulated by FDA; excessive use may be neurotoxic in humans

HEXACHLOROPHENE

hexacyanoferrate(II): $[Fe(CN)_6]^{4-}$; iron complex used to prepare ferrocyanate pigments

hexafluoropropylene: HFP; $CF_2=CFCF_3$; produced as by-product when $CHClF_2$ (CFC 22) is heated to 700–900°C; this reaction is used to manufacture $CF_2=CF_2$ (TFE) which is the monomer of polytetrafluoroethylene (PTFE); use: intermediate for manufacture of Teflon; monomer of the vulcanizable elastomer hexafluoropropene/vinylidene; *see* fluorocarbon polymers

hexafluorosilicate: ore; raw material for manufacture of sodium aluminum fluoride (cryolite)

hexaglycerol: *see* trimethylolpropane

hexahydrocyclohexane: *see* inositol

hexametapol: *see* hexamethylphosphoric triamide

hexamethylenediamine: $H_2N(CH_2)_6NH_2$, HMDA; produced from adiponitrile by high pressure catalytic (Co-Cu or Fe) hydrogenation; used to manufacture nylon 66 by condensation with adipic acid; *see* nylon 6,6, interfacial polymerization

hexamethylene 1,6-diisocyanate: OCN

$(CH_2)_6NCO$; HMDI; produced by phosgenation of hexamethylenediamine (HMDA); used for production of urethanes and urethane polymers

hexamethylene glycol: *see* 1,6-hexanediol

hexamethylenetetramine: HMTA, hexa; hexamine; formed from ammonia and formaldehyde; use: curing agent for phenolic and urea resins, accelerator for slow curing rubber vulcanizates, starting material for manufacture of the chelating agent nitrilotriacetic acid (NTA) and its salts, intermediate in the pharmaceutical industry; during World War II its nitrate, cyclotrimethylene-trinitramine, was an ingredient of the cyclonite explosives that were used in blockbuster bombs

HEXAMETHYLENETETRAMINE

hexamethylphosphoramide: *see* hexamethylphosphoric triamide

hexamethylphosphoric triamide: hexamethylphosphoramide; hexametapol; HMPA; HMPT; $[N(CH_3)_2]_3PO$; use: solvent, deicing additive for jet fuel, UV absorber in PVC compounds; potential carcinogen

hexamine: *see* hexamethylenetetramine

1,6-hexanediol: $HO(CH_2)_6OH$; produced by hydrogenation of caprolactam, omega-hydroxycaproic acid or adipic acid; use: plasticizer, raw material for manufacture of hexamethylenediamine (HMDA), polyesters and polyurethanes

1-hexanol: *n*-hexyl alcohol; $C_6H_{13}OH$; produced by reduction of ethyl caproate with Na in absolute alcohol; use: for chemical synthesis

hexose: any sugar containing six-carbon atoms

hexylene glycol: $(CH_3)_2C(OH)CH_2CH$ $(OH)CH_3$; produced by hydrogenation of diacetone alcohol; use: in cosmetics, as coupling agent to castor oil in hydraulic brake fluids

n-hexyl trimellitate: plasticizer for vinyls and PVC/ABS blends; helps to resist fogging, extraction, and migration

HF heating: *see* high frequency heating

high Btu: high fuel or heating value, about 1000 Btu per cubic foot

high-ceiling diuretic: term used for the most effective diuretics such as furosemide

high-density polyethylene: HDPE; linear polyethylenes with densities >0.94 g/mL; produced by polymerization at low pressure using Ziegler catalysts; high density results from long chains with few side branches; strong, hard rigid material, highly crystalline, with good chemical resistance and high softening temperature

high explosive: detonating explosive; explosive with high reaction rate; characterized by a pressure rise to 100,000–5,000,000 psi within microseconds; the resulting shock wave (detonation wave) propagates at 1–7 miles/ sec; classification distinguishes between primary explosives, which are used as detonators, and secondary explosives, which usually require detonators to initiate reaction

high-flash: *see* naphtha

high-frequency heating: HF heating; material is placed between two electrodes that create an electric field between 3 and 200 mc/sec; rapid heating occurs due to dielectric loss

high-frequency welding: *see* high-frequency heating

high fructose corn syrup: HFCS; sweetener prepared by enzymatic digestion of starch; fructose is obtained from glucose using immobilized glucose isomerase; fructose content on a dry basis ranges from 42–95%

high fructose syrup: HFS; *see* high fructose corn syrup

high-gloss inks: inks having low pigment content and low paper penetration

high-gradient magnetic separation: HGMS; a fine-coal cleaning technology using large magnets to remove small, weakly magnetic materials such as pyrite from the coal

high-load melt index: rate of flow of molten resin through an orifice of 0.0825 in diameter when subjected to a force of 21.6 kg at 190°C (ASTM)

highly oriented yarn: HOY; *see* poly(ethylene terephthalate)

high-performance liquid chromatography: HPLC; separation method for liquid samples; choice of which particular liquid column chromatography (LCC) to select depends on the molecular weight of the sample: >2000 daltons one uses exclusion chromatography (EC) with an aqueous or nonaqueous phase; <2000 daltons the decision depends on whether the sample is ionic or nonionic; with acidic samples the solid phase is an anion exchange/ion pair, with basic samples it is a cation exchange/ion pair; with nonpolar samples the best choice is reverse phase chromatography with an aqueous mobile phase; *see* chromatography; liquid column chromatography; liquid–liquid chromatography; liquid–solid chromatography

high-performance liquid crystalline polymers: *see* aramids, aromatic copolyesters; liquid crystals

high-performance polymers: materials resistant to high stress such as high temperature and pressure; polymers with high modulus, tensile and shear strength

high polymer: macromolecule with a molecular weight greater than 10,000 dalton

high-pressure injection molding: HIPM; *see* reaction injection molding

high-pressure laminates: laminates molded and cured at pressures of 1200–2000 psi

high-pressure molding: molding or laminating process using pressures >200 psi

high-pressure polyethylene process: pressures of 15,000–50,000 psi are used to obtain a highly branched, high molecular weight polyethylene (LDPE, low density polyethylene) with specific gravities of 0.91–0.92

high-pressure powder molding: in the plastics technology powdered polymers are molded by high-pressure compaction at room temperature and then heated to complete cure or polymerization; limited to polymers that do not release vapors upon heating; examples are polypropylene (PP), polypropylene oxide (PPO), and polytetrafluoroethylene (PTFE)

high-pressure spot: in reinforced plastics, an area containing little resin

high-shrink fiber: produced by drawing synthetic fibers without stabilizing them after stretching; use: to produce bulky yarns by blending with low-shrink (stabilized) fibers followed by additional heat setting; *see* low-shrink fibers

high-test molasses: in sugar purification, the fraction that contains 22–27% sucrose and 50–55% invert sugars

hindered isocyanate: *see* isocyanate generator

HIPS: high-impact polystyrene; *see* styrene plastics

histamine: substance that causes allergic symptoms; it is liberated when cells are damaged or by antigen–antibody reaction; extreme reaction can be fatal; antihistamines that alleviate the symptoms by competition with histamine for receptor sites are called H_1 blockers, e.g., histapyrrodine, chlorpheniramine, promethazine

H
N
N

NH$_2$CH$_2$CH$_2$

HISTAMINE

histapyrrodine: antihistamine

C$_6$H$_5$

CH$_2$CH$_2$NCH$_2$C$_6$H$_5$

N

HISTAPYRRODINE

histidine: amino acid

H
N
N

NH$_2$

HOOCCHCH$_2$

HISTIDINE

histones: small proteins tightly bound to DNA; histones are rich in the basic amino acids lysine and arginine; histones and DNA form a consolidated structure (chromatin) during cell division; cell division, which is one step in the cell cycle, can be followed using a microscope and observing the changes in the chromosomes that contain the chromatin; *see* cell cycle

hob: master model of hardened steel that is pressed into a block of softer material to form a mold cavity; this action is called hobbing

Hock process: oxidative cleavage of cumene to phenol and acetone; *see* phenol

Hock's elasticity tester: falling ball apparatus to determine resilience of rubber; ball falls at an angle onto the test piece; the rebound angle from the vertical is a measure of the resilience

Hoekstra plastometer: a parallel plate apparatus for determining the plasticity of small test pieces

Hoffman process: manufacture of hydrazine by treating aqueous urea with NaOCl

hog: heavy machine used for reducing particle size and cutting up scrap

hold-down groove: in molding technology, a small groove is cut into side wall of a mold to assist in holding the molded article in that member while the mold opens

hollow fibers: synthetic filaments made with hollow centers by extruding through special spinning jets; use: as semipermeable membranes in purification and extraction systems

holocellulose: natural fibrous material consisting of high molecular weight α-cellulose and low molecular weight hemicellulose

holofol process: process for making articles from unvulcanized, calendered sheets; after both sides of the sheet are partially vulcanized they are treated with a swelling agent that causes them to split; after the swelling agent evaporates the hollow article is fully vulcanized

holography: the interference pattern of light waves surrounding the object to be preserved with a reference wave of coherent light (laser) is recorded on a photographic emulsion or other recording materials such as photopolymers or photoresists; when this record of the hologram is illuminated with a replica of the coherent beam it can be displayed by reflection on a surface where it appears as a three dimensional image; the technique can be used to record electromagnetic waves outside the visible spectrum and to display them in the visible range

homo-: prefix denoting same, e.g., a catalyst that is dissolved in the reaction medium is termed a homogeneous catalyst; members of the homologous series of alkanes differ only by the number of their carbon atoms; proteins are said to be homologous if they contain identical sections of amino acids

homologous temperature: the ratio of the

temperature of a material to its melting temperature

homolysis: symmetrical breaking of a covalent bond often leading to free radicals

homopolymer: macromolecule consisting of a single type of repeating unit, i.e, one type of monomer; examples: polyethylene, polyglycine

Hookean elasticity: ideal elasticity; elasticity that conforms with Hooke's law: $T = E(L-L_0)/L_0$ (T is the imposed tensile stress, $L-L_0$ is the change in length under stress, L_0 is the original length of the specimen); the proportionality constant, E, is called Young's modulus or modulus of elasticity

Hooke's Law: *see* Hookean elasticity

hoop stress: circumferential stress due to external or internal pressure of cylindrical material

hopper: container holding material before it is being fed into a processing machine; for example, a hopper dryer is a feeding device for molding of thermoplastics in which dry, hot air is forced through the feed pellets; a hopper loader (hopper filler) automatically feeds materials into a processing machine using mechanical or pneumatic means

hormesis: in medicine, a stimulatory effect of low doses of agents that are toxic at high doses; the effect is believed to be due to stimulation of defensive mechanisms

hormone: biologically active substance released by glands directly into the bloodstream, e.g., insulin, estradiol, cortisone; glands that secrete hormones are called endocrine glands

hotbench test: in plastics technology, a method to determine (1) the gelation properties of plastisols; and (2) the softening temperature of a plastic, by casting a film onto a temparature gradient plate

hot gas welding: the use of hot air or nitrogen to weld plastics that cannot be exposed to an open flame

hot melts: thermoplastics that are solid at room temperature; use: for adhesives and coatings

hot mixing: in plastics manufacture, a process that raises mixing temperature to 150–180°C by increasing the speed of the rotors; use: to increase plasticity and facilitate the incorporation of fillers

hot-short: describing a hot material that is inelastic, nonstretchable, and easily broken under tension

Hottenroth number: titration of the viscose solution with NH_4Cl solution to check the ripening of the viscose, i.e., its ease of coagulation in the wet spinning process; the ripening process stretches over several days, modifying the xanthate distribution; the properties of the spun viscose rayon strongly depend on the correct coagulation rate

HOY: highly oriented yarn; *see* poly(ethylene terephthalate)

HT-1: original designation of an experimental fiber made from para-phenylenediamine and terephthalic acid; *see* aramids

hull: in textile technology, a dark speck of matter that appears to be in the fabric of a fabric-base laminated sheet

humectants: agents that increase the ability of moisture to adhere to a material; use: antistatic agents, coatings

humic acid: *see* humus

humin: *see* humus

humus: organic matter present in peat, soils, sediments, and water; it is a complex mixture of substances that have varying composition; these "humic substances" are fulvic acid, which is water soluble, humic acid, which is soluble only in alkaline solution, humin, which is insoluble at any pH; use: fertilizer, growing medium, chemical feedstock, and for soil amendment; in the form of peat it is used as fuel

hybrid composites: materials with two or more reinforcements each of which provide desired properties, e.g., alumina-silica fibers for high-temperature insulation and carbon fiber for strength

hybridoma: *see* monoclonal antibodies

Hydeal process: catalytic (Cr, Mo, or Co oxide) hydrodealkylation of alkylaromatics at 500–650°C and 30–50 bar; higher temperature and pressure are required when reaction is carried out without catalyst; for example, the production of benzene from toluene

Hydrar process: manufacture of cyclohexane by catalytic (Ni/support) gas phase hydrogenation of benzene; this UOP process is run in a series of three reactors with increasing temperature (400–600°C) at 30 bar; the short residence time prevents isomerization to methylcyclopentane

hydrated silica: $SiO_2 \cdot nH_2O$; semisolid, gelatinous material that contains 28% solids, becomes fluid on addition of water and gels again on standing; produced by treating sodium silicates with mineral acids; use: paper and textile coatings

hydrazides: $RCONHNH_2$; metal scavengers; *see* additives

hydrazine: H_2NNH_2; produced (1) by a continuous liquid-phase process from ammonia and sodium hypochlorite via formation of chloramine (NH_2Cl); the initial production of NaOCl takes place <30°C, the final reaction of chloramine and ammonia leads to product and is run at 130°C (Raschig [Olin] process); use of primary or secondary amines leads to substituted hydrazines; (2) from urea by reaction with NaOCl at ~100°C (Hoffman urea process); (3) from methyl ethyl ketone and ammonia in the presence of hydrogen peroxide at ~50°C (Ugine

Kuhlmann process); and 4. with chlorine instead of hydrogenperoxide (Bergbau or Bayer process, not yet commercial, similar to [3]); use: intermediate in the manufacture of agricultural chemicals, blowing agents, high-energy fuels, and pharmaceuticals; also used for treating boiler water where it acts as oxygen scavenger

hydrazones: $RR'C=NNH_2$; metal scavengers; *see* additives

hydroabietic acid: the dihydro and tetrahydroabietic acids are obtained by hydrogenation of abietic acid, which is derived from pine rosin; use: emulsifier

hydroabietyl alcohol: $C_{19}H_{31}CH_2OH$, plasticizer for cellulose nitrate, ethyl cellulose, and PVC

hydrocarbon plastics: plastics consisting of monomers containing only carbon and hydrogen; in the plastics industry hydrocarbon resins are considered to be low molecular weight resins made from impure monomers derived from coal-tar, cracked petroleum distillates, and turpentine; use: binders in asphalt flooring, processing aids, and coating additives

hydrocarbonylation: *see* hydroformylation; oxo process; Reppe processes

hydrocarboxylation: *see* Reppe processes

hydrochloric acid: HCl; muriatic acid; hydrogen chloride; produced mainly as (1) byproduct from the chlorination of hydrocarbons, dehydrochlorinations, and incineration of chlorinated wastes; (2) by reaction of NaCl and H_2SO_4; (3) very pure HCl is produced by burning H_2 and Cl_2; use: pickling and cleaning metal parts, widely used in chemical manufacturing

hydrochlorothiazide: thiazide; diuretic agent that increases flow of urine from kidneys and thus acts as hypertensive (lowers blood pressure) agent

169

HYDROCHLOROTHIAZIDE

hydrocodone: dihydrocodeinone; produced from codeine by heating in the presence of Pt or Pd; analgesic

hydrocracking: catalytic cracking of petroleum in the presence of hydrogen gas at pressures up to 3000 psi; catalysts used are metals such as Ni, Co, W, S, Mo, Pt, or Pd; catalyst supports are acidic oxides or metal oxides; process converts S–, N–, and O– containing compounds to H_2S, NH_3, and H_2O

hydrocyanic acid: HCN; hydrogen cyanide, prussic acid, formonitrile; produced by (1) ammoxidation of methane with a brief residence time at 1000–1200°C, yields HCN and water (Andrussow process); (2) ammoxidation of propylene (Sohio process); (3) ammonodehydration of methane in the absence of air in Pt-, Ru-, or Al-coated corundum tubes at 1200–1300°C (Degussa process); (4) ammono deoxidation of light petroleum in a fluidized bed of fine coke at 1300–1600°C Shawinigan process); also (5) as undesirable by-product in the Sohio acrylonitrile process (*see* acrylonitrile); use: as disinfectant, fumigant, military war gas; in the cyanide process for mining and metallurgy; in chemical manufacture to produce N-containing groups

hydrodealkylation: dealkylation of alkylaromatics in the presence of H_2; *see* Hydeal process

hydroformylation: oxo synthesis; manufacture of aldehydes, alcohols, and acids from olefins; the reaction that leads to "oxo" aldehydes consists of the catalytic addition of H_2 and CO to terminal olefins at elevated temperature and pressure: $RCH=CH_2 + CO + H_2 \rightarrow RCH_2CH_2CHO$; the aldehydes serve as feedstock for "oxo" carboxylic acids and alcohols; industrial processes use C2 to C20 linear and branched olefins; addition to the terminal double bond occurs more readily at position 1 than at position 2, thus leading to a mixture richer in *n*-aldehydes than *iso*-aldehydes; most industrial processes use Co and Rh phosphine catalysts

hydrogasification: conversion of coal to gaseous and liquid fuel by hydrogenation; *see* coal gasification

hydrogels: ionically cross-linked hydrophilic polymers; the most widely used polymer is poly(hydroxyethyl methacrylate); use: for biomedical devices, contact lenses

hydrogen: H; element; H_2 gas is mainly produced (1) by steam reforming, i.e, reaction of refinery hydrocarbons and natural gas with steam; (2) from water gas (CO + H_2) by fractionation using liquid air at low pressure (Linde–Fränkl–Caro process) or by cooling the gases by expansion from 20 atmospheres to lower pressure (Claude process); (3) by the electrolysis of water; (4) as by-product in the electrolytic production of NaOH; and (5) in several fermentation processes; use: for chemical synthesis including that of ammonia, HCl, and methanol, for coal gasification and liquefaction, as chemical-reducing agent, for low-temperature welding; gelled hydrogen (liquid hydrogen thickened with silica powder) is used as rocket fuel; H_2 gas is expected to become an important energy source in the future; *see* steam reforming

hydrogenated methylabietate: produced from rosin; use: plasticizer for cellulose nitrate, ethyl cellulose, acrylic and vinyl resins, and polystyrene; *see* abietic acid

hydrogenation: catalyzed reductions using elemental hydrogen; use: coal liquefaction, in the chemical industry, metallurgy

hydrogen bonding: electrostatic bond between H and the highly electronegative elements O, N, and F

hydrogen chloride: *see* hydrochloric acid

hydrogen cyanide: *see* hydrocyanic acid

hydrogenolysis: bond cleavage in the presence of H$_2$, e.g., reductive cleavage of esters to alcohols (RCOOR' → RCH$_2$OH + R'OH), cleavage of ethers to alcohols (R–O–R' → ROH + R'H); also *see* Hydeal process

hydrogen peroxide: H$_2$O$_2$; produced by reduction of alkylanthraquinone (2-ethyl- or *t*-butyl-) to the corresponding quinone that undergoes rapid autoxidation with evolution of hydrogen peroxide; use: oxidizing and reducing agent, raw material for manufacture of organic peroxides, disinfectant, bleach

hydrol: molasses obtained from corn starch hydrolysis

hydroliquefaction: *see* hydrogasification

hydrolysate: product of hydrolysis reaction

hydrolysis: decomposition of a compound by reaction with water, e.g., catalyzed reaction of water with esters yields an acid and an alcohol; catalytic reaction of nitriles and water leads to acids

hydrophilic: having an affinity to water; hydrophilic substances are lipophobic, i.e, have little or no affinity to fatty materials; hydrophilic substances may be ionic or contain polar molecules; examples are cellulosics (cotton, rayon, paper), proteins (wool), soaps; *see* dipole moment, polar molecules

hydrophobic: having little or no affinity to water; hydrophobic substances are nonpolar and have an affinity to fatty materials, i.e, they are lipophilic (oils, high molecular weight fatty acids, acrylics, and polyesters); materials made from hydrophobic fibers are less comfortable than those made from hydrophilic fibers because they do not take up moisture from the body; can be made more comfortable with a hydrophilic finish or by blending them with cotton, wool or rayon; *see* dipole moment; polar molecules

hydroquinone: *p*-dihydroxybenzene; *p*-hydroxyphenol; produced by oxidative cleavage of *p*-diisopropylbenzene (*see* Hock process), oxidation of phenol with H$_2$O$_2$ and H$_2$SO$_4$ (Rhone-Poulenc, Ube), catalytic (Ru and Re) hydrocarbonylation of acetylene (Reppe synthesis) used by duPont; use: photographic developer, polymerization inhibitor, antioxidant, dye intermediate

hydrorotator: *see* hydroseparator

hydrorubber: colorless mass, molecular weight 30,000–150,00 daltons; produced by catalytic hydrogenation of rubber; use: interlayer in laminated safety glass, impregnating agents, addition to high-pressure lubricants

hydroseparator: hydrotator; a device for separating solids, primarily separating fines; the container receives more fluid than its volume provides, the overflow removes most of the water and some fine particles; use: for primary dewatering or cleaning

hydrosol: in the plastics industry, a suspension of a resin such as PVC or nylon in water

hydrosulfite: *see* sodium dithionite

hydrotreating: in petroleum refining, the catalytic hydrogenation of petroleum to release low sulfur liquids and H$_2$S from sulfur-rich hydrocarbons and ammonia from nitrogen-containing hydrocarbons; use: to produce low-sulfur and low-nitrogen fuels; *see* hydrogenation

hydrotrope: water-soluble chemicals that increase the water solubility of slightly soluble organic compounds by increasing their dispersion in water; common hydrotropes are ammonium and alkali salts of aromatic sulfonic acids (derived from toluene, xylene, and cumene); use: to solubilize insoluble ingredients in liquid detergents and thus decreasing their viscosity, permitting more free-flow

hydroxyapatite: tricalcium phosphate; Ca$_5$ (PO$_4$)$_3$OH; main constituent of bones and teeth; use: suspension agent when combined with surfactant

hydroxyethylmethyl methacrylate: rigid hydrophilic polymer that becomes flexible when saturated with water; used for soft contact lenses, biomedical devices, masonry coatings

hydroxyl amine: NH_2OH; produced (1) by reduction of ammonium nitrite with SO_2 to $HON(SO_3NH_4)_2$ followed by hydrolysis (Raschig process); (2) by catalytic (Pd) reduction of NO_3^- in the presence of phosphate buffer (a step in DSM's HPO caprolactam process); (3) by catalyzed (Pt) reduction of pure NO in dilute acid (BASF, Inventa processes); use: raw material for production of cyclohexanone oxime; *see* caprolactam

hydroxyl value: a measure of the OH groups in an organic material; in plasticizers, large hydroxyl values indicate that the plasticizer may become incompatible during aging due to the many unesterified alcohol groups; in urethane technology the hydroxyl number is important in the selection of polyols

p-**hydroxyphenol**: *see* hydroquinone

hydroxyproline: amino acid derivative

HYDROXYPROLINE

Hygas process: production of synthesis natural gas from coal; manufacture takes place in·a single vessel containing four separate vertically stacked contacting zones: on top of the reactor is a drying zone, followed by two hydrogasification zones; the bottom zone converts char into synthesis gas that serves as the hydrogen source in the two higher zones

hygroscopic: absorbing water from air; examples of hygroscopic substances are $MgSO_4$, $CaCl_2$, silica gel

hyper-: prefix denoting super, e.g., hypertension means high blood pressure

hypnotics: agents that reduce central nervous system (CNS) activity; used as sedatives

hyperfiltration: *see* reverse osmosis

hypo (1) synonym for sodium thiosulfate; (2) prefix denoting less, e.g., hypoallergenic is a term applied to cosmetics that elicit little or no allergic reactions

hypochlorites: salts of hypochlorous acid, HClO; OCl⁻; salts decompose to release oxygen; use: oxidant, household bleach; LiOCl is added to dishwasher detergent

hypocholesteremics: agents that lower blood cholesterol, e.g., clofibrate

hypoglycemic agents: drugs that lower blood sugar; term is usually limited to drugs that are taken by mouth

hysteresis: time lag exhibited by a material on reacting to changes imposed on it: for example, the retention of induction in iron and steel when the magnetic force is cyclically changed or the cyclic noncoincidence of elastic loading and unloading curves that one observes under cyclic stress; when plotting stress versus strain the loading and unloading curves enclose an area (hysteresis loop) that represents the amount of mechanical energy that has been lost as heat

I

ibuprofen: p-$(CH_3)_2CHCH_2C_6H_4CH(CH_3)$ COOH; antiinflammatory drug

ideal elasticity: *see* Hookean elasticity

Ignifluid stoker: device that burns low grade coal in the fluidized mode

ignition element: component of fuse caps in commercial detonators

ilmenite: $FeTiO_3$; mineral; a minor source of titanium; used to produce TiO_2: the dried, ground ore is digested with sulfuric acid, the iron is reduced and removed as $FeSO_4$; the impure TiO_2 is purified by washing with H_2SO_4 and is calcined at 1650°F to increase its crystallinity and thus raise the refractive index; *see* rutile

imazapir: imidazolinone herbicide

2-imidazolidone: 2-imidazolidinone, ethyleneurea; prepared from ethylenediamine and CO_2 under pressure; use: insecticide, raw material for polymers, finishing agents for textiles and leather, additives for adhesives, lacquers and plastics

2-IMIDAZOLIDONE

imides: conpounds containing the group $RN[C(O)-]_2$, where R may be a hydrogen or an alkyl or aryl group

imines: compounds containing the group $>C=NH$

immediate set: deformation that can be measured immediately after removing the load that caused the deformation

immittance spectroscopy: *see* impedance spectroscopy

immobilization: technique used to restrict the mobility of catalysts or reagents in a reaction mixture; the method is used for synthesis and analysis; immobilization facilitates product separation and catalyst or reagent recycling; immobilization may be achieved by encapsulation, adsorption, inclusion in gels, zeolites or liposomes, or by chemical tethering to solid supports; in biotechnology, whole cells to act through their own immobi-

lized enzymes; immobilized enzymes are used for analytical probes and sensors; *see* affinity chromatography; solid phase peptide synthesis

immunized cotton: cotton in which the cellulose has been esterified (usually with acetic anhydride) so that it resists cotton dyes

immunoassays: highly sensitive tests for the presence of small amounts of biologically active materials such as proteins, hormones, and drugs; tests are based on the specific reaction of a substance (antibody) with the chemical (antigen) to be detected; the antibody–antigen complex can be identified by various techniques: radioimmunoassay (RIA) uses a radioactive label such as ^{125}I, fluoroimmunoassay uses a fluorescent label on the antigen, spin immunoassay uses ESR to identify a free-radical-labeled antigen, enzyme immunoassay (EIA) uses enzyme-labeled antibodies or antigens that are detected by testing for enzyme activity; enzyme-multiplied immunoassay technique (EMIT) uses an enzyme–antibody complex that inhibits enzyme activity; heterogeneous immunoassays use a second immobilized antibody leading to an insoluble antibody$_1$–antigen–antibody$_2$ complex (e.g., enzyme-linked immunosorbent assay [ELISA]); automation of immunoassays permits their inclusion in standard clinical laboratory tests; one of the uses for immunoassays is testing for drug abuse

immunoglobulins: class of antibody proteins isolated from blood serum, e.g., IgG, gamma globulin

impact modifier: additive, usually an elastomer or plastic incorporated in a plastic compound to improve impact resistance

impact resistance: relative susceptibility of materials to fracture under stresses applied at high speeds

impact strength: measure of toughness; the ability of a material to withstand a blow without fracturing; two common instruments to test for impact strength are the *Izod* (ASTM D 256) and the *Charpy* testers; the Izod impact strength is determined by the difference in height of a swinging pendulum before and after it breaks a notched specimen that is held vertically; in the Charpy test the specimen is held horizontally between two supports and impact strength is measured by the energy (foot-lb or kg/cm) needed to break the specimen; other tests are the *drop-weight test* in which weights are dropped from different heights, *free falling dart test* (ASTM D 1709) for testing films by dropping darts with a rounded head onto 10 specimens and recording the percent failure, and the *tensile impact test* (ASTM D 1822-61T) in which the specimen is attached to a pendulum, strikes an anvil and the height difference after breakage measures the impact strength

impedance spectroscopy: IS; modulus spectroscopy; dielectric spectroscopy; immittance spectroscopy; technique to determine physicochemical properties of solids and of dispersions of conductive materials in nonconductive fluids; the method analyzes the electrical responses to a series of small amplitude AC signals

impreg: phenol–formaldehyde–wood composite that contains resin bonded to the internal surfaces of the wood cell walls that due to the resin concentration keeps the wood in an expanded condition; this wood composite does not swell when exposed to water

impregnation: process of filling the interstices of a porous material with a substance such as paint or resin

impulse sealing: joining thermoplastic sheets by first pressing them together, then applying a short pulse of heat and cooling them immediately

indan: dihydroindene; *see* indene

indanthrene: vat dye based on anthraquinone; colorfast, used for cotton

INDANTHRENE

indene: isolated from coal tar; use: chemical intermediate, for production of copolymers with benzofuran and with cumarone in the presence of BF_3

INDENE

index of refraction: *see* refractive index

India ink: heavy drawing ink colored with lampblack or other mineral pigments

India rubber: natural rubber

indigo blue: indigo, Vat blue 1; a blue vat dye that is produced by air oxidation of indoxyl in the presence of base; obtained in antiquity from plants (*Indigofera tinctora, Isatis tinctoria*); mode of dyeing: the water-soluble sulfonate of reduced indigo is applied to material and then is reverted to the insoluble dye by oxidation with sodium nitrite or bichromate; used to dye denim for blue jeans

INDIGO BLUE

indigo carmine: disodium-5,5'-indigo disulfonate; dye used to analyze for oxidizing agents, as dye in kidney testing

indium: In; metallic element; use: in bear-

174

ing alloys, dental alloys; indium compounds, such as the arsenide, phosphide, selenide, and telluride, are used for semiconductors p-n junctions

indole: produced from coaltar and synthetically from o-toluidine; use: chemical precursor, ingredient in perfume; indole-3-acetic acid is a plant growth regulator; indole-3-butyric acid is a plant rooting agent; 3-methyl-indole (skatole) is a perfume ingredient

INDOLE

indomethacin: indocin; an antiinflammatory drug

INDOMETHACIN

in-duct sorbent injection: for pollution abatement lime is sprayed into the ductwork between the boiler and smokestack to absorb SO_2

induration: heat hardening of solid particles; used for agglomeration

industrial gases: the most important are acetylene, air, ammonia, argon, carbon dioxide, carbon monoxide, chlorine, fluorine, fluorocarbons, helium, hydrogen, krypton, methane, neon, nitrogen, nitrous oxide, oxygen, propane, sulfur dioxide, xenon

industrial waste: as distinguished from municipal or agricultural waste, industrial waste

requires special treatment depending on the industry from which it derives

inert gases: gaseous elements in group 8; see helium; neon; argon

infiltration/inflow: in waste water treatment the total quantity of water entering the sewer system

infra-red spectroscopy: I.R.; an analytical technique that uses the molecular absorption of infrared radiation; energy absorbed in infrared (2.5–50 μm, 4000–200 cm^{-1}) is due to stretching, bending, and twisting vibrational modes uniquely characteristic of the structure of each compound; the absorption bands appear at specific wavelengths allowing identification of bonds and detailed structural features; I.R. bands provide information on vibrations due to changes in dipole moment and reveal the configuration, conformation, symmetry, and H-bonding; a continuous infra-red light source provides the radiation that is analyzed with single or double beam spectrophotometers or I.R. Fourier transform interferometers; the sample holders need infra-red transmitting cell windows (NaCl, KBr, AgCl, CaF$_2$, CsBr, etc.); applications: analytical tool and process control instrumentation

infusorial earth: see diatomaceous earth

ingrain dyes: insoluble dyes that are produced by reaction with fabric, e.g., azoic dyes, copper phthalocyanines

inhibitor: agent that prevents or retards a chemical reaction; opposite of accelerator; use: to prevent undesirable effects such as mildew, corrosion, polymerization

initiators: agents or devices used to start a reaction; in polymerization technology initiators are free radicals (i.e, peroxides, azo-, or diazo-compounds), UV light, or Lewis acids/bases (BF$_3$); in explosives technology initiators may be electrical or mechanical impulses

injection blow molding: process in which an extruded hollow material is put into a blow mold for final shaping; *see* blow molding

injection molding, thermoplastics: the major industrial method of converting thermoplastic materials into finished molded products; powdered or granular polymers from a hopper are fed into a heated chamber where the polymer is softened; the softened material is then moved into a mold with the help of a screw or a ram; injection molding pressure (ASTM D 883-65T) is maintained until the polymer has hardened after which the article is removed from the mold; *see* injection molding, thermosets; reaction injection molding

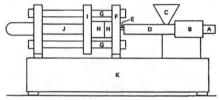

INJECTION-MOLDING MACHINE
(A) hydrolic motor for turning the screw; (B) hydrolic cylinder and piston that allows the screw to reciprocate about three diameters; (C) hopper; (D) injection cylinder (a single-screw extruder); (E) nozzle; (F) fixed platen; (G) tie rods; (H) mold; (I) movable platen; (J) hydraulic cyliner and piston. These are used to move the movable platen and to supply the force needed to keep the mold closed. (K) machine base.
[Copyright Kirk-Othmer. *Encyclopedia of Chemical Technology,* 3d ed., John Wiley & Sons, Ltd.]

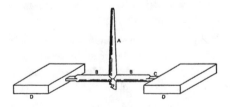

A FULL SHOT
(A) sprue; (B) runner; (C) gate; (D) part.
[Copyright Kirk-Othmer. *Encyclopedia of Chemical Technology,* 3d ed., John Wiley & Sons, Ltd.]

injection molding, thermosets: thermoset plastics production was a slow, multistep process with none of the advantages of thermoplastic injection molding; screw injection molding of thermosets (direct screw transfer, DST) is a process for modified thermoset resins that have a long liquid time at ~200°F and rapid cure in the mold at 350–400°F; the exact working time temperature relationships have to be known for satisfactory operation

ink donor film: IDF; *see* thermal printing

ink jet printing: IJP; nonimpact printing in which ink is forced through small nozzle(s) onto a surface; patterns may be produced by spraying dye from movable jets, by directing the emerging drops electrostatically or by controlling the velocity of the emerging drops so that only drops that are needed are ejected

inks: colored coatings comprised of a pigment, a vehicle (e.g., water, oil, polymer) and additives (e.g., driers, waxes); writing inks are water solutions of pigments; printing inks range from liquids to heavy pastes in which the pigment is suspended in drying oils or other organic media; flow is controlled by the addition of lubricants such as polyethyleneglycol; depending on the printing process, drying results from absorption, oxidation, evaporation, precipitation, quick setting (absorption and coagulation) or cold setting by use of plasticized waxes with melting points of 150–200°F. Magnetic inks for mechanical processing of bank checks contain >50% magnetic powder. Lithographic inks are highly viscous, thus cannot mix with the water phase and are only deposited on selected hydrophobic surfaces. Flexographic inks are formulated so that they do not attack the rubber plates in the rotary letterpresses. Rotogravure and intaglio printing inks can be based on organic solvents because these printing processes are done directly from copper plates

innerlayer: *see* tires

inorganic fibers: *see* fibers, inorganic

inorganic polymers: *see* polymers, inorganic

5-inosinic acid: IMP; a flavor enhancer produced from meat extracts and by enzymatic deamination of muscle adenylic acid

5-INOSINIC ACID

inositol: $C_6H_6(OH)_6$; hexahydroxycyclohexane; member of vitamin B complex

insecticides: pesticides for insect control; these toxins act on the reproductive or nervous system of larval or adult insects; their toxicity to mammals varies; use of insecticides is controlled by the FDA

in situ: phrase meaning in its original place; thus in situ foaming is the insertion of a liquid foamable plastic such as epoxy, polystyrene, or urethane into a cavity before foaming (used for insulating brick work)

Instron machine: instrument to measure tensile properties

insulation: acoustic, electric, thermal, vibration, chemical (acid, base, corrosion), light

insulation board: noncompressed fiberboard

insulin: polypeptide hormone regulating sugar metabolism; mammalian and human insulins differ slightly in amino acid composition so that insulin from pig pancreas has been used for diabetes therapy; human insulin has been expressed in *Escherida coli* through genetic engineering

intaglio printing: process using engraved plates; *see* inks; printing

integrated gasification combined cycle: IGCC; in power generation a system in which coal is gasified and then used to drive a generator

inter-: prefix denoting between, such as intermolecular (between molecules); *see* intra-

intercalation: process of placing atoms or ions within crystals without disrupting the crystal structure; the process is reversible; in chemotherapy, intercalation of molecules, such as cisplatin, between the bases in the DNA double helix disrupts the function of the DNA and thus the growth of cancer cells

interceptor sewers: in waste water treatment, sewers used to collect sewage for delivery to treatment plant

interface: the surface between two solids, two immiscible liquids or two immiscible phases

interfacial polymerization: polymerization occurring at or near the interface between two immiscible solutions, e.g., nylon rope from sebacyl chloride in CCl_4 and hexamethylenediamine in the water phase; production of various condensation polymers using di-acid chlorides in a nonpolar organic solvent and diamines or glycols in an aqueous system that contain an acid acceptor (sodium carbonate or other base); the two solutions are metered into a vessel with stirring and the polymer precipitates immediately at the interface; allows easy synthesis of small experimental amounts of a wide variety of condensation polymers; commercial use for production of polymers that are thermally unstable under normal melt condensation conditions

interference: in electromagnetic systems, the effect produced by two or more superimposed waves; these waves may offset each other by being out of phase and/or they may have different frequencies and intensities; use: for spectroscopy, holography, radio and video transmission

interferon: group of antiviral proteins secreted by killer T cells; secretion is stimulated by interleukin-2, which in turn is secreted by helper T cells in response to antigens; interferons act against viruses and stimulate macrophage activity

interleukins: IL1, IL2, IL3; vital components of the immune system; proteins that regulate the activation and proliferation of white blood cells such as macrophages, T and B cells; IL1 is secreted by macrophages in response to foreign proteins (antigens); IL1 stimulates helper T cells to secrete IL2 and stimulates B cells to secrete agents that cause the growth and differentiation of cells; IL2 when secreted by helper T cells, stimulates killer T cells that secrete several substances including interferon; the acquired immunodeficiency syndrome (AIDS) virus disrupts part of this system

internal lubricant: in plastics technology, a lubricant that is incorporated into the material before processing; such lubricants include waxes, and fatty acids and their salts

internally heated-retort process: wood distillation in which the retort is heated by internally circulating gases

internal mixers: machines with contoured blades rotating in a closed chamber, e.g., Banbury mixer, Gordon plasticizer, Werner-Pfeiderer mixer, and Shaw Intermix

internal plasticizers: agents that are incorporated during polymerization rather than during compounding

interpenetrating networks; IPN: in polymer technology, a combination of two polymers in network form, at least one of which is synthesized and/or cross-linked in the immediate presence of the other; sequential IPNs are formed when a netweork of polymer 1 is swollen with polymer 2, plus its cross-linker and initiator, and polymerized in situ; simultaneous interpenetrating networks (SINs) are formed when both polymerizations are carried out simultaneously; semi-IPNs have one cross-linked and one linear polymer; unlike blends, IPNs swell but do not dissolve in solvents; creep and flow of the material are suppressed

interpolymer: copolymer in which the two monomer units are so intimately distributed that the copolymer can be considered to be homogeneous

intra-: prefix denoting within; i.e, intramolecular (within a molecule); *see* inter-

introfier: an agent that will change a colloidal solution to a true solution by changing the wetting properties and fluidities

intumescent coatings: when exposed to flames or intense heat these coatings bubble and foam thus protecting the substrate by cutting off the oxygen supply, e.g., aromatic sulfonates, magnesium stearate/silicone/talc, cycloaliphatic chlorine/antimony oxide

inulin: polysaccharide isolated from plants such as the Jerusalem artichoke and dahlias; used to isolate fructose

Inverformer: paper manufacturing device similar to fourdrinier machine

inverse electrodialysis: unit operation in which diffusion of ions through a semipermeable membrane causes an electric potential that can be used to perform work; formerly known as reverse electrodialysis

invert sugar: equimolar mixture of glucose and fructose obtained by hydrolysis of cane sugar solution; sweeter than sucrose; used in food products and confectionary as sweetener and to hold moisture (humectant)

invertase: saccharase, β-fructosidase; enzyme isolated from yeast; used commercially to catalyze the reaction sucrose $\rightarrow$ glucose + fructose

inverter: device that converts DC current to AC current

inviscid melt spinning: IMS; method to make high-quality fibers from low-viscosity melts; pressurized melt is forced through a pinhole die in a controlled atmosphere of oxygen and propane; the gas hardens the fiber surface thus preventing droplet formation or breakage; also *see* redrawn inviscid fiber spinning; RIMS

iodine: I; element; produced from Chilean nitrate-bearing earth (caliche) and from seaweed; use: manufacture of iodine containing compounds, germicides, antiseptics, X-ray opaque diagnostic agents, catalyst in organic synthesis, for antigoiter therapy

iodine value: measure of the degree of unsaturation of fats and oils; iodine value equals centigrams of iodine absorbed by 1 g of fat or oil

ion exchange: reversible chemical process in which a solid and a solution containing acid or basic groups can exchange their ions; the resins for this are usually cross-linked solids containing sulfonic, carboxyl, or amino groups; the resin can be a solid or a film; use: for water softeners in which Na^+ are exchanged for Ca^{++} in the solution, for the collection of metallic ions, removal of organic acids and bases (formic acid, nicotine, strepto- mycin), for purification; *see* chromatography

ionic polymerization: addition polymerization, involving carbonium or carbanions (cationic or anionic polymerization); a stable cation or anion complex consisting of monomer and catalyst start the reaction; cationic catalysts are Lewis acids (Friedel–Craft catalysts), anionic catalysts include alkali metals, alkoxides, hydroxides; high rates of polymerization at low temperatures (-100 to $-70°C$) are characteristic for the ionic reactions; termination of the polymer chain is usually by transfer reaction

ionization foaming: in plastics manufacture, foaming due to release of hydrogen from polyethylene by exposure to ionizing radiation

ionomers: polymers that contain intermolecular ionic bonds and intramolecular covalent bonds; transparent, tough thermoplastics and compatibilizing agents

ion scattering spectroscopy: ISS; microanalytical technique to detect the elemental composition of a solid by bombarding its surface with low-energy ions; sensitivity extends to the outermost atom layer and the lateral resolution is a few millimeters

iprodione: *see* fungicides

IR: *see* isoprene rubber

I.R.: *see* infrared spectroscopy

iron oxides: produced by air oxidation of $Fe(OH)_2$ and by reduction of nitrobenzene with iron; minerals vary in color, heat resistance and chemical reactivity; Fe_2O_3, red iron oxide pigment is heat resistant to 1000°C; FeOOH, yellow iron oxide is heat resistant to 189°C; Fe_3O_4, black iron oxide pigment is resistant to 180°C

iron pentacarbonyl: $Fe(CO)_5$; made from iron ore and CO; use: antiknock agent for automotive fuels, catalyst, raw material for finely divided iron (for powdered iron cores in high frequency coils); hazard: lung irritant, affects nervous system, causes liver and kidney damage

iron pyrite: pyrite; FeS_2; "fool's gold"; iron ore; formerly mined for sulfur; its presence in coal requires its removal before using coal for fuel

irregular polymer: a polymer whose molecules can not be described by only one species of constitutional unit in a single sequential arrangement (IUPAC definition)

isoamyl alcohol: *see* isopentyl alcohol

isoamylenes: isopentenes; C_5H_{10}; C5 olefins; use: chemical feedstock; *see* isoprene

isoamyl esters: $RCOO(CH_2)_2CH(CH_3)_2$; use: solvents, in perfumery as odorants, e.g., pear flavor (isoamyl acetate), apple flavor (isoamyl isovalerate), for masking unpleasant odors, in organic synthesis, as plasticizer (isoamyl phthalate)

isoascorbic acid: erythorbic acid; *d*-erythro-ascorbic acid; use as antioxidant in food products to retain color and flavor, as reducing agent for photography industry

isobutanol: $(CH_3)_2CHCH_2OH$; produced from propylene by hydroformylation via isobutyraldehyde; use: solvent, organic reagent

isobutene: *see* isobutylene

isobutylene: isobutene; $[CH_3]_2C=CH_2$; isolated from C4 refinery stream, also as coproduct in oxirane process; use: raw material for production of *t*-butanol, butyl rubber, isoprene, methacrylic acid, methyl-*t*-butyl ether (MTBE), oil additives, pivalic acid, polyisobutene

isobutyraldehyde: $[CH_3]_2CHCHO$; manufactured by oxo process from propylene (CO and H_2 at 130–160°C and 1500–3000 psi over Co catalyst); use: organic synthesis

isobutyric acid: $(CH_3)_2CHCOOH$; produced by oxidation of isobutyraldehyde; use: raw material for methacrylic acid

"iso-$C_{12}H_{24}$": *see* tetrapropene

isocil: 5-Br-3-isopropyl-6–CH_3-uracil; herbicide for noncrop areas

isocyanate: *see* isocyanates

isocyanate generator: raw material for polyurethane resins, a mixture of isocyanate, a polyester, and a phenol; the mixture is stable at room temperature and reacts to form a polyurethane resin at 160°C

isocyanates: compounds having the general formula $R(N=C=O)_n$; produced by phosgenation of amines; reactive compounds, particularly the aromatic *p*-diisocyanates; they react with alcohols to form carbamic acid esters or urethanes; use: chemical synthesis, e.g., production of polyurethanes, bonding agents, pesticides

isocyanurates: produced by trimerization of isocyanates; use: specialty monomers for flame retardant, resilient, rigid polymers

ISOCYANURATES

isoelectric point: pH at which the charge on a molecule in solution is zero; colloidal particles flocculate at the isoelectric point, amino acids become zwitterions and exhibit minimum viscosity and conductivity

isofluorphate: $([CH_3]_2CHO)_2P(F)O$; cholinesterase inactivator; war gas, highly toxic

isoflurane: $CF_3CHClOCHF_2$; inhalation anesthetic

Isoforming process: isomerization of *m*-xylene to *p*-xylene using $Al_2O_3 \cdot SiO_2$ catalyst (Exxon process)

isolac: condensation product of rubber and β-naphthol; use: reinforcement for shoe soles and uppers, for wire insulation

Isomar process: isomerization of *m*-xylene to *p*-xylene using $Al_2O_3 \cdot SiO_2$ and Pt/H_2 catalysts (UOP process)

isomerization: catalytic industrial processes to obtain commercially useful isomers; for example, the conversion of unbranched hydrocarbons to branched hydrocarbons during gasoline reforming; the conversion of phthalic to terephthalic acid; the isomerization of maleic acid to fumaric acid; the formation of cyclohexane from methylcyclopentane; the conversion of *m*-xylene to *p*-xylene

isomers: different compounds having the same empirical formula; in complex molecules or ions bearing different ligands their spacial arrangement may differ (e.g., *cis*- and *trans*- $PtCl_2(NH_3)_2$ [the anticancer drug cisplatin and its inactive geometric isomer]); in organic compounds the difference may be in the carbon skeleton (e.g., pentane and isopentane), in the position of substituents (e.g., amyl alcohol and isoamyl alcohol), in the spacial arrangement of groups (e.g., *cis*- pentene-2 and *trans*-pentene-2 or *o*-, *m*- and *p*-xylene), in spacial arrangement that leads to chiral compounds that can rotate plane polarized light, i.e, optical activity (e.g., D- and L-isoamyl alcohol); isomers have different physical, chemical, and/or biological properties

isoniazid: 4-pyridinecarboxylic acid hydrazide; made by reacting isonicotinic acid ethyl ester with hydrazine; antituberculosis agent

isonicotinic acid: 4-pyridine carboxylic acid; *see* isoniazid

iso-octane: $[CH_3]_3CCH_2CH[CH_3]_2$; highly branched hydrocarbon; use: as standard for antiknocking properties of gasoline; iso-octane has an octane number 100; *see* octane number

isooctanol: *see* 2-ethylhexanol

isopentyl alcohol: isoamyl alcohol, $[CH_3]_2 CHCH_2CH_2OH$; made by hydroformylation of isobutene; use: solvent, extractant, for organic synthesis, odorant in the perfume industry; hazard: central nervous system depressant

isophthalic acid: 1,3-dicarboxybenzene; meta-phthalic acid; produced by air oxidation of *m*-xylene using Co and Mn acetate catalyst (Amaco process); use: manufacture of plasticizers, poly(benzimidazole) (PBI), *m*-phenylene-isophthalamide, isophthaloyl chloride (raw material for the aramid fiber Nomex)

isophthaloyl chloride: *see* isophthalic acid

isoprene: $CH_2=C(CH_3)CH=CH_2$; isolated from the C5 stream of petroleum cracking, also prepared by dehydrogenation of isoamylenes, by reaction of isobutylene with formaldehyde (*see* hydroformylation), or by

the catalytic reaction of acetone with acetylene; polymerizes to form polyisoprene rubber (IR); hazard: skin irritant, narcotic in high concentrations

isoprene rubber: *cis*-1,4-polyisoprene; natural rubber

isopropanol: *see* isopropyl alcohol

isopropyl acetate: $CH_3COOCH[CH_3]_2$; produced by esterification of isopropyl alcohol or by acid catalyzed addition of acetic acid to propylene; use: solvent, for soil consolidation by mixing with ethyl acetate and water glass

isopropyl alcohol: $(CH_3)_2CHOH$, IPA, isopropanol; prepared by catalytic ($WO_3 \cdot ZnO/SiO_2$) hydration of propylene (ICI high pressure process) or using H_3PO_4/SiO_2 (Vega medium pressure process); use: as rubbing alcohol, solvent, antifreeze, antiseptic; also used to manufacture acetone, isopropyl acetate; hazard: inhalation or ingestion of large quantities may cause flushing, headache, vomiting, narcosis, coma; 100 mL can be fatal

isopropyl benzene: *see* cumene

isopropyl carbanilate: IPC, INPC, isoPPC, Propham; $C_6H_5NHCOOCH(CH_3)_2$; herbicide

isopropylidene acetone: *see* mesityl oxide

isoquinoline: used in synthesis of dyes, insecticides, antimalarials, rubber accelerators

ISOQUINOLINE

Isosiv process: separation of paraffins by adsorption using molecular sieves (UCC process)

isosorbide: diuretic

isosorbide dinitrate: dilator of coronary (heart) blood vessels

ISOSORBIDE DINITRATE

isotactic polymers: regular polymers in which the configurational repeating units are all identical with the configurational base unit

ISOTACTIC POLYMERS

isotropy: tendency of a material to have similar properties in all directions; materials may be isotropic for one property but not for others, e.g., an unoriented polymeric structure has isotropic mechanical properties; an oriented structure has unisotropic properties: for example, a molded plastic versus oriented fiber or film; *see* anisotropy

itaconic acid: $CH_2=C(COOH)CH_2COOH$; produced by batch fermentation of molasses using *Aspergillus terreus*; use: copolymer in acrylic and methacrylic resins to permit better acceptance of printing dyes and inks, plasticizer, lube oil additive

Izod Impact Strength: *see* impact strength

J

J acid: $3-SO_3H-6-NH_2-\alpha$-naphthol; starting material for dyestuff synyhesis

Jade Green: vat dye

jet: in fiber technology, a spinneret for extruding polymer melts or solutions to make filaments or filament yarns

jet dyeing: in textile manufacture, pressure jets move a fabric loop through dye bath

jet molding: modification of injection molding designed for thermosets

jet printing: *see* ink jet printing

jig: clamping device; in coal cleaning, a device that induces pulsating motions through a coal/water slurry thereby loosening dirt particles from the coal; in textile technology, a device for dyeing by repeatedly rolling a full width fabric under tension through a dye bath; *see* beck

jig welding: welding of thermoplastic materials between suitably shaped clamping devices (jigs)

jojoba oil: derived from the seeds of the jojoba plant, which grows wild in the southwestern United States; contains mostly liquid wax that consists of esters of monounsaturated C20 to C22 acids and monounsaturated C20 to 22 alcohols, the oil is chemically almost equivalent to sperm whale oil whose use as lubricant additive, in high-quality waxes and for cosmetics is being phased out

Joule-Thompson expansion cycle: process for liquefaction of gases using an expansion valve or nozzle

jute: a cellulosic fiber from the plant *Corchorus* of India and Pakistan; use: as cordage, coarse textiles or fabric in resin reinforcement

K

K acid: 1-amino-8-naphthol-4,6-disulfonic acid; azo dye intermediate

K.A.: (ketone/alcohol); process to manufacture adipic acid from cyclohexane by a catalytic oxidation to a mixture of cyclohexanone and cyclohexanol and from there to the acid

kanamycin: antibiotic complex used against meningococci and tuberculosis

kaolin: China clay, Chinese white alumina, porcelain clay; aluminum silicate with a high fusion point; remains white upon firing; use: for ceramics manufacture and as filler for cosmetics, paints, paper, rubber, and textiles

kaolinite: $Al_2O_3 \cdot SiO_2 \cdot H_2O$; main ingredient of kaolin

Karl Fischer reagent: methanol solution of I_2, SO_2, and pyridine; used to determine water content in resins

kauri: hard resin obtained from the kauri pine in New Zealand; use: varnishes, paints

Kaysam process: in the plastics industry, process used to produce hollow articles, such as shoes and toys, by rotating a heat-sensitive latex in two directions in a hollow, heated mold; after gel hardens on the mold surface the object is removed and vulcanized

KBW Coal Gasification process: entrained flow gasifier; *see* coal gasifiers

Kel-Chlor process: coproduction of chlorine from HCl during oxychlorination of ethylene in the presence of NO_x catalyst and H_2SO_4

Kel-F: *see* fluorocarbon polymers

kelp: *see* algae

keratin: protein constituent of feathers, hair (wool), hoofs, horns, nails, skin; generally harder than other proteins due to S-S cross-linking; obtained from animal sources by calcination; use: filler in rubber and plastics, coatings for pills that dissolve only in the intestines

kermes: natural red dye stuff produced from dried insects; oldest dye stuff known

kerogen: solid hydrocarbon material in oil shale

kerosene: (kerosine) a crude oil fraction distilling above gasoline (hydrocarbon fraction between C10 and C16), it is refined to jet fuel and range oil

ketene: $H_2C=C=O$; produced by pyrolysis of acetic acid or acetone at 500–600°C using CS_2 as catalyst; use: for production of acetic anhydride, as acetylating agent, for production of diketene, sorbic acid, and propiolactone, for cross-linking polymers

2-ketogluconic acid: isoascorbic acid; erythorbic acid; $HOOCCO(CHOH)_3CH_2OH$; intermediate in ascorbic acid manufacture; produced by fermentation of glucose; use: antioxidant, preservative

ketohexamethylene: *see* cyclohexanone

ketone resins: condensation products of ketones and formaldehyde; additives for chlorinated rubber

kettle soap: *see* full-boiled kettle process

Kevlar: *see* aramids

Keyes process: production of 100% ethanol by distillation of 95% aqueous ethanol in the presence of benzene

kier: in textile manufacture, a large metal tank used for scouring cotton material by boiling in alkaline solution (kiering); also used for dyeing or bleaching

kieselguhr: *see* diatomaceous earth

killed steel: steel deoxidized by addition of Al and ferrosilicon at the melting temperature until bubbling ceases

kinases: enzymes that catalyze transfer of phosphate, e.g., from adenosine triphosphate (ATP) to an acceptor

kinematic viscosity: ratio of viscosity to density; measured in Stoke (cgs unit)

kinetic molecular weight: M_k; *see* molecular weight

king's green: copper acetoarsenite

kirksite: aluminum-zinc alloy used for molds for blow molding; has a high thermal conductivity, thus accelerates cooling cycles

kiss-roll coating: process for applying thin coatings to substrates such as paper; one roller is immersed in a coating fluid and transfers a small amount of the fluid to a second roller from which coating is applied to substrate

kitazin: $([CH_3]_2CHO)_2P(O)SCH_2C_6H_5$; fungicide used primarily for rice

kling test: method to determine the relative degree of fusion of thin sections of plastics; a folded specimen is immersed in a solvent and the time in which disintegration starts is measured; the preferred solvent system is one in which disintegration of the fully fused system starts within 5–10 min

kneaded rubber: an artist's eraser made from unvulcanized rubber with high softener content; used for charcoal drawings, cannot erase hard pencil markings.

kneaders: pair of intermeshing blades used to mix semidry or rubbery plastic masses

knock: ignition of a portion of the gasoline in the cylinder head due to spontaneous oxidation; the extra pressure results in a sound, called knock; straight chain hydrocarbons cause more knocking than branched chain hydrocarbons and aromatics; *see* octane number

knockout: part of a mold used to eject the molded product, e.g., knockout pin, knockout plate; in agrochemistry, a measure of rate of insecticide action

Knoop microhardness: determined by the indentation caused by a diamond pyramid; the Knoop hardness number (KHN) equals the load times 14.2 divided by the square of the indentation

Koch acids: tertiary α carboxylic acids; produced from olefins by reaction with mineral acid followed by carbonylation and hydrolysis; Koch reaction is based on stability of tertiary carbonium ions; these hindered acids are heat resistant, the esters resist hydrolysis; used as components of oils, for manufacture of paints and resins

Koch reaction: *see* Koch acids

KOH number: in rubber technology, a term used to denote the amount of ammonium salts in natural latex: the KOH number equals the grams KOH needed to decompose ammonium salts contained in 100 g of dry solids in latex; KOH number indicates processing properties of latex

Koji fermentation: manufacture of citric acid using *Aspergillus niger* on moist bran; a solid-state process, only used in Japan

Kopper's process: production of butadiene by pyrolysis of cyclohexane

Koppers-Totzek: coal gasification with oxygen and steam at atmospheric pressure and 1400–1600°C using an entrained flow process; *see* coal gasifiers

korax: 1-chloro-1-nitropropane; fungicide

kraft process: production of strong, relatively inexpensive paper by digesting wood with a mixture of caustic soda, sodium carbonate, and sodium sulfide

Krebs unit: a measure of viscosity used for the Stormer viscometer; weight in grams that will rotate the viscometer paddle 100 revolutions in 30 seconds

Kroll process: production of metallic titanium by reduction of $TiCl_4$ with Mg at red heat and atmospheric pressure in a helium or argon atmosphere

krypton: Kr; element #36; inert gas; obtained by fractionation of air and from ammonia manufacture; krypton-85 is a radioactive isotope isolated from irradiated nuclear fuel and is used for detecting leaks and activating phosphors in self-luminous markers

L

L-: prefix denoting an optically active compound whose stereochemical structure can be related to L-(-)-glyceraldehyde

lac: *see* shellac

lac dye: a natural orange or crimson dye stuff extracted from *Coccus lacca*; this insect is an important source of shellac

lacquers: glossy coatings that dry rapidly by evaporation of solvents without oxidation or polymerization; lacquers are based on high molecular weight polymers such as cellulose esters, acrylics, polyurethanes, and vinyls; they usually redissolve in the original solvent; unlike varnishes, lacquers are usually pigmented or opaque; use: for automobile, furniture, metal, plastic, paper and textile finishes

lactam: cyclic amide; ε-caprolactam and lauryl lactam are industrially important; penicillins are β lactams

lactase: enzyme that catalyzes the hydrolysis of lactose to glucose and galactose; microbial sources for industrial hydrolysis are *Kluyveromyces fragilis* and *K. lactis, Lactobacillus* sp., and *Saccharomyces lactis*; *see* enzymes

lactic acid: $CH_3CH(OH)COOH$; produced from milk by fermentation of sugar, or from acetaldehyde via the nitrile; use: food additive, chemical raw material

Lactobacillus brevis: microorganism used for the industrial fermentation of glucose to malic acid

Lactobacillus delbruckii: microorganism used for the industrial fermentation of glucose to lactic acid

lactofen: diphenylether herbicide

lactone: cyclic ester derived from a hydroxycarboxylic acid, e.g., caprolactone that serves as raw material for polyester

ladder polymers: double-stranded chain polymers that are more resistant to degradation; for example, on heating polyacrylonitrile at 300–400°C it cyclizes to form a series aromatic rings (black orlon); cellulose that is built up of glucose chains and supported by a parallel hydrogen bond chain is a natural ladder polymer; asbestos minerals known as amphiboles consist of two parallel silicate chains joined by oxygen atoms

lagoon: in waste water treatment, a shallow pond where sunlight, bacteria, and oxygen interact to purify the water; *see* stabilization pond

lake: water-insoluble salt of cationic or anionic dyes

Lambiotte process: internally heated retort process for distillation of hardwood

lamella: plate-like molecular or crystalline structure; lamellae are present in many different materials ranging from the mineral mica to most polymers; in polymers lamellae are the basic crystalline unit and are 5–20 nm thick; the polymer chain direction is normal to the lamellar surfaces so that the chains must be folded; deformation extends some of the folds when polymers are drawn so that some of the chains unravel and form interconnections between the lamellae; thus, the polymer does not flake in the way crystalline mica does nor do the lamellae slide against each other, the way graphite layers do, but even highly drawn materials have partially folded crystalline units; only extremely high strength materials or very stiff polymeric fibers have few chain-folded structures (e.g., aramids, ultra high tenacity poyethylene)

lamella thickener: settling device that has inclined plates for continuous removal of solids

laminar flow: movement of one layer of fluid over another

lamina: thin plate, layer, or sheet; the dimensions of laminae are larger than those of lamellae; laminates are structures consisting of multiple layers (laminae), such as laminated fabrics, glass, plastics, paper, or wood; produced by heat sealing or with adhesives; used to provide strength, fire resistance, permeability or nonpermeability

laminate: *see* lamina

laminated glass: *see* safety glass

lampblack: pure carbon obtained from burning carbon-containing substances in insufficient air; use: pigment

lanolin: grease obtained from wool; complex mixture of esters of higher alcohols (including cholesterol) and long chain fatty acids; used in cosmetics, medicinals, paints

laser: *l*ight *a*mplification by *s*timulated *e*mission of *r*adiation; light amplification is achieved in a gas, liquid, or solid by an energy inversion so that more particles in the laser are in an excited state than in the ground state; the result is that the excited atoms or molecules pass the excess energy to the unexcited particles thereby amplifying their energy emission; use: light sources for spectroscopic measurements in infra-red, visible (Raman), and ultraviolet spectra, material processing: welding, cutting, drilling; measurement of physical parameters, holography, surgical procedures, optical fiber telecommunication systems, nonlinear optics, printing, military weapon

laser xerography: a nonimpact printing method that can produce impact letter quality using dot matrix printing; *see* xerography

latex: stable water dispersions of synthetic or natural polymers; in particular, the name for the white sticky colloidal emulsion derived from the rubber tree

latex paint: water emulsions of polyvinyl acetate, polyacrylics, and polystyrene-butadiene

latex thread: extrusion of rubber latex to form elastomeric fibers

latices: plural of latex

laughing gas: *see* nitrous oxide

lauric acid: C12 saturated fatty acid; used for fine toilet soap manufacture

lauryl lactam: produced by acid catalyzed Beckmann rearrangement from lauryl oxime, which in turn is derived from cyclododecanone; use: for the manufacture of nylon 12

LAURYL LACTAM

laxatives: cathartics; purging agents that fall into four classes (1) bulk forming (bran, methyl cellulose); (2) lubricants (mineral oil); (3) saline (milk of magnesia, sodium sulfate); and (4) stimulants (castor oil, phenolphthalein, senna)

LD_{50}: measure of toxicity; amount of a substance that causes death in 50% of exposed animals or humans

leachate: liquid that has percolated through some medium such as soil or solid waste and contains extracted dissolved or suspended material

lead-acid battery: *see* batteries

lead azide: $Pb(N_3)_2$; a common primary explosive in blasting caps; prepared by adding 2% aqueous NaN_3 to 5% aqueous $PbAc_2$ solution that contains a small amount of dextrin; more sensitive than nitroglycerin, less than mercury fulmanate; requires a strong igniter for detonation

lead glass: *see* glass

lead pigments: colors range from white to red: basic white lead ($2PbCO_3 \cdot Pb(OH)_2$), sublimed white lead (75% $PbSO_4$, 20% PbO, 5% ZnO), Leipzig yellow or chrome yellow ($PbCrO_4$), basic lead chromate (red paint used for protection of steel, $2PbO \cdot CrO_3$), litharge (PbO, yellow)

lead styphanate: lead trinitroresorcinate; a primary explosive; serves as igniter for lead azide; extremely sensitive to spark, static electricity, and flame; has good thermal stability

Lebedev process: one-step catalytic ($MgO \cdot SiO_2$) synthesis of butadiene from ethyl alcohol at 400°C

Leblanc process: former process for manufacture of Na_2CO_3 (soda ash)

lecithin: mixture of glycerol phosphatides having the general formula $CH_2(OOCR')CH$ $(OOCR'')CH_2OPO_4NR_3$; produced primarily by extraction from soy beans, but is also present in other vegetable oil seeds and in animals; composition varies according to the source; used widely in the food, plastic, paint, pharmaceutical, and textile industries as emulsifier, antioxidant, release agent, stabilizer, and surfactant; in the rubber industry as softener and curing accelerator

Leclanché system: *see* batteries

lectin: plant protein that binds to saccharides

Leipzig yellow: *see* lead pigments

letterpress ink: ink with high penetration power for absorbent paper

letterpress printing: *see* printing

letterset printing: indirect printing using raised type; *see* printing

levodopa-carbidopa mixture: levodopa: $R-CH(NH_2)COOH$, carbidopa: $R-C(CH_3)$ $(NHNH_2)COOH$, where R is $3,4-(OH)_2C_6H_3$ CH_2-; the L-forms are used for Parkinson's disease and are unique in that carbidopa helps levodopa to cross the blood–brain barrier after which it is decarboxylated to pro-

duce the neurotransmitter dopamine; carbidopa that does not cross the blood–brain barrier inhibits the decarboxylation of levodopa in the blood; therefore small amounts of levodopa are sufficient

levopropoxyphene: *see* propoxyphene

levulose: name for fructose; name indicates that fructose is a ketone and that it rotates plane polarized light counterclockwise

Lewis acid: electron pair acceptor such as $AlCl_3$ or BF_3

Lewis base: electron pair donor such as NH_3

ligand: molecule, ion or atom that donates an electron pair to a central metal atom thus forming a complex; NH_3 donates one electron pair to a complex, e.g., $Cu(NH_3)_4^{2+}$ and is called a unidentate ligand; porphyrins (chlorophyll, heme, and phthalocynanine) that contain 4 nitrogen atoms are tetradentate ligands: monodentate, bidentate, and tridentate phosphines play important roles in the design of modern catalysts; catalysts containing chiral phosphine ligands are used for synthesis of chiral products; *see* chelates

ligand exchange chromatography: LEC; *see* chromatography

light Arabian crude oil: a low sulfur petroleum

light-density soap products: *see* fluff process

light gas oil: petroleum fraction that boils above kerosene (200–350°C); use: diesel fuel (lower boiling fraction), No. 2 fuel oil (home heating oil); *see* diesel oil

light naphtha: second petroleum fraction that boils at 20–150°C; straight run gasoline; it distils off after removal of methane, ethane, propane and butane; contains aliphatic and cycloaliphatic hydrocarbons; use: for gasoline production, fuel, solvent

light (soda) ash: sodium carbonate product

from Solvay process; hydration of light ash followed by calcination converts the low-density carbonate to dense ash, which is preferred by the glass industry; dense ash has about twice the bulk density of light ash

light water reactor: a nuclear reactor that uses boron containing water as coolant

lightwood: the resinous portion of wood after the decay of sapwood

lignin: complex nonlinear polymer consisting of phenolic rings; the major noncellulosic polymeric constituent of wood; potential biomass material; use: extender in phenolic resins, low gravity filler in rubber

lignite: lowest rank of coal; has a heating value of 4000–8300 Btu/lb; in the United States lignite is found mainly in the Gulf Coast and northern Great Plains

lignosulfonates: in the sulfite process of pulping, the soluble product of the reaction of lignin with HSO_3^-

lime: *see* calcium oxide

lime soap: soap curd resulting in hard water from insoluble calcium and magnesium fatty acid salts

limestone: *see* calcium carbonate

limestone injection multistage burners: LIMB; in power generation, a system in which lime is injected into a boiler above low NO_x burners to absorb SO_2

lime sulfur: calcium polysulfide; a historic pesticide

limonite: iron ore; *see* ocher

lindane: BHC; 1,2,3,4,5,6-hexachlorocyclohexane; use: insecticide; hazardous chemical waste (EPA)

Linde process: *see* air-liquide process

linear alkyl benzene: LAB; biodegradable, anionic surfactant; use: laundry product

linear alkyl sulfonates: LAS; biodegrad-able, high sudsing, anionic surfactants (e.g., sodium dodecylbenzene sulfonate); nonlinear alkylbenzene sulfonates (ABS) are not biodegradable; laundry products

linear high-density polyethylene: LHDPE; *see* low-pressure polyethylene processes

linear low-density polyethylene: LLDPE; strong, largely linear polyethylene with some branching (comonomers); produced in the gas phase by the catalytic Unipol process that is run at 100°C and pressure of 100–300 lbs/in.^{2}; LLDPE is a thermoplastic with a density of 0.90–0.94 g/mL; use: pipe, film, molding, wire, and cable coating

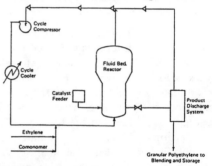

UNION CARBIDE'S "UNIPOL" PROCESS FOR LLDPE
[Copyright *Riegel's Handbook of Industrial Chemistry*, 8th ed., Van Nostrand-Reinhold Co., Inc.]

linear polymer: unbranched polymer

linen: a cellulosic fiber material from the bast tissue of the flax plant, *Linum usitatissimum;* fibers are very fine, flexible, strong; oldest known textile fiber.

linoleic acid: 9,12-octadecadienoic acid; *see* linseed oil

linolenic acid: 9,12,15-octadecatrienoic acid; *see* linseed oil

linseed oil: drying oil derived from flax seeds; consists of linoleic, linolenic, oleic, and saturated fatty acids

linuron: 3,4-dichloro-$C_6H_3NHCON(CH_3)$ OCH_3; urea herbicide

lipases: enzymes that hydrolize fats and oils; microorganisms that produce the enzyme are used in waste water treatment (*Candida lipolytica*), as digestive aids (*Aspergillus niger*), in the tanning industry (*Mucor javanicus*), and for improvement of flavor (*Rhizopus arrhizus*); lipases are effective when used in nonpolar solvents such as hexane; *see* enzymes

lipids: animal and plant fats, oils, and waxes

liposomes: lipid vesicles in which aqueous compartments are enclosed by a lipid layer

liquefaction: conversion of a solid into a liquid by chemical reaction

liquefied natural gas: LNG; a fuel; bp, −162°C; prepared by cooling natural gas with a mixture of liquid nitrogen and hydrocarbons using a cascade system; LNG is shipped by ocean vessels and by trucks

liquefied petroleum gas: LPG; petroleum fraction consisting of propane and butane; marketed as liquefied gas

liquid–column chromatography: LCC; separation process in which the mobile phase is a liquid; used for materials that cannot be vaporized, e.g., polymers, ionic or labile compounds; the mobile phase flow is controlled by pressure; *see* chromatography, liquid–liquid chromatography; liquid–solid chromatography

liquid crystals: l.c.; organic compounds with a physical structure intermediate between the crystalline (three-dimensional) solid state and the melt; these intermediate phases exist in the presence of solvents (lyotropic l.c.) or as function of temperature (thermotropic l.c.); liquid crystals consist of long anisotropic molecules, usually containing two or more benzene rings in *para* position together with intermediate substituents: R-$(C_6H_4)_n$-X-C_6H_4-R', where n = 1,2; X = −CH=N−, −N=N(O)−, −CH=N(O)−, −CH=CH−, −C≡C−, −O(C=O)−, R and R' are

alkyl, alkoxy or acyl groups; these molecules are packed so as to lead to three different molecular structures (1) nematic in which the rod-like molecules are arranged parallel to one another; (2) smectic in which the molecules form sheet-like units; (3) cholesteric in which the molecules have rod-like arrangements as in nematic l.c., but in addition the rods change direction from layer to layer; not only small molecules but polymers can have liquid crystalline properties; polymer l.c. fall into two groups (1) main chain l.c. (aramids and aromatic polyesters) in which the *p*-aromatic units are in the polymer backbone; (2) side chain l.c. in which the *p*-phenylene chains are attached to polyacrylate, polymethacrylate, or polysiloxane backbones by flexible spacers; here the backbone determines the temperature ranges of the liquid crystalline transitions; use: watch and calculator displays, optical data storage, wave guides, optical switches, holography

liquid detergents: heavy-duty laundry products that contain 50–60% water; careful blending maintains their viscosity, clarity and low freezing point; formulations include surfactants, builders, anticorrosion agents, antiredeposition agents, foam regulators, whiteners, solubilizers, hydrotropes, and sometimes antistatic and softening agents

liquid injection molding: LIM; *see* injection molding; thermosets

liquid–liquid chromatography: LLC; partition chromatography; separation process in which the stationary and the mobile phases are liquids that greatly differ in polarity; in normal phase LLC the stationary phase is polar or hydrophilic, in reverse phase LLC the stationary phase is nonpolar; *see* chromatography, liquid column chromatography

liquid membranes: double emulsions produced by gentle dispersion of a water-in-oil (w/o) emulsion in a second aqueous phase; the internal emulsified phase is separated from the external aqueous phase by a hydrocarbon layer so that the three layers form a liquid membrane

liquid metal fast breeder: LMFBR; *see* sodium cooled breeder reactor

liquid reaction molding: LRM; *see* reaction injection molding

liquid–solid chromatography: LSC; separation absorption chromatography; process used for samples that are soluble in nonpolar or moderately polar solvents; *see* chromatography; liquid column chromatography

liquid sugars: sugars dissolved in water; prepared by dissolution of purified sugars or by purification of in-process liquors; economically important for the food industry

litharge: yellow lead oxide, PbO, massicot; made by heating lead in air; use: in storage batteries, ceramics, fluxes, glazes, glasses, match heads, pigments; as rubber accelerator, raw material for red lead, when mixed with glycerin it is used as plumber's cement

lithium: Li, metallic element; produced by electrolysis of molten LiCl/KCl mixture; use: in aerospace alloys, in batteries with high-power densities, as catalyst (e.g., production of *cis*-polyisoprene from isoprene), as raw material for manufacture of organolithium reducing agents; *see* batteries

lithium aluminum hydride: $LiAlH_4$; produced from lithium hydride; reducing agent for organic synthesis

lithium carbonate: Li_2CO_3; produced from lithium ores; use: during aluminum manufacture, as flux in the glass and ceramic industries, as drug for manic–depressive patients

lithium chloride: LiCl; used in some solvent systems for spinning fibers, as electrolyte in dry batteries, and for the manufacture of metallic lithium

lithium hydride: LiH; produced by reaction of lithium and hydrogen at 700°C; use: reducing agent

lithium hydroxide: LiOH; produced by reaction of lithium carbonate and calcium hydroxide; use: to manufacture greases, i.e, lithium soaps, lithium stearate

lithographic inks: high viscosity inks usually dissolved in linseed oil and an aliphatic solvent

lithographic printing: widely used process in which ink is transferred from a planar surface; important method for fabrication of semiconductor devices; patterns are transferred onto chips by light exposure and are then developed; *see* printing

lithol: group of azo pigments used in printing inks, industrial enamels, paints, plastics, rubber, and cosmetics, e.g., lithol rubine:

LITHOL RUBINE

lithopone: mixture of $BaSO_4$ and ZnS; white pigment; produced by heating a dried mixture of BaS and $ZnSO_4$ and plunging it into cold water; use: in manufacture of paints, inks, oilcloth, linoleum, and rubber materials

"living polymers": produced in highly purified anionic polymerization systems that lack a termination reaction thus maintaining a "living" polymer system; the resulting polymers have narrow molecular weight distributions; this system is used for the commercial production of very accurately defined block copolymers of polystyrene and polybutadiene; the systems are terminated by destroying the catalyst

long-oil varnish: varnish that contains ≥30 gallons of oil per 100 lb of varnish

long ton: 2240 lb

loop diuretic: the most potent and rapidly acting group of diuretics; these drugs act directly on the part of the kidney that controls the transfer of Na^+ (the loop of Henley); the most commonly prescribed member of this group is furosemide

loss modulus: dissipation of energy when a material is deformed

low alloy steels: steels containing <5% of alloying elements

low Btu gas: *see* producer gas

low density polyethylene: LDPE; branched, tough thermoplastic having densities of 0.91–0.93 g/mL; produced by free radical polymerization under high pressure; use: injection molding, household goods, packaging, blow molding bottles, films; they melt near 100°C and thus cannot be used for equipment that has to be sterilized

Low E-glass: low emissivity glass; glass coated with a transparent fluorine-doped SnO_2 film retains heat supplied by solar energy or from heated rooms; use: energy conserving window panes

low explosives: *see* deflagrating explosive; propellant explosive

low-pressure molding: molding or laminating at pressures of ≤200 psi (ASTM D 883-75a)

low-pressure polyethylene processes: heterogeneous catalytic polymerization of ethylene at ~450 lb/in.^{2} or less leads to high-density polyethylene (HDPE); catalysts used are transition metal oxides (usually Cr or Mo oxide on silica–alumina supports) or Ziegler–Natta catalysts; polymerization proceeds only in the absence of water, oxygen, or polar materials; use: compression and injection molding, films, polyethylene applications where the higher strength, stiffness, and temperature resistance are essential, e.g., sterilization of medical supplies; *see* PHILLIPS PARTICLE FORM POLYETHYLENE PROCESS

low-shrink fibers: stabilized, oriented acrylic staples; use: production of bulky yarns by blending low-shrink and high-shrink (unstabilized) fiber staple; during dyeing in hot water the unstabilized fibers shrink and deform the stable ones thus producing a more voluminous structure; *see* high-shrink fiber

LPG: *see* liquefied petroleum gas

lubricant: additive that reduces the coefficient of friction; substance that prevents adhesion (e.g., graphite, waxes, fluorohydrocarbons)

lubricating oil: produced from the petroleum fraction boiling >350°C; contains C20 to C70 aromatic and aliphatic hydrocarbons; refined asphalts, resins, and waxes are separated as by-products during lubricating oil manufacture; used to provide thin films between moving machine parts; formulations contain additives such as antioxidants, corrosion, and rust inhibitors, viscosity index improvers, pour-point-temperature depressants, detergents, dispersants, foam depressants

lucite: poly(methylmethacrylate)

luminous paints: paints containing radioactive elements that emit α and β rays; phosphorescent or fluorescent paints that become activated by light, e.g., ZnS (green), CdS (yellow and orange)

Lurgi gasifier: fixed bed coal gasifier; *see* coal gasifier

Lycra: polyurethane elastomer; *see* Spandex

lysine: $H_2N(CH_2)_4CH(NH_2)COOH$; essential amino acid; produced by fermentation, by hydrolysis of soy proteins, and by cya-

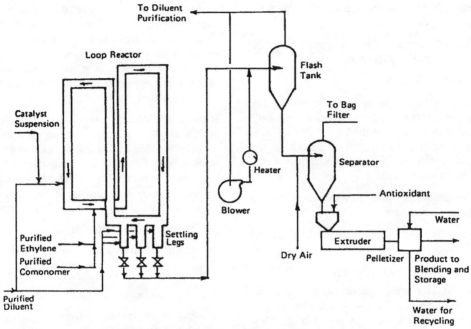

PHILLIPS PARTICLE FORM POLYETHYLENE PROCESS
[Copyright *Riegel's Handbook of Industrial Chemistry,* 8th ed., Van Nostrand-Reinhold Co. Inc.]

noethylation of acrylonitrile with HCN, CO_2, and NH_3; use: food and feed additive, pharmaceutical

lymphokines: hormones that control the nature and extent of all immune responses including antibody formation; they have molecular weights of 15,000–40,000 daltons; interleukins are lymphokines

lyophilic: ability to form colloidal suspensions easily

lyophilization: *see* freeze-drying

lyotropic: ability to form liquid crystalline phases in solution

lysis: in biotechnology, refers to destruction of cells

-lysis: suffix denoting breaking down, decomposition, e.g., pyrolysis, cytolysis; the ending for the adverb is -lytic, i.e, pyrolytic, cytolytic

M

macrolides: antibacterials characterized by large rings, e.g., erythromycin; isolated from fermentation processes; used against organisms that have developed resistance to penicillin, streptomycin and tetracyclins

macromolecules: high polymers; molecules with high molecular weights (to >1,000,000 daltons) that are built up from low molecular weight units (monomers); naturally occurring polymers include cellulose and proteins, synthetic polymers include polyvinyl compounds, polyesters, nylons, and polyurethanes; *also see* homopolymer, copolymers, block polymer or copolymer, Spandex

macrophage: large white blood cell, part of the immune system; *see* interferon, interleukins

madder: ground root of *Rubia tinctorum,*

which has been stored long enough to develop the red-orange pigment alizarin (dihydroxyanthraquinone)

madder lake: alkaline solution of red anthraquinone dye; *see* lake

Madison process: *see* wood hydrolysis

magic angle spinning: to achieve NMR spectra of solids the sample must be spun rapidly about an axis oriented at an angle of about 54.7° (the magic angle) with respect to the external magnetic field; this spinning leads to an averaging of the orientation of the molecules similar to that present in a solution; *see* nuclear magnetic resonance

magnesia: MgO; magnesium oxide; produced by calcining dolomite, magnesite, or magnesium chloride obtained from seawater; use: lining of refractory ovens, additive to give scorch resistance to rubbers, fertilizer, adsorbent, chemical reagent

magnesium: Mg; metallic element; produced by electrolysis of molten $MgCl_2$ or by thermal reduction of dolomite; use: in alloys (mainly with Al, Be, Mn, Si and Zn), as cathode protection against corrosion, as reducing agent and for steel desulfurization

magnesium carbonate: $MgCO_3$; mineral; dolomite; use: raw material for MgO production

magnesium chloride: $MgCl_2$; produced from seawater and brine; use: for electrolytic production of magnesium

magnesium oxide: *see* magnesia

magnesium silicate: white pigment used for paints

magnesium sulfate: $MgSO_4$; mineral used for fertilizer; anhydrous salt is used as drying agent

magnetic ink: inks that can be magnetized after printing for recognition by electronic reading equipment

magnetic resonance imaging: *see* nuclear magnetic resonance

magnetohydrodynamics: MHD; method of electric power generation in which hot ionized gases (plasma) are passed through a magnetic field to create a current; thus electricity is directly produced from thermal energy; in closed cycle MHD the conducting medium (working fluid) is recycled; in open systems the hot nonionized gas is used to generate steam and is then discarded

major tranquilizers: neuroleptics; *see* antianxiety agents

malachite green: Basic Green 4; a triarylmethine dye; used in acrylics, inks, lacquers, leather, and paints, as indicator for pH 0.5–1.5

malathion: $(CH_3O)_2P(S)SCH(CH_2COOC_2H_5)COOC_2H_5$; organophosphate insecticide

maleic acid: *see* maleic anhydride

maleic anhydride: MA; a key chemical intermediate; produced originally by air oxidation of benzene over V_2O_5 at 450°C and 2–5 bar (Alusuisse–UCB process), as byproduct during the manufacture of phthalic anhydride from naphthalene, by catalytic (V_2O_5) oxidation of butylene and now from butane by fluidized bed oxydation; use: feedstock for manufacture of 1,4-butanediol, tetrahydrofuran, succinic anhydride, γ-butyrolactone, fumaric and malic acids; end products include polyester resins, alkyd resin modifiers, pesticides, plasticizers; to replace phthalic anhydride in alkyd resins, to increase hardness for enamels and resist yellowing

MALEIC ANHYDRIDE

maltose: disaccharide that yielding glucose

on hydrolysis; used in confectionery as extender of sucrose; *see* saccharides

mandrel: form to mold the inner surface of a hollow molded object

maneb: $(S_2CNHCH_2CH_2NHCS_2)^=Mn^{++}$; fungicide

manifold: a pipe with several outlets

manila: natural resin

Mannheim process: production of HCl by treating NaCl with H_2SO_4; salt cake (sodium sulfate) is a by-product

mannomethylose: 6-deoxymannose; *see* arabic gum

mannose: a six-membered sugar; an isomer of glucose

margarine: oleomargerine; butter substitute made from ~80% vegetable oils and 20% milk with addition of vitamins A and D and artificial coloring; the vegetable oils are hydrogenated to achieve the desired hardness; the melting point of margarine depends on the degree of fatty acid unsaturation; the higher the unsaturation the lower the melting point

marijuana: *see* hemp

marine oils: unsaturated oils containing 5 to 6 nonconjugated double bonds; believed to be beneficial for reduction of blood cholesterol; the hydrogenated oils are used for shortening

marsh gas: *see* methane

mask: *see* photoresist

mass dyed: mass colored; *see* melt dyeing

massecuite: in sugar refining, a mixture of crystals and mother liquor obtained during sugar crystallization

massicot: *see* litharge

mass polymerization: *see* bulk polymerization

mass spectrometry: the mass spectrometer provides quantitative and qualitative information about the atomic composition of materials; the sample is fragmented by electrical energy and the resulting charged gaseous fragments are deflected by a magnetic field within the spectrometer; the deflection is a function of ionic charge and particle mass and the position of the impact of the ions is recorded; the heaviest particle represents the weight of the original molecule, the lighter ones, the fragments into which it has dissociated; the resulting spectra are characteristic of the molecular structure; desorption techniques permit ionization and desorption of samples from solutions, solids, or solid films without the need for vaporization: desorption is achieved by bombardment with high-energy particles whose energy of impact is rapidly dissipated through the solution or solid

masterbatch: in rubber technology, a compound with a high concentration of additives; small amounts of masterbatch are added to batches during compounding

mastic gum: natural solid resin used in adhesives and lacquers

matrix: the continuous phase of a composite plastic

maximum therapeutic index: MTI; the ratio of tolerated dose to minimum effective dose

Mazzoni SCN process: an automated continuous saponification of fats

mechanical pulp: wood that is ground mechanically; groundwood

medium-oil varnish: varnish containing 20–30 gallons of oil per 100 lb resin

melamine: cyanuramide; 2,4,6-triamino-s-triazine; manufactured by heating urea to 350–400°C in the presence of Al_2O_3 in a fluidized bed reactor; used to make melamine (formaldehyde) resins; *see* triazine

melamine (formaldehyde) resins: thermosetting resins; lower molecular weight resins are used for impregnating paper and for laminating; for textile finishing where they provide soil and wrinkle resistance; high molecular weight resins are used for tableware

melamine/phenolic resins: mixture that has the stability of phenolic resins and the coloring properties of melamines

melt dyeing: in the plastics and textile industry, dye is incorporated into the polymer before the material is molded or spun

melt extrusion: *see* melt spinning

melt index: in polymer technology, the number of grams of thermoplastic resin that can be forced through an orifice (0.0825 = in. diameter) when subjected to a force of 2.16 kg in 10 min at 190°C (ASTM D 1238)

melt spinning: in fiber technology, process of forcing molten polymer through holes of a spinneret and cooling the resulting filaments; *see* PROCESS FOR MELT SPINNING, spinning

melt strength: strength of the molten plastic; important for molding, extruding, and drawing molten resins

membranes: thin films that can act as semipermeable barriers; membrane material selection depends on properties such as hydrophilicity or hydrophobicity, resistance to pressure, temperature and chemicals, sterilizability, dimensional stability; most membranes lack tensile and tear strength so that they need supports (woven and nonwoven nylon or polyester, silica and sand castings, glass beads, screens and perforated plates, sintered materials, plastic grids); membranes may be formed directly on the supports or attached to the support after formation; used for separation (microfiltration [MF], osmosis, reverse osmosis [RO], dialysis, ultrafiltration [UF], electromembrane processes), chromatography, electrodes, collectors, controlled release devices, sludge purification, batteries, fuel cells

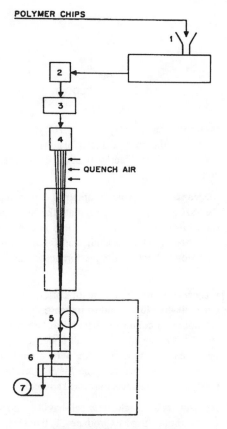

PROCESS FOR MELT SPINNING
(1) melt extruder; (2) pump; (3) sand filter; (4) spinnerette; (5) finish wheel; (6) godets; (7) take-up.
[Courtesy D.M. Considine, P.E.]

membranes, biological: cellular and intracellular membranes control the activity and mobility of ions and molecules; modern drugs, such as calcium blockers, are based on affecting transmembrane passage

menhaden: North Atlantic coastal fish whose highly unsaturated oil is used for shortening upon hydrogenation

mepasine: the C14 to C18 paraffin fraction of hydrocarbons during Fischer–Tropsch synthesis

meperidine: analgesic; narcotic; addictive

CH₃
|
N

C₆H₅   COOC₂H₅

MEPERIDINE

meprobamate: $(H_2NCOOCH_2)_2C(CH_3)$ (C_3H_7); first tranquilizer, formerly called "Miltown"; less effective than the diazepams; antianxiety drug

mercaptans: organic compounds containing the -SH group; used to regulate the molecular weight of synthetic rubbers and other polymers by acting as chain transfer agents in emulsion polymerization; odorants for natural gas

mercerization: treatment of cotton to produce silk-like luster and to increase dyeability: cotton (a cellulose fiber) fabric or yarn is swollen by concentrated alkali under tension, becomes stronger, rounder with a smooth surface, which increases the luster; mercerized cotton thread has improved strength

mercury cell: electrolytic cell having a mercury cathode; used to produce Cl_2 from brine

mercury fulminate: $Hg(OCN)_2$; primary high explosive; highly sensitive; prepared in small batches by gradually adding 1 lb Hg to ~8 lb HNO_3; this solution of mercuric nitrate is refluxed with 10 lb 95% ethanol; solid mercury fulminate separates as gray crystals that are washed with cold water, drained, and stored in bags under water; use: in fuse caps when mixed with $KClO_3$; in military shells; easily detonated by spark, heat, or friction

Merrifield resin: polystyrene-p-(chloromethyl)styrene; use: support to fix the carboxy-terminal of the growing peptide in solid state peptide synthesis

mersolate: sodium salt of mepasine sulfonate; *see* mepasine

mesh: the opening size of a sieve; mesh size is used as a measure of the size of granules and powders

mesityl oxide: $(CH_3)_2C=CHCOCH_3$; methyl isobutenyl ketone; isopropylidene acetone; produced from acetone via aldolization to diacetone alcohol followed by water elimination (using H_2SO_4 or H_3PO_4) at ~100°C; use: raw material for manufacture of methylisobutyl ketone (MIBK) and methyl isobutyl carbinol (MIBC)

messenger RNA: mRNA; *see* ribonucleic acid

mestranol: estrogen; oral contraceptive

H₃C   OH

C≡CH

CH₃O

MESTRANOL

meta-anthracite: highest rank of anthracite coal

metaldehyde: cyclic tetramer of acetaldehyde; use: cooking fuel, snail poison

metal-matrix composites: MMC; high strength engineering materials; called fiber-reinforced metals (FRM) in Japan; *see* composites

metal vapor deposition: MVD; production of metal coatings using sublimation of the metal; the technique permits the production of monoatomic layers

metallic pigments: metallic powders such as Al, Au, Zn, and bronze; made by compounding metal powder with lubricant into a paste; the most important metallic pigment is made from aluminum; use: in organic coatings and inks

metallizing: coating plastics with metals

metallocene: *see* ferrocene

metallurgical coal: met coal; grades of coal suitable for coke production needed to reduce iron in steel manufacture; typified by high Btu and low ash content

metastable state: condition of a system that exists above the stable energy state; addition of a small amount of energy causes change to a stable condition

metathesis: reaction of the type AX + BY → AY + BX; *see* olefin metathesis

met coal: *see* metallurgical coal

methacrolein: $CH_2=C(CH_3)CHO$; methacrylaldehyde; produced as by-product of propylene hydroformylation; use: for production of resins and copolymers; toxic

methacrylaldehyde: *see* methacrolein

methacrylamide: $CH_2=C(CH_3)CONH_2$; use: monomer for acrylic resins

methacrylic acid: $CH_2=C(CH_3)COOH$; produced (1) from acetone by base-catalyzed addition of HCN via the acetone cyanohydrin $[(CH_3)_2C(OH)CN]$ that is reacted with dilute H_2SO_4 to eliminate water and hydrolyze the nitrile; (2) from isobutylene by oxidation; use: primarily to prepare methyl methacrylate

methacrylic ester: *see* methylmethacrylate

methadone: $C_2H_5COC(C_6H_5)_2CH_2CH(CH_3)$ $N(CH_3)_2$; analgesic; narcotic; addictive

methanal: *see* formaldehyde

methanation: formation of methane by reduction of CO and CO_2 with H_2; used to remove these oxides from synthesis gas

methane: CH_4; marsh gas; fire damp; principal ingredient of natural gas, averaging 75 wt%, range 23–99 wt%; atmospheric methane hinders heat radiation from the earth and is a more effective "greenhouse" gas than CO_2; use: raw material for synthesis gas, hydrogen cyanide, acetylene; fuel

methane reforming: *see* reforming

methanol: CH_3OH, methyl alcohol, wood alcohol; manufactured by (1) catalytic conversion of synthesis gas having a 2:1 molar ratio of H_2/CO; processes use various conditions (high pressure: 300–350 bar, ZnO/Cr_2O_3 catalyst, 320–380°C [BASF, UK–Wesseling]; medium pressure, 50–250 bar, catalyst combinations of CuO, ZnO, Cr_2O_3, and Al_2O_3, 200–350°C [BASF, Catalyst and Chemical, Inc., Nissui-Topsoe, Pritchard, Vulcan]); low pressure: 40–100 bar, $Cu/ZnO/Al_2O_3$ catalyst, 250°C [ICI, Lurgi]); (2) by oxidation of a propane–butane mixture followed by distillation; use: raw material for organic synthesis, solvent, fuel, automotive fuel, feedstock for production of gasoline (*see* methanol-to-gasoline process), possible source for single cell protein (SCP)

methanol-to-gasoline process: Mobil process, MTG; manufacture of gasoline using a fluidized zeolite catalyst, ZSM-5

methionine: $CH_3SCH_2CH_2CH(NH_2)COOH$; natural amino acid; the industrial production from acrolein and CH_3SH leads to the *d,l* racemic mixture; use: supplement in animal fodder

methomyl: $CH_3SC(CH_3)=NOCONHCH_3$; carbamate pesticide; contact insecticide

methotrexate: anticancer drug; chemotherapeutic

METHOTREXATE

3-methoxybutanol: $CH_3CH(OCH_3)CH_2$ CH_2OH; produced from crotonaldehyde and methanol via 3-methylbutanal; use: component of hydraulic fluids; acetate is used as solvent for paints

methoxychlor: $p-CH_3OC_6H_4CH(pC_6H_4)$ CCl_3; insecticide

methoxyflurane: *see* anesthetics

methyl alcohol: *see* methanol

methylamines: monomethylamine, MMA, CH_3NH_2; dimethylamine, DMA, $(CH_3)_2NH$; trimethylamine, TMA, $(CH_3)_3N$; manufactured by catalytic alkylation of ammonia with methanol or methyl ether, by reaction of CO, NH_3, and H_2 at high pressure, and reduction of HCN or of formamide; use: raw materials for rocket fuel, pesticides, and additives; MMA is used as raw material for surfactants and photographic developer; DMA is used to manufacture industrial solvents, rubber accelerators, water treatment agents, and additives for petrochemicals; TMA is used for the manufacture of choline chloride, $[(CH_3)_3N^+ CH_2CH_2OH]Cl^-$, which is used in animal feed and medicine

N-**methylaminopropionic acid**: $CH_3NH CH_2CH_2COOH$; use: to remove H_2S from synthesis gas (Alkazid process)

methylammonium nitrate: MAN; component of formulations for slurry explosives

methylamyl alcohol: *see* methyl isobutyl carbinol

methyl bromide: CH_3Br; produced from methanol by bromination in the presence of phosphorus; use: soil fumigant, for organic synthesis

methyl *t*-butyl ether: MTBE; $CH_3OC (CH_3)_3$; produced by reacting methanol with isobutylene using an acidic catalyst such as an ion exchanger in a fixed-bed reactor; it has a high octane number of 115–135; use: additive in unleaded motor fuel to increase octane number

methyl cellulose: MC; cellulose methyl ether; produced by treating cellulose with alkali followed by reaction with a methylating agent (~1.8 of 3 available hydroxyl groups are substituted); use: emulsifying and thickening agent, laxative, substitute for water soluble gums; in formulations (e.g., paints, adhesives, textile sizing, soil conditioners, soaps)

methyl chloride: CH_3Cl; chloromethane; produced by chlorination of methane at 400–450°C; chlorination processes vary mainly by the mode of heat removal from the highly exothermic reaction; methyl chloride is separated from the other chlorination products by distillation; in the reaction of methane in air with molten $CuCl_2/KCl$ the reagent acts as catalyst and provides chlorine (oxychlorination); also produced from methanol by treatment with HCl at 100–150°C in the presence of $ZnCl_2$ or at 300–380°C and 2–3 bar in the presence of Al_2O_3; use: raw material for production of higher chloromethanes, general methylating agent, catalyst in rubber processing, refrigerant, in high-temperature thermometers; hazardous chemical waste (EPA)

methyl cyanide: *see* acetonitrile

methyldopa: blood pressure lowering agent; acts by competition for the enzyme L-dopa decarboxylase that is needed for the synthesis of the neurotransmitter L-norepinephrine; norepinephrine causes the constriction of blood vessels and thus stimulates increase of blood pressure

$$COOH$$
$$CH_2C(CH_3)NH_2$$

HO

OH

METHYLDOPA

methyl ethyl ketone: MEK; $CH_3COC_2H_5$; produced by gas phase dehydrogenation of 2-butanol or by oxidation of 1-butene using ZnO or Cu-Zn alloy catalyst; use: solvent for paints, oils, resins, nitro- and acetyl cellulose, for production of methyl ethyl ketone peroxide

methyl ethyl ketone peroxide: MEK hydroperoxide; an isomeric mixture produced by treating methyl ethyl ketone with H_2O_2; use: polymerization initiator, for curing unsaturated polyester resins

methylenedianiline: MDA; raw material for production of 4,4'-diphenylenemethane diisocyanate

methylene-di*para*phenylene isocyanate: MDI; *see* 4,4'-diphenylmethane diisocyanate

2-methyl-5-ethylpyridine: MEP; produced by reaction of ammonia with acetaldehyde or paraldehyde; liquid phase production uses NH_4Ac catalyst at 220–280°C and 100–200 bar; use: raw material for manufacture of nicotinic acid and its derivatives

methylformate: $HCOOCH_3$; *see* *N*-methyl formamide

methyl isobutenyl ketone: *see* mesityl oxide

methyl isobutyl carbinol: MIBC; methylamylalcohol; $(CH_3)_2CHCH_2CH(CH_3)OH$; produced by catalytic (Cu or Ni) reduction of mesityl oxide; use: solvent, for flotation in coal purification, as feedstock for chemical synthesis

methyl isobutyl ketone: MIBK; $(CH_3)_2CH$ CH_2COCH_3; produced by (1) catalytic (Cu or Ni) reduction of mesityl oxide; (2) from acetone by a single step process using Pd catalyst with zeolite (DEA process) or zirconium phosphate (Tokuyama Soda process); use: solvent and extractant, for flotation in ore recovery, feedstock for chemical synthesis

methyl mercaptan: CH_3SH; produced by reaction of methanol with H_2S; used for manufacture of dimethylsulfoxide

methylmethacrylate: $CH_2=C(CH_3)COO$ CH_3, MMA; produced from acetone via the cyanohydrin or by oxidation of isobutylene to methacrylic acid; also produced from ethylene by hydroformylation to propionaldehyde followed by condensation with formaldehyde; use: monomer for polymethylmethacrylates

2-methyl-4-nitroaniline: *see* nonlinear optics

methyloamides: *see* alkanol amides

methylolation: introduction of a methylol group; *see* hydroformylation

methylol group: $-CH_2OH$

methylolphenol: o-$HOC_6H_4CH_2OH$; prepared by condensing phenol with formaldehyde; raw material for phenolic resin coatings

methyl parathion: $(CH_3O)_2P(S)O$-p-(C_6H_4) NO_2; insecticide; cholinesterase inhibitor

2-methyl-1-pentene: 2MP1; produced by (1) dimerization of propene using $(C_3H_7)_3Al$ catalyst at 200°C and 200 bar (Goodyear Scientific Design process); (2) by continuous codimerization of propene and butene using trialkylaluminum and a nickel salt as catalyst (Dimersol process); use: organic precursor, monomer for resins, starting material for isoprene

4-methyl-1-pentene: 4MP1; produced by dimerization of propene using Na/K_2CO_3 catalyst; use: monomer of poly-4MP1

methyl phenylcarbinol: $C_6H_5CH(OH)CH_3$; produced by oxidation of ethylbenzene; intermediate in the Halcon production of styrene; use: cosmetic, dye, and food industries

***N*-methyl formamide**: $HCON[CH_3]_2$; dimethylformamide; produced by reaction of methyl formate with dimethylamine or by action of CO on dimethylamine in methanol at 80–100°C and 100–300 bar; use: solvent, particularly for polar polymers; implicated in causing testicular cancer

***N*-methylpyrrolidone**: NMP; produced by reaction of γ-butyrolactone and methylamine; use: solvent, e.g., in synthesis gas purification (Purisol process), for extraction of aromatics from BTX stock, for acetylene purification, for isoprene extraction from C5 petroleum fraction

N-METHYLPYRROLIDINE

methyl rubber: the first synthetic rubber (developed in 1900)

methylsilicone rubber: produced from dimethyldichlorosilane by addition of water to form $(CH_3)_2Si(OH)_2$ followed by condensation in the presence of iron chloride and sulfuric acid; chain length can be controlled with trimethyl chlorosilane, cross-linking is achieved with methyltrichlorosilane

α-methyl styrene: $C_6H_5C(CH_3)=CH_2$; produced as by-product in cumene manufacture; use: monomer

4-methylthio-2-hydroxybutyric acid: MHA; $CH_3SCH_2CH_2CH(OH)COOH$, methioninehydroxy analogue; use: substitute for methionine in fodder supplements

methyl violet: a triphenylmethine pigment; produced by chlorination of nitromethane to chloropicrin (Cl_3CNO_2) and oxidative condensation with aromatic amines; used in paints and inks for typewriter ribbons and permanent markers

Met-L-X: *see* fire-extinguishing agents

metoprolol: p–$CH_3OCH_2CH_2C_6H_4$ OCH_2 $CH(OH)CH_2NHCH(CH_3)_2$; a β-blocker; lowers blood pressure, relieves angina, used for heart attacks

metribuzin: preemergence herbicide for crops

METRIBUZIN

MeV: million electronvolt; *see* electronvolt

mevinphos: organophosphate insecticide;

only the *trans* isomer is active

MEVINPHOS

M-glass: glass with high beryllia content

MHD: *see* magnetohydrodynamics

MIBC: *see* methyl isobutyl carbinol

mica: crystalline silicates that have a perfect basal cleavage; used as fillers in thermosetting resins

micelle: a colloid aggregate composed of surface-active molecules; micelles form when the solubility limit for single molecules is reached; in a micelle the molecules are oriented, e.g., in a nonpolar solvent all hydrophilic ends point away from the solvent, whereas in a polar solvent the hydrophobic ends point away; spherical micelles have diameters between 20 and 30 Å; when reactants are only soluble in the micelles, one observes an increase in reaction rate (micellar catalysis); micelles unlike vesicles break upon dilution

Michaelis–Menten model: an expression for enzyme kinetics in which the enzyme (E) reacts reversibly with the substrate (S) to form a transition state (ES) which decomposes irreversibly to give product (P)

$$E + S \underset{k_{-1}}{\overset{k_{+1}}{\rightleftarrows}} ES \overset{k_{+2}}{\rightarrow} E + P$$

the model permits the calculation of the reaction velocity (v) in terms of the maximum observable velocity (v_{max}); the model only holds for enzymes that have a single active site

Michler's ketone: $[(CH_3)_2N\text{-}p\text{-}C_6H_4]_2CO$; produced by heating N,N-dimethylaniline with phosgene; intermediate in the production of triphenylmethine dyes

micrinite: constituent of coal

microanalytical techniques: for nondestructive methods that give information on the near-atomic level; *see* individual techniques: analytical electron microscopy (AEM), Auger electron spectroscopy (AES), diffuse reflectance IR Fourier transform spectroscopy (DRIFT), electron spectroscopy for chemical analysis (ESCA), extended X-ray absorption fine structure spectroscopy (EXAFS), infra-red spectroscopy (IR), ion scattering spectroscopy (ISS), nuclear backscattering spectroscopy (NBS), Raman spectroscopy, scanning electron microscopy (SEM), scanning tunneling microscopy (STM), secondary ion mass spectroscopy (SIMS), transmission electron spectroscopy (TEM), ultraviolet photoelectron spectroscopy (UPS), X-ray photoelectron spectroscopy (XPS)

microballoons: hollow spheres (diameter 0.05–0.15 mm) filled with an inert gas; use: low density filler, for syntactic foam; to reduce evaporation of liquids by floating microballoons on the surface

microencapsulation: enclosure of gas, liquid or solid materials within semipermeable membranes; use: slow release pharmaceuticas and agricultural materials

microfiltration: process using filters having pore sizes 0.1–10 μm in diameter (microfilters); for high-temperature filtrations ceramic membranes made mainly from stabilized α-alumina are being developed; *see* membranes

micrometer: micron; 10^{-6} meter; μm

micron: micrometer, 10^{-6} meter; μm

microorganism: living organisms not visible to the naked eye; bacteria, fungi

microspheres: hollow spheres of glass or plastic with 4–20 μm diameters; they are incorporated into materials to impart strength, reduce density, and improve resistance to crazing

microwave heating: heating with electromagnetic radiation having frequencies of 10^9–10^{10} cps (radar range)

migration: in polymer and rubber technology the spontaneous transfer of molecules or ions within a formulated compound or across the interface between solids

mil: 0.001 in.

mild steel: low carbon steel that has low strength and good conductibility

milk of lime: aqueous calcium hydroxide suspension

milled soap: produced by calendering soap chips into a continuous strip while incorporating additives such as color and perfume

millimicron: nm; nanometer; 10^{-9} m, 10 Å

mills: devices to do the grinding and pulverization from about 1 in. size to 150 mesh or finer; attrition mills have one or two rotating disks at an adjustable distance; hammer mills and roller mills are smaller versions of crushing machines; ball and pebble mills are rotating drums partially filled with a mobile grinding medium harder than the material to be ground; *see* beneficiation

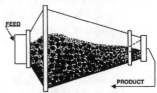

Highly schematic representation of operating principle of a cone-type pebble or ball mill.
[Courtesy D. M. Considine, P.E.]

Mills process: a continuous process for making high-grade toilet soap in which saponification is catalyzed by ZnO and the product is purified by vacuum distillation

Milori blue: *see* ferrocyanate pigments

mineral black: pigments made from ground coal

mineral fibers: inorganic fibers such as asbestos, carbon, glass, metals

mineral oil: refined petroleum oil; used in pharmaceutical and textile industry; as a laxative

mineral rubber: solid bituminous materials such as asphalt or petroleum residues

mingler: in sugar manufacture, a ∪-shaped trough containing a rotating agitator

minimum filming temperature: MTF; in polymer technology, the lowest temperature at which a polymer will coalesce and form a continuous film

minus sieve: the portion of a powder sample that passes through a standard sieve

mirex: polychlorinated pesticide; used primarily against fire ants

"mixed oil": mixture of cyclohexanone and cyclohexanol obtained catalytic (Co[II]) air oxidation of cyclohexane; oxidation of "mixed oil" with nitric acid yields adipic acid

mixers: devices used to intimately intermingle two or more materials; they include mills, blenders, agitators, kneaders, and mixers

modacrylic fiber: a synthetic polymer that contains 35–85 wt% acrylonitrile and usually 20% vinylchloride or vinylidenechloride; use: for carpets, industrial filters, winter ware

moderator: in nuclear technology, material needed to cool the reactor and absorb neutrons; examples: water, heavy water, metallic sodium, and carbon dioxide

modified cellulose: *see* mercerization

modified wood: wood that has been subjected to chemical or heat treatment

modulus: measure of resistance against any sort of deformation, such as the modulus of elasticity (Young's modulus), modulus of flexure, modulus in compression; in textile technology, the stiffness of a fiber as measured from the initial slope of a stress–strain curve; the ratio of the applied force (stress) to elongation or bend (strain); *see* tensile properties

Moessbauer spectroscopy: NGR, nuclear γ ray spectroscopy; technique that measures the recoilless emission and resonant absorption of γ rays (<150 kEv) emitted from radioactive elements; used to measure oxidation states of metals and the number and nature of ligands in complex ions and complex biological molecules

Mohs hardness scale: *see* hardness

moisture absorption: absorption of water from the atmosphere in contrast to water absorption that occurs during immersion

moisture content: moisture regain; percentage of water in a material after exposure to moisture under standard moisture and temperature conditions compared with its weight in the dry state; *see* standard moisture regain

moisture regain: *see* moisture content

moisture-set inks: inks composed of a water-insoluble resin that are dissolved in a water-miscible solvent; when the printing is exposed to steam or water mist the resin precipitates and binds the pigment to the surface

molasses: the residual mother liquor in sugar extraction from sugar cane, sugar beets and starch hydrolysis; use: food and feed additive, nutrient in fermentation processes

molding compound: molding powder; premix of polymer and additives ready for molding and curing

molding powder: *see* molding compounds

mold release: effected by external lubricants or by additives that rise to the surface during molding to speed release

molecular sieves: porous crystalline materials with pore sizes 4–5Å; used to separate small molecules from solutions and suspensions; industrial uses include the separation of olefins in the gas phase (Parex and Isosiv processes) and in the liquid phase (Molex process); *see* zeolites

molecular weight: the sum of the atomic weights in a molecule referred to the standard atomic weight of the carbon-12 isotope (12.); for polymers several different molecular weight measurements are of importance: M_n, the number average molecular weight that is measured by osmotic pressure; M_v, the viscosity average molecular weight that is based on viscosity measurements; M_w, the weight average molecular weight that is measured with light scattering or with the ultracentrifuge, M_z also from ultracentrifuge measurements correlates with melt elasticity data, and M_k, the kinetic molecular weight that is based on polymer yield divided by the moles of initiator used in the charge

molecular weight distribution: MWD; in polymer technology, the weight average molecular weight divided by number average molecular weight; M_w/M_n; the MWD of polymers depends on methods of polymerization and affects processing and physical properties

molten salt breeder: MSBR; in nuclear technology, an experimental breeder using uranium and thorium fluoride in molten ^7LiF-BeF_2

molybdate orange: $PbCrO_4$ modified with $PbMoO_4$; use: pigment

monel: Ni-Cu alloy

monoamine oxidase: MAO; *see* monoamine oxidase inhibitors

monoamine oxidase inhibitors: MAO inhibitors; antidepressant drugs; monoamine oxidase deaminates neurotransmitters such as epinephrine and thereby removes their effects, which include increased heart beat, raised blood pressure, as well as feelings of euphoria; MAO inhibitors reduce MAO ac-

tivity and thus affect the mood of the patient; example of MAO inhibitors is phenelzine $(C_6H_5CH_2CH_2NHNH_2)$

monoammonium phosphate: MAP; granular MAP is produced by ammoniating phosphoric acid to a stoichiometric ratio and drying the product above 200°F; use: fertilizer

monoazo dyes: colored compounds that contain a single –N=N– linkage; *see* azo dyes

monocalcium phosphate: $Ca(H_2PO_4)_2$; use: fertilizer, animal feed supplement

monochloroacetic acid: $ClCH_2COOH$; produced by liquid phase chlorination of acetic acid; use: intermediate for dyes, pesticides, and pharmaceuticals; for manufacture of carboxymethylcellulose

monochromatic: in reference to radiation, light composed of a single wavelength, e.g., a monochromatic X-ray beam

monoclonal antibodies: MAbs; produced by fusing specialized mouse spleen or human lymph cells with long-lived ("eternal") cancer cells; the resulting cells are called hybridoma and when cultured they produce a single type of antibody; the name monoclonal antibodies refers to the fact that the antibodies are derived from a single cell clone; each MAb has unique properties of isoelectric point, solubility, and tolerance to heat, pH, and salt conditions; they can be tagged with various chemicals such as I^{131} and with biologicals; use: in sensitive bioassays, vaccine production, and therapy

monoethanol amide: MEA; surfactant, use: to produce emulsifiers, foam boosters and dispersants

monodentate: *see* ligands

monofil: *see* monofilament

monofilament: monofil; single polymer filament of indeterminate length usually prepared by melt extrusion

monoglycerides: minor constituent of fats

monomers: small molecules that can react to form large molecules (macromolecules) consisting of many monomer units, i.e, polymers; the following are major industrially important monomers: acrylic esters, acrylonitrile, bisphenol A, 1,4-butanediol, butadiene, caprolactam, epichlorohydrin, ethylene, ethylene glycol, ethylene oxide, formaldehyde, hexamethylene diamine, isocyanates, isoprene, methyl methacrylate, phenol, phthalic anhydride, propylene, propylene glycol, propylene oxide, styrene, terephthalic acid, vinyl acetate, and vinyl chloride

monomethyl amine: *see* methylamines

Monsavon process: in soap manufacture, continuous saponification in which a hot emulsion of fat and caustic is fed into a kettle where saponification is completed under agitation; the crude soap is removed and washed countercurrent with hot caustic; washed soap is continuously removed from the top of the tower, washed and pumped into a settling tank

monuron: p-$ClC_6H_4NHCON(CH_3)_2$; pre-emergence urea herbicide

mordant: in textile manufacture; chemical that is applied to the fabric so that dye will adhere; mordants are usually acids or salts, mordant dyes are usually lakes

Morehouse mill: a high-speed stone mill

morphine: analgesic; narcotic; addictive drug; *see* alkaloids

MORPHINE

morphology: physical and structural characteristics of a substance; the macroproperties (thermal and mechanical) of a material depend on its chemistry and the physical arrangement of its structural features; i.e, in polymer technology, the chemical structure determines the temperature range of use and the physical structure determines its mechanical properties; elucidation of the morphology requires experimental methods (i.e, X-ray diffraction, spectroscopy, microscopy); the aim of these measurements is to develop understanding of the molecular structure that will permit prediction of the thermal and mechanical properties of a product; similar concepts apply in the study of alloys, blends of polymers, composites, bioproducts; in each case the morphology of the molecular structure that correlates with the macroproperties of a material is being established

mortar: a mixture of cement, lime, sand, and water that is plastic and workable for several hours before it settles into a solid mass

MOS: metal oxide semiconductor

motionless mixers: tubular devices in which the motion of the entering liquid mixtures is controlled by the internal structure

motor octane number: MON; *see* research octane number

mountain top removal mining: surface process restricted to steep mountains in which mined areas are restored for other use

mRNA: messenger RNA; one of the groups of ribonucleic acids (RNA) involved in the biosynthesis of proteins by translation of the genetic information encoded in the DNA; mRNA is the template that copies the sequence of the bases in the DNA; an mRNA molecule is produced for each gene or groups of genes; *see* RNA; tRNA; genetic code

multidentate: *see* ligands

multifilament: yarn consisting of many single filaments

muriatic acid: *see* hydrochloric acid

mutagen: agent that causes genetic mutations

mycelium: filamentary body of fungi

N

nacreous: pertaining to mother-of-pearl

naled: $(CH_3O)_2P(O)OCH(Br)CBrCl_2$; nonpersistent insecticide

nano-: n; prefix denoting 10^{-9}, e.g., nanometer, nm, 10^{-9} m

naphtha: crude petroleum fractions containing mainly C10 or lower hydrocarbons; also obtained from distillation of coal tar or shale oil and as by-product from carbonization of coal; naphtha fractions divide approximately into light naphtha (casing head, straight run gasoline), which distils at 50–200°F; solvent naphtha, which distils below 320°F; heavy naphtha (high-flash), which distils at 200–400°F; use: solvent, fuel, chemical feedstock; *see* petroleum; petroleum refining; reforming; cracking

naphthalene: produced from coal tar by fractional distillation and from oil raffinates by liquid–liquid extraction or absorption; pure naphthalene is made from a mixture of alkyl homologues by hydrodealkylation at 550–750°C and 7–70 bar (Unidak process); use: chemical intermediate, precursor of phthalic anhydride, β-naphthol

NAPHTHALENE

naphthene: nonaromatic cyclic hydrocarbons isolated from petroleum; they are converted to aromatics during catalytic reforming

naphthenic acids: saturated higher fatty acids isolated from petroleum by extraction with aqueous NaOH; their salts, naphthenates, are used as drying agents for paint and cellulose preservatives

naphthol: mono-hydroxynaphthalene; existing as two isomers: α-naphthol (1-hydroxy naphthalene) and β-naphthol (2-hydroxy naphthalene); produced by sulfonating naphthalene and fusing the resulting sodium naphthalene sulfonate with caustic soda; use: precursor of azo dyes, perfumes and in chemical synthesis

naphthol AS: the water-soluble compound and its derivatives are coupling agents for the production of water-insoluble azodyes

NAPHTHOL AS

naphthylimide: brightening agent for laundry detergents

Natta catalysts: catalysts that lead to stereospecific polymerizations; made from $TiCl_3$ and aluminum alkyls; *see also* Ziegler–Natta catalysts

natural fibers: cotton, flax, silk, and wool; cotton and flax are based on cellulose, silk and wool on proteins; silk is the only filament ("endless", very long) fiber, the other three are staple yarns; flax is the raw material for linen fabrics

natural gas: gaseous petroleum fraction consisting mostly of methane but also includes C2 to C4 hydrocarbons, usually needs clean-up to remove moisture and sulfur compounds; use: fuel, chemical feedstock and for reforming

natural rubber: caoutchouc; *cis*-1,4-polyisoprene

naval stores: mixture of gum and oleoresin produced (1) by tapping pine trees; (2) by extraction or steam distillation of pine wood;

or (3) from sulfate pulping of pine wood; a source of turpentines (mixture of terpene hydrocarbons) and rosin (abietic acid and its isomers); turpentine is used as solvent; turpentine and/or some of the isolated hydrocarbons have many uses including mineral flotation, textile processing, odorants, adhesives, paper sizing, chewing gum, softening agents in rubber formulations; rosin is used as chemical feedstock (including for manufacture of styrene-butadiene rubber), for preparation of emulsifiers, paper size, adhesives, protective coatings, varnish

neat soap: product of batch saponification process; consists of 65% sodium soap, 35% water and traces of alkali, glycerol, and salt; use: feedstock for toilet and laundry soaps

necking: necking down, drawing down; in the textile industry, a sharp change in diameter of an undrawn filament or yarn during extension; *see* drawing

necking down: *see* necking

negative catalyst: an inhibitor

negative dyes: *see* acid dyes

nematic crystals: liquid crystalline phase of materials in which rod-like molecules are arranged in parallel; *see* liquid crystals

nematocide: pesticide against worms (nematodes)

neo acids: branched carboxylic acids produced by the Koch reaction

neohexene: 3,3-dimethyl-1-butene; produced from 2,4,4-trimethyl-2-pentene and ethylene by metathesis; use: antiknock additive to motor fuel

neon: Ne; element; obtained from air: at liquid nitrogen temperature H_2, He, and Ne remain in gas form; H_2 is removed by oxidation to water, Ne and He are separated by distillation; use: in lamps, as an inert atmosphere

neopentyl glycol: 2,2-dimethyl-1,3-propanediol; produced by aldol condensation of isobutyraldehyde and formaldehyde; use: for synthesis of polyesters, resin paints, lubricants and plasticizers

neoprene: generic name for chloroprene polymers; *see* poly(chloroprene)

neuroleptics: major tranquilizers; *see* antianxiety drugs

neurotoxin: toxic substance that attacks the nervous system; use: pesticides, war gas

neutral sulfite semimechanical process: NSSC; pulping of hardwoods in which sulfite liquor is buffered to maintain a slightly alkaline pH; NSSC pulp is the best available for corrugating material; *see* sulfite process

never press fabrics: *see* ease-of-care textiles

New Drug Application: NDA; application is submitted to the U. S. Food and Drug Administration (FDA) following clinical trials; FDA approval is the last step before commercial drug production

news ink: colored pigments or carbon black suspended in mineral oil

newsprint: paper consisting of 80% groundwood and 20% chemical fiber

niacin: nicotinic acid; a member of the vitamin B complex; antipellagra vitamin

NIACIN

niacinamide: nicotinic acid amide; member of the vitamin B complex; *see* niacin

nicotine: *see* alkaloids

nicotinic acid: *see* niacin

nicotinic acid amide: *see* niacin

nigre: in soap manufacture, a dark-colored layer occurring between the neat soap and the caustic solution; nigre contains some soap, salts and impurities

nigrosine dyes: black dyes formed by reaction of nitrobenzene with aniline in the presence Fe^{2+} and HCl; *see* azine dyes

niter: *see* potassium nitrate

nitramine: *see* tetryl

nitre: *see* potassium nitrate

nitric acid: HNO_3; a strong acid, manufactured by oxidation of ammonia over a Pt/Rh catalyst; use: as acid in the laboratory and industry; for manufacture of ammonium nitrate fertilizer, dyes, plastics, chemicals, and industrial explosives; for pickling stainless steel and for recovery of uranium from ion exchange resins

nitric phosphates: fertilizers produced by acidification of phosphate rock with nitric acid (Odda process) or with mixtures of acids (nitric and sulfuric; nitric and phosphoric acids); unlike the manufacture of ammonium phosphates no sulfur is needed for manufacture of nitric phosphates

nitride process: reaction of metals such as aluminum with atmospheric nitrogen to form metal nitrides; the nitrides yield ammonia when treated with water; use: for nitrogen fixation

nitrides: binary nitrogen compounds; industrially important nitrides include the ceramic materials boron nitride, BN, silicon nitride, Si_3N_4, and aluminum nitride (AlN); these materials are nonconducting, resistant to chemicals, and stable at high temperatures but are not stable in oxidizing environments; their high strength and hardness leads to applications where good mechanical properties are required at high temperatures

nitriding: hardening of steel by exposure to NH_3 at high temperature

nitrile: RCN; organic compounds containing the -CN group, e.g., acrylonitrile, ace-

tonitrile, succinonitrile; use: intermediates for the manufacture of carboxylic acids, amines, polymers

nitrile barrier resins: copolymers of acrylonitrile (>60%) and acrylates, butadiene, or styrene; impervious to gases and water vapor

nitrile rubber: NBR; arylonitrile-butadiene copolymers prepared by free radical copolymerization; applications include oil-resistant parts such as well parts, fuel hose, gaskets

nitrilotriacetic acid: NTA; $(N(CH_2COOH)_3$; chelating agent, formerly used to replace phosphates in detergents but this use has been discontinued because NTA might cause birth defects

nitrobenzene: $C_6H_5NO_2$; produced by nitration of benzene with nitric and sulfuric acid; batch process operates at 50°C, continuous process uses three nitrating vessels with temperature increasing from 35–40°C to 50°C to 55–60°C; waste sulfuric acid is recycled; use: for manufacture of aniline, as chemical intermediate, oxidizing agent, solvent (especially for Friedel–Crafts reactions), raw material for organic synthesis, additive

nitrocellulose: nitrocotton; produced by nitrating cellulose (e.g., wood pulp, cotton linters) with a mixture of concentrated nitric and sulfuric acid; highly flammable; dinitrocellulose is used for plastics (celluloid is nitrocellulose with camphor as plasticizer), lacquers, and adhesives; higher nitrates are explosive and are used for manufacture of dynamite (nitroglycerin + nitrocellulose) and as primary explosives; as historical interest: the first synthetic fiber spun (by deChardonnet) used nitrocellulose in an ether and alcohol solution, denitrating it subsequently to rayon for safety reasons

nitrocellulose cotton: *see* pyroxylin

nitro chalk: in the fertilizer industry, a mixture of limestone and ammonium nitrate

nitrocotton: *see* nitrocellulose

nitrocyclohexanone: produced by nitration of cyclohexanone; intermediate in the Techni-Chem process for manufacture of caprolactam; *see* caprolactam

nitro dyes: compounds characterized by a nitro group α to the chromophore; a small group of low cost, colorfast dyes; *see* dyes

nitrogen: N; element; N_2 gas is isolated from air during production of oxygen; used as inert atmosphere; *see* nitrogen fixation

nitrogen fixation: reaction by which atmospheric nitrogen enters the synthesis of proteins, nucleic acids, and other biological materials; the conversion is achieved by *rhizobium* bacteria that invade certain plant roots to form nodules; nonorganic nitrogen fixation can be achieved by forming metal nitrides

nitroglycerin: $C_3H_5(ONO_2)_3$; nitroglycerol; glyceryl trinitrate; NG; produced by slowly adding >99.9% pure glycerin to a mixture of concentrated sulfuric and nitric acid keeping the temperature to 2–3°C; the oil is then separated from the spent acid, washed with aqueous Na_2CO_3, followed by concentrated NaCl, and stored in lead lined tanks; a colorless oil, highly sensitive to shock, especially when air bubbles are present—almost a primary explosive; gelatinized by nitrocellulose; use: in dynamites; nitroglycerin relaxes involuntary muscles, particularly those of the blood vessels near the heart; by increasing blood flow to the heart it acts to relieve pain caused by oxygen lack to the heart muscle in angina pectoris

nitroglycerol: *see* nitroglycerin

nitromannite: mannitol hexanitrate; primary high explosive; more sensitive to shock than nitroglycerin; use: in composition caps and fuses

nitroso rubber: tetrafluorethylene-trifluoronitrosomethane copolymer; produced by free radical copolymerization; incombustible, resistant to chemicals and solvents; its low T_g (–51°C) makes it useful for low-temperature applications

nitrosoureas: alkylating agents; *see* anticancer drugs

nitrosyl chloride: NOCl; produced by action of Cl_2 on $NaNO_3$ or from nitrosylsulfuric acid ($NOHSO_4$) and HCl; oxidizing agent, catalyst, organic synthesis; decomposed in situ to NO and Cl during manufacture of cyclohexyloxime [$(C_6H_{11}NO)$]; highly toxic

nitrotoluenes: *o*-, *p*-, and *m*-nitrotoluenes are produced by continuous nitration of toluene under milder conditions than those needed for nitrobenzene; use: intermediates for dyes, perfumes, pharmaceuticals, dinitrotoluenes, and trinitrotoluene

nitrous oxide: N_2O; laughing gas; prepared by heating NH_4NO_3 at 200°C in aluminum vessels; liquefies at 100 atm pressure; use: anesthetic

noble gases: *see* inert gases

Nomex fiber: *see* aramids

noncellulosics: polymers and fibers not based on cellulose

nonionic surfactants: most of these low sudsing detergents are polyoxyethylene [$-(CH_2CH_2O)_nH$] and polyoxypropylene [$-(CH_2CH_2CH_2O)_nH$] derivatives of hydrophobes such as alcohols, alkylphenols, fatty acids, and amides; produced by catalytic reaction of ethylene or propylene oxide with the substrate at high temperature and pressure; reaction is stopped by cooling when the desired amount of the oxide has reacted; examples are alcohol ethoxylates (AE) and alkylphenol ethoxylates (APE); use: active ingredients of heavy-duty laundry detergents

noniron fabrics: never press fabrics; *see* ease-of-care fabrics

nonlinear alkylbenzene sulfonates: ABS; *see* linear alkyl sulfonates

nonlinear optics: phenomenon caused by the interaction of a light beam with the material through which it passes; when a second beam enters before the molecules or atoms have time to relax, the emerging beam may be changed in intensity and/or frequency from the incident beam; materials having strong nonlinear optical properties include inorganic substances such as crystalline GaAs and Si, CS_2, fused silica, KH_2PO_4, and $LiNbO_3$; organic molecules with nonlinear properties include 2-methyl-4-nitroaniline (MNA) and 4-*N*,*N*-dimethylamino-4'-nitrostilbene (DANS); nonlinear optical materials are used for devices such as switches and waveguides

nonlinear polymers: *see* branched polymer

nonwoven fabrics: products intermediate between woven textiles and paper; fiber staples or filaments are held together by entanglement (felts), fusion (bonded fabrics), or with the aid of an adhesive; produced with textile (dry-laid nonwovens) or paper (wet-laid nonwovens)-making equipment; thermoplastic synthetics such as nylon, polyester, and polyolefins can be spin-bonded using the filaments that emerge from spinerettes; spin-bonding does not require additional adhesives; use: in coverings for automobile interiors and roofs, diapers, hospital gowns

norethynodrel: steroidal contraceptive

NORETHYNODREL

normal phase liquid–liquid chromatography: *see* liquid–liquid chromatography

normal superphosphate: *see* ordinary superphosphate

noruron: urea herbicide

$(CH_3)_2NCONH$

NORURON

notatin: glucose oxidase; this enzyme is derived from *Penicillium notatum:* use: as oxygen scavenger in the food industry, with peroxidase for quantitative determination of glucose, with catalase for removal of glucose or oxygen

note: in flavor-testing the individual components of complex aromas, e.g., a woody note is present in vanilla and in pepper

novaculite: fine-grained quartz; use: filler for silicone rubber, epoxies, urethane foam, and polyvinylchloride

novalac: *see* phenol-formaldehyde resin

NPDES: national pollutant discharge elimination system; standard for pollution control

NPK fertilizers: formulations that contain nitrogen (ammonia, urea, ammonium salts), phosphorus (phosphates, phosphoric oxide), and potassium (potassium salts, potassium potash); *see* fertilizers

nuclear backscattering spectroscopy: NBS; a microanalytical technique to determine the elemental composition of a solid surface to a depth of 10–20 nm with a 1–2 mm lateral resolution; determination is made by bombardment with high-energy ions

nuclear breeder: reactor that converts uranium into plutonium, which can then be used for fuel to generate electricity

nuclear γ ray spectroscopy: *see* Moessbauer spectroscopy

nuclear magnetic resonance: NMR; analytical method for determination of the detailed spacial arrangement of atoms in a molecule and their interactions with one another; the measurements are done in a strong external magnetic field modified by a weaker radio frequency field; well-resolved resonances with the radio frequency field are given by nuclei with a spin of 1/2 such as ^1H, ^{13}C, ^{19}F, ^{29}Si, ^{31}P; the exact frequency and intensity depends on the environment and the concentration of each atom while the area under each peak relates to the number of atoms giving the signal; more detailed data can be obtained by using several of the nuclei (such as ^1H and ^{13}C) simultaneously; in spectra taken of liquids or solutions the signals originate from the tumbling molecules; solid materials (polymers, powders) require a special sample spinning procedure known as magic angle spinning to obtain a high resolution spectrum similar to those available for liquids and solutions; MRI (magnetic resonance imaging) is a medical application of NMR used for diagnosis; *see* magic angle spinning

nuclear reactors: electric energy generators using the nuclear fission of uranium, thorium, or plutonium; the reactor core consists of fuel rods and a moderator or coolant to absorb the neutrons released during fission. *see* NUCLEAR REACTOR

nucleation: initiation of crystallization within a material by a small ordered region (homogeneous nucleation) or by a foreign particle (heterogeneous nucleation); the crystallite size in a polymer can be controlled by the addition of foreign nuclei

nucleophilic antibiotics: *see* anticancer drugs

nuclide: a proton or neutron; particle in an atomic nucleus

number average molecular weight: M_n; polymer molecular weight determined by a method that counts the number of molecules (e.g., osmometry); *see* molecular weight

nutgall: *see* gall

nylon: $[-CO(CH_2)_nCONH(CH_2)_mNH]_x$ or $[-CO(CH_2)_nNH]_x$; generic term for synthetic

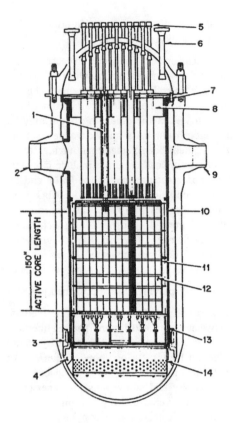

NUCLEAR REACTOR
(1) Control-element assembly fully withdrawn; (2) 42-in.-ID outlet nozzle; (5) control-element drive mechanism; (6) instrumentation nozzle; (9) 30-in.-ID inlet nozzle; (12) fuel assembly; (2) and (9) are outlet and inlet for pressurized water used as a coolant; (a) fuel rods contain UO_2 pellets in containers with low neutron absorption characteristics; (b) control rods are in selected fuel assemblies, they hold materials with high neutron absorption characteristics for regulation of the rate of power generation; (c) pressurized water is the coolant in the primary cooling system and produces steam for energy production in a heat exchanger.
[Courtesy D. M. Considine, P.E.]

aliphatic polyamides; for aromatic polyamides, *see* aramids; use: for manufacture of synthetic fibers and plastics

nylon nomenclature: based on monomer from which the nylon is derived: single numbers give the number of carbon atoms in the straight-chain amino acid monomeric unit (e.g., nylon 6 is made from caprolactam); in double numbers the first one is the number of carbon atoms of the diamine and second the number of carbon atoms of the dibasic acid from which the polymer is derived (e.g., nylon 6,6 is made from hexamethylene diamine and adipic acid)

nylon 6: $(-NH(CH_2)_5CO)_n$; polycaprolactam; made from caprolactam by condensation polymerization in an autoclave or continuous reactor at elevated temperatures; at the melt, the polymer is in equilibrium with ~10% monomer, which is removed before melt-spinning; fibers and plastics are melt-extruded; spin orientation at high take-up velocities leads to a different crystal structure than that obtained on drawing

nylon 6,6: $(-NH(CH_2)_6NHCO(CH_2)_4CO)_n$; polyhexamethylene adipamide; produced from the salt of hexamethylenediamine and adipic acid by melt condensation; the salt provides the required stoichiometric ratio for a high molecular weight polymer; the molecular weight can be controlled by using a slight excess of adipic acid or adding 1% acetic acid; polycondensation of the salt takes place in water solution under pressure while the temperature is raised to 270–280°C, then pressure is reduced, vacuum applied, and the nylon melt extruded as a ribbon from the bottom of the vessel and cut into molding material; use: fibers, plastics by melt extrusion

nylon 6,10: polyamide manufactured by condensation polymerization from hexamethylenediamine and sebacic acid

nylon 6,12: polyamide manufactured by condensation polymerization from hexamethylenediamine and 1,12-dodecanedioic acid

nylon 11: polyamide manufactured by condensation polymerization from ε-aminoundecanoic acid

nylon 12: polyamide made by condensation polymerization from lauryl lactam

nystatin: macrocyclic antifungal agent. *see* NYSTATIN

nytril: generic term for synthetics consisting mainly of vinylidene dinitrile

O

obsidian: volcanic glass that changes to pumice on melting; flash roasting causes a 15-fold expansion; the resulting low-density perlite is used for acoustic tile, insulation, wall board, and lightweight aggregate in cement

ocher: ochre; white to dark brown earth

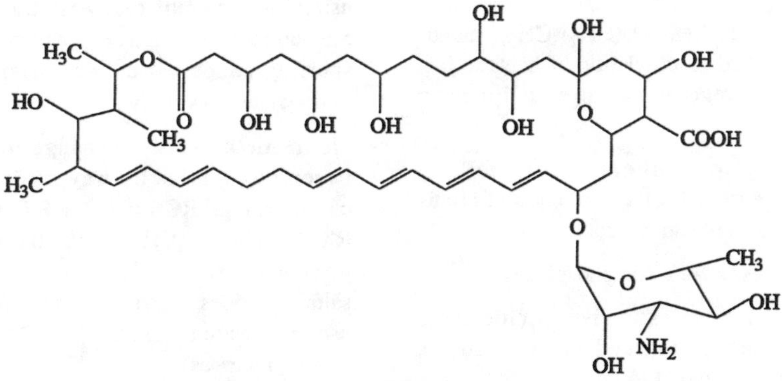

NYSTATIN

pigment colored with hydrated iron oxides; made by washing and grinding in oil; color varies with additives and geographic origin: white is ordinary clay, golden ocher contains chrome yellow, sienna is found in Italy (calcining yields "burnt sienna"), van Dyke brown comes from Germany, yellow and brown ochre are limonite, umber comes from Italy and contains manganese oxide, burnt umber is more red and is made by calcining raw umber

Octafining process: isomerization of m-xylene to p-xylene using $Pt/Al_2O_3 \cdot SiO_2$ and H_2 at 400–500°C and 10–25 bar; Pt/H_2 decreases dehydrogenation at this temperature, the alumina-silicate promotes isomerization

octane number: measure of an automotive fuel's ability to avoid knocking; the scale ranges from 0 (n-heptane) to 100 (iso-octane, 2,2,4-trimethyl pentane); research octane number (RON) is determined with a fixed spark advance, an air inlet temperature of 125°F and engine speed of 600 rpm; motor octane number (MON) is determined with variable spark timing at 300°F and an engine speed of 900 rpm; gasoline is usually labeled with an arithmetic mean of the two octane numbers, R+M/2; antiknock agents are additives that improve octane number; they include aromatics, cumene, t-butanol, diisobutene, methanol, and methyl t-butyl ether (MTBE)

2-octanol: $CH_3CH(OH)(CH_2)_5CH_3$; manufactured by hydroformylation of heptene followed by hydrogenation; use: antifoaming agent

octylphenyl phosphoric acid: OPPA; added to kerosene for the extraction of uranium oxide from phosphate rock

Odda process: *see* nitric phosphates

odorants: substances to give specific odors or to disguise unpleasant odors; for example, mercaptans are added to odorless natural gas for detection of gas leaks; fruit and flower aromas are used in cosmetics and for aesthetic purposes

off-set printing: lithography; also known as set-off; *see* printing

off-set yield strength: in polymer technology, stress at which the strain exceeds the normal yield point and yield strength by a specified amount (the offset); at the yield point the initial portion of the stress–strain curve deviates from linearity; *also see* physical testing

oil extended rubber: easily processable rubber made from a highly viscous latex that contains up to 62.5% oil, e.g., oil extended butadiene-styrene copolymer (OP rubber)

oil of vitriol: sulfuric acid having a density of 66° Bé

oil paints: pigments for which linseed oil serves as the major binder

oils: *see* fats

oil shales: petroleum solids; they contain organic material (kerogen) that yields shale oil on heating; shale oil can be refined to conventional petroleum products but at 1990 petroleum prices it is not economically competitive

olefinic thermoplastic elastomers: copolymers of propylene and ethylene, prepared by using Ziegler–Natta catalysts; use: for flexible automobile parts such as bumper guards, sporting goods, hose, wire insulation, and tubing; *see* polyolefins

olefin metathesis: exchange reaction between two olefins by way of double bond cleavage, e.g., RCH=CHR + R'CH=CHR' → R'CH=CHR + RCH=CHR'; the reaction is reversible and is catalyzed by Mo, W, or Rhe salts or oxides; several industrial processes use the reaction (e.g., the SHOP and the Triolefin processes)

olefin polymers: *see* polyolefins

olefins: unsaturated hydrocarbons; produced from petroleum by catalytic cracking; use: for special manufacturing methods; raw materials for chemical synthesis; *see* individual olefins

Olefin-siv process: separation of isobutene from the C4 olefin fraction with the help of molecular sieves

oleic acid: $CH_3(CH_2)_6CH=CH(CH_2)_6COOH$; obtained from animal fat (lard) and plants (soybean); a basic nutrient; also used as raw material for manufacture of soap, textile finishes, and chemicals

oleomargarine: *see* margarine

oleoresins: balsams; mixture of vegetable plant resin and essential oil; use: mixing vehicles for varnishes, aluminum paint

oleum: fuming sulfuric acid

oligo-: prefix denoting few

oligomer: polymer having few (~2–10) repeating units

oligopeptide: peptide made up of 10 or less amino acids, e.g., "thyrotropin releasing factor" (3 amino acids) and bradykinin (9 amino acids)

oligosaccharide: saccharide consisting of 2–10 simple sugars; class of antibiotics that is active against gram-positive and anaerobic bacteria

oligotrophic lakes: deep lakes with low supply of nutrients resulting in little organic matter

olive oil: produced from the pulp of olives; virgin oil is unrefined and used for its flavor; pure olive oil is refined and deodorized; low in unsaturates and high in oleic acid; use: nutrient

opacifiers: additives that reduce translucence, e.g., titanium dioxide (anatase and rutile)

open-cell foamed plastic: cellular plastic in which gases can travel freely between cells

open-pan salt: *see* Grainer process

optical bleach: *see* optical brighteners

optical brighteners: (optical bleaches, fluorescent whitening agents, FWA) agents that absorb UV radiation and reflect visible blue-white radiation thereby masking the yellowing of fabrics

optical fibers: extremely pure glass fibers, free of dirt or voids; the transparent material has a higher refractive index than the outside thus retaining the light injected at the end all along the fiber; optical fibers are usually coated for protection against surface damage, various types of glass are used, some with lateral refractive index gradient, fiber diameter is <100 μm; use: telecommunications, replacing less efficient metal cables; medical applications for internal body investigations and operations

optical storage devices: *see* electronic chemicals

orange peel: in the plastics industry, an uneven surface resulting from overpolishing, overheating, or overcarburizing in the mold

ordinary superphosphate: OSP; normal superphosphate; contains 16–18% P_2O_5; produced by treating phosphate rock with sulfuric acid; use: as fertilizer or for bulk blending for manufacture of NPK fertilizers

organelles: small structures within cells that serve a specific biological purpose, e.g., ribosomes are the locus for protein synthesis

organophosphates: the most widely used class of insecticides (e.g., parathion, malathion); toxic to humans

organosol: suspension of resin in plasticizer together with at least 5% of volatile organic liquid; prepared from plastisols by reducing its viscosity

organotin compounds: in polymer technology, efficient, compatible stabilizers for polyvinyl chloride (PVC) that impart clarity, e.g., tin-alkyl(aryl) sulfides and oxides, organotin carboxylic acid salts, mercaptides, and trialkyl(aryl) tin alcoholates; in agriculture, a family of pesticides, e.g., triphenyltin acetate (fentin acetate, $[C_6H_5]_3SnOCOCH_3$), triphenyltin hydroxide (fentin hydroxide, $[C_6H_5]_3SnOH$), triphenyltin chloride (fentin chloride, $[C_6H_5]_3SnCl$), bis(trineophyltin)-oxide (fenbutatin oxide, $[(C_6H_5)C(CH_3)_2 CH_2)_3Sn]_2O$), tricyclohexyltin hydroxide (cyhexatin, $[C_6H_{11}]_3SnOH$), and tricyclohexyltin-1,2,4-triazole (azocyclotin)

orientation: in polymer technology, the process of stretching or drawing a polymer to align its molecular chains and thus improve properties; stretching in one direction produces uniaxial orientation, in two directions biaxial orientation; oriented polymers shrink in the directions of orientation upon heating unless they are heat-set or annealed; fibers have unidirectional orientation, films require two-dimensional orientation and stabilization; see birefringence

orientation release stress: the stress (force/unit area) exerted by the reheated oriented polymer in returning to its preoriented condition

orphan drug: medication for rare diseases; orphan drugs while important do not have wide use

orthene: $CH_3O(CH_3S)P(O)NHC(O)CH_3$; acephate; insecticide

orthophosphates: salts of phosphoric acid

orthophthalate plasticizers: a family of widely used plasticizers that includes dioctylphthalate (DOP), di-isodecyl phthalate (DIDP), and ditridecyl phthalate (DTDP)

osmometry: measurement of osmotic pressure; used for molecular weight determination, particularly for large molecules (e.g., for polymers it determines the number average molecular weight)

osmosis: solvent flow through a semiperme-able membrane to equalize the concentration of the solute on both sides of the membrane; the pressure needed to stop the flow of solvent (osmotic pressure) is proportional to the concentration of the solute and can be used to determine the molecular weight of the solute; also see reverse osmosis

Ostromislenski process: manufacture of butadiene by cracking an ethanol–acetaldehyde mixture over copper or tantalum oxide catalyst; the butadiene fraction is extracted from the reaction mixture with 2,2'-dichloroethyl ether

outfall: in waste water treatment, the exit of a sewer or drain where effluent is discharged into receiving waters

outgassing: under vacuum, the evaporation of a volatile substance (e.g., a solvent) leading to an improvement of the vacuum

oxalacetic acid: $HOOCCOCH_2COOH$; intermediate in citric acid manufacture from glucose

oxalic acid: $HOOCCOOH$; ethane diacid; produced from molten sodium formate; by nitric acid oxidation of sugar or by reaction of strong alkali with sawdust; use: reducing agent for iron compounds, rust stain remover, metal polish, bleaching agent and mordant for dyes

oxamide: $H_2NOCCONH_2$; produced by one-step oxidative dimerization of HCN in the presence Cu(II) catalyst; use: slow-release fertilizer, stabilizer for nitrocellulose, flame-proofing agent, wood preservative

oxazine dye: group of xanthene dyes in which the chromophore contains a six-membered 1-N, 4–O heterocycle

O-X-D process: see oxydehydrogenation

oxetane: $(CH_2)_3O$; propyleneoxide, trimethylene oxide; see polyoxetane

oxidation dyes: dyes that are applied to a fiber in colorless form and are then oxidized, e.g., aniline is applied to cotton and oxidized to aniline black using hydrogen peroxide or sodium bichromate in the presence of a metal catalyst

oxidation pond: in waste water treatment, a pond in which organic matter is being oxidized by bacteria while oxygen is sometimes bubbled into the water; *see* stabilization pond

oxidative coupling: term used in polymer chemistry denoting the polymerization of two hydrogen-containing monomers by oxidation resulting in loss of water; used extensively in the polymerization of phenols

oxirane: *see* ethylene oxide

oxirane process: production of propylene oxide by oxidation of propylene with hydroperoxides; the economics of the process depends on the by-products that are derived from the hydroperoxide; thus, use of C_6H_5 $CH(OOH)CH_3$ leads to styrene, $(CH_3)_3COOH$ leads to isobutene, and $(CH_3)_2C(OH)OOH$ gives isopropanol

oxirane value: percent oxygen absorbed by an olefin during epoxidation; a high oxirane value indicates a large number of oxide bonds in epoxy resins

oxo process: hydroformylation; manufacture of aldehydes from olefins, CO, and H_2. *see* THE COMMERCIAL OXO PROCESS

oxo products: alcohols, aldehydes, and acids produced by hydroformylation; the primary reaction product, "oxo" aldehyde can be converted to "oxo" alcohols and "oxo" acids

oxyacetylene torch: used for welding and cutting metals; oxyacetylene flame produces the highest temperature of any gas; *see* acetylene

4,4'-oxy-*bis*(benzene-sulfonylhydrazide): foaming agent; synthesis from diphenyl ether and chlorosulfonic acid ($ClSO_3H$) yields the dichlorosulfone ether that is then reacted with hydrazine and ammonium hydroxide to give the product; use: for foamed polyethylene insulation; *see* blowing agent

oxychlorination: catalytic chlorination of hydrocarbons using HCl in the presence of air, e.g., production of 1,2-dichloroethane from ethylene, mono- and di-chlorobenzene from benzene, propene to allyl chloride; *see* Kel–Chlor process

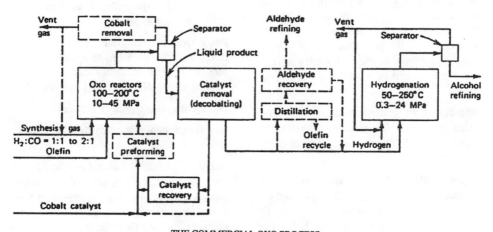

THE COMMERCIAL OXO PROCESS
To convert MPa to atm, divide by 0.101.
[Copyright Kirk-Othmer. *Encyclopedia of Chemical Technology*, 3d ed., John Wiley & Sons, Ltd.]

oxydehydrogenation: formation of C=C in the presence of oxygen; used in the following manufacturing processes: butadiene from *n*-butene (O-X-D process), formaldehyde from methanol in the presence of Ag or Cu, acetone from isopropanol in the presence of ZnO, Cu, or Cu-Zn, methacrylic acid from isobutyraldehyde, and styrene from ethylbenzene in the presence of metal oxides

oxyfliorofen: diphenyl ether herbicide

oxygen: O; element; O_2 (dioxygen) gas is produced by liquefaction of air followed by separation from nitrogen and argon using fractional distillation; the air is first purified by filtration to remove atmospheric dirt, CO_2 is removed by passing the air through towers filled with coke and washing it with NaOH solution; cooling of the compressed air proceeds either by Joule-Thompson expansion at constant enthalpy or by the Claude expansion cycle, which proceeds at constant entropy; the boiling point of oxygen is 13°C higher than that of nitrogen, therefore, fractional distillation is needed for final separation and purification; oxygen is used in metallurgical and chemical processes where an oxygen-enriched environment increases the reaction rate, shortens the cycle time, and leads to lower cost; use: for cracking natural gas by partial oxidation, as raw material for oxidations, for pollution control as oxidizer of biological waste, in rocketry as support for fuel, and in medicine as breathing aid

oxygen bleach: granular bleaches containing sodium perborate; less powerful than chlorine bleaches; use: for presoaking and cleanser formulations

Oxygen Index: OI; the volume percent of oxygen/(oxygen + nitrogen) that will support combustion of a test material; because air contains 21% oxygen the OI of air is 21; the higher the OI of a material the more fire resistant

P

package dyeing: process in which yarns wound on perforated cylinders are placed into a dye bath and the solution is forced alternately from the inside and outside of the package

package spinning: pot spinning; one of several methods to collect rayon yarn after the spin-bath operation; the yarn is thrown against the wall by centrifugal force in a fast rotating pot (Topham box, named after the inventor), is washed and dried, and is then ready for further textile operations; other methods wash the yarn on a bobbin or have a continuous washing procedure before wind-up; *see* viscose

packed tower: for air pollution control, a tower filled with crushed rock or wood chips; the polluted air is forced up the tower while a liquid is sprayed downward; the pollutants dissolve in or react with the liquid

packing fraction: in an unconsolidated or bulk solid the volume occupied by the particles or agglomerates

packing volume: in an unconsolidated or bulk solid the space that is unoccupied by particles or agglomerates

Pacol Olex process: gas phase dehydration of C6 to C19 *n*-paraffins to a statistical mixture of linear mono-olefins using a fixed-bed catalyst at 400–600°C (UOP process)

pad dyeing: *see* dyeing processes

pad steam continuous dyeing: *see* dyeing processes

paints: pigments suspended in a liquid medium; properties for which paint films are judged include stability, resistance to water and environmental stress, gloss, color, opacity, hardness, and flexibility; paint formulations consist of (1) finely ground pigment; (2) a volatile solvent (e.g., water, hydrocarbons); (3) binder (resins, drying oils, latexes); and (4) additives (e.g., stabilizers, defoamers, pesticides, dispersants, fire retardants, UV absorbers, coalescing agents); the original binders were drying oils, modern binders are resins that include alkyds, acrylics, and epoxies that provide stable insoluble protective coatings after solvent evaporation; in water dispersions the binders consist of latexes; air pollution considerations have increased the use of water-based latex paints; in the past, paint has been applied only by brush or roller; modern methods include use of spray equipment in which paint is atomized by pressure, steam or heat; electrostatic spraying in which the paint is attracted to a conductive surface; electrodeposition in which the paint is deposited onto a conductive surface from a water bath; powder coating in which dry pigment is applied to the object and then fused onto the surface; use: for decorative effects and protective coatings

paint vehicle: liquid medium in which pigment is suspended; consists of solvent and binder; *see* paints

palamoll: copolymers of butadiene and diethylfumarate

palmitic acid: $CH_3(CH_2)_{14}COOH$; saturated unbranched fatty acid having 16-carbon atoms; use: soap manufacture, food

palm oil: derived from palm kernels; this oil has a high palmitic, lauric, and myristic acid content; use: cooking oil and confectionary fat

panning and pressing: in soap manufacture an old process for the fractional crystallization of fatty acids; molten fatty acid is run into pans, chilled and pressed to remove the unsaturated fatty acids; the process is limited to fatty acid mixtures that solidify easily

paper fibers: fibers, such as cotton, rayon, and synthetics, that are not suitable for textile operations are blended with normal paper raw materials such as shredded rags

paper maker's alum: *see* alum

paraffin hydrocarbons: C_2H_{2n+2}; paraffins; alkanes; aliphatic hydrocarbons; saturated linear or branched hydrocarbons; at room temperature, C1 to C4 alkanes are gases, C5 to C16 are liquid, >C16 they are solids (paraffin wax); derived from petroleum by fractionation; separation of branched and linear hydrocarbons is achieved by adsorption with molecular sieves or extractive crystallization with urea (used mainly for C15 to C30 hydrocarbons); paraffins are converted to terminal olefins by cracking, and to internal olefins by catalytic dehydrogenation (*see* Pacol Olex process) and chlorodehydrogenation (i.e, chlorination followed by removal of HCl); sulfonation leads to alkylsulfonates that are used as detergents; new methods to activate C-H bonds catalytically promise the use of paraffins in chemical synthesis; the hydrocarbons per se are now used as fuel, illuminating gas, lubricants, solvents, laxatives, plasticizer, additives, candles, sizing

paraffins: *see* paraffin hydrocarbons

paraffin wax: *see* paraffin hydrocarbons

paraform: *see* paraformaldehyde

paraformaldehyde: $(CH_2O)_n$; paraform; low molecular polymer of formaldehyde; produced by evaporation of aqueous formaldehyde; a white powder that regenerates formaldehyde on heating; use: instead of formaldehyde where water is not desired, catalyst and hardener for synthetic resins, disinfectant; *also see* trioxane

parahydroxy benzoic acid: PHB; p-HOC$_6$H$_4$COOH; monomer for aromatic polyesters

paraldehyde: 2,4,6-trimethyl-1,3,5-trioxan; cyclic trimer of acetaldehyde; use: substitute for acetaldehyde when its higher boiling point and lower reactivity are of value, promoter in the manufacture of terephthalic acid from p-xylene, antioxidant, depressant; flammable

PARALDEHYDE

paramagnetism: solid state property due to the random alignment of the spins of the orbital electrons; paramagnetic substances are weakly magnetic

paraphthalate plasticizers: mostly solids derived from alcohols and terephthalic acid; they are better than orthophthalates with respect to low-temperature flexibility and lacquer marring

paraquat: postemergence herbicide; produced by dimerization of pyridine with Na/NH_3 and oxidation followed by methylation with CH_3Br; use: pretreatment of fields before planting crops, eradication of illegal marijuana plantings, controlling weeds in lakes and rivers

$(CH_3OSO_3^-)_2$

PARAQUAT

parathion: $(C_2H_5O)_2P(S)O p\text{-}C_6H_4NO_2$; general insecticide

parenteral: refers to administration of drugs by injection into the body cavity

Parex process: recovery of p-xylene from C8 mixture by continuous selective adsorption on a fixed bed of solid adsorbent (UOP process)

pargyline: $C_6H_5CH_2N(CH_3)CH_2C\equiv CH$ antidepressant and blood pressure lowering drug; monoamine oxidase (MAO) inhibitor

Paris green: $Cu(CH_3COO)_2\cdot 3Cu(AsO_2)$; general insecticide

parison: in blow molding the hollow tube of thermoplastic material that is inflated inside the mold during the molding process

Parkes' process: cold vulcanization of rubber

partially oriented yarns: POY; *see* poly (ethylene terephthalate)

particle: in solid materials, a convenient identifiable subdivision, such as grains, prills, small agglomerates, crystals

particle board: particles of wood bonded together by a resin adhesive

particle size: in materials, size of the particles expressed in particle diameter, mesh or sieve size

parting agent: release agent, lubricant

partition chromatography: *see* liquid–liquid chromatography

parylenes: group of thermoplastics based on poly-p-xylene and its halogen derivatives

pascal: S.I. unit for pressure; 1 Newton/m^2

Pasteur effect: in fermentation of a carbohydrate biomass the suppression of ethanol production due to the presence of air

pathogenic: disease producing

PBI: *see* poly(benzimidazoles)

PCB: *see* polychlorinated biphenyls

PDU: process development unit; *see* pilot plant

peanut oil: by-product of low grade peanuts; high in oleic acid; use: cooking oil

pearlescence: mother-of-pearl-like appearance

pearlescent pigments: thin colored platelets having a high refractive index; these crystals are imbedded into a solid material matrix or dispersed in emulsions; only a portion of light reflects from the crystal planes, the rest of the light passes through; fish scales supply natural pearlescents, which consist primarily of guanine crystals; synthetic pearlescents are made from TiO_2-coated crystalline mica, crystalline Bi or Pb compounds

pearl polymerization: *see* suspension polymerization

peat: highly organic soil derived from dried marshy regions; use: a fuel

pebble mill: used to reduce coarse materials to powders; *see* mill

pectin: acidic polysaccharides with an average molecular weight of 50,000–180,000; the main constituent of pectins is D-galacturonic acid and its methyl ester; natural ingredient of fruit and plants, it provides them with turgidity; use: gelling agent and thickener in acidic foods such as jams and jellies

pectinase: enzyme derived from *Aspergillus niger*; use: for clarification of fruit juice, pectin decomposition, and viscosity reduction of fruit pulp

peel ply: in laminates, the outer layer that is discarded to improve bonding of the other layers

pellet: solid particles produced by agglomeration or by extrusion using pellet mills that have hole diameters from 1/16 to ~1 in.; used for compacting coal, clays, plastics, catalysts; *see* agglomeration

pendulum impact strength: *see* hardness

penicillin acylase: penicillin amidase; enzyme used to produce 6-aminopenicillanic acid from penicillin G

penicillin amidase: *see* penicillin acylase

penicillins: group of antibiotics originally derived from the mold *Penicillium notatum* by surface fermentation; modern industrial submerged aerobic fermentation uses *P. chrysogenum* (Wisconsin 49-133 strain) and achieves concentrations of 20–30 g per L of broth; this batch fermentation is followed by removal of fungus mycelium, extraction of the penicillin, purification with solvents such as methyl isobutyl ketone or amyl acetate; penicillin is usually recovered as sodium or potassium salt; semisynthetic penicillins are

made by use of enzymes such as penicillin acylase (isolated from *Bacillus megaterium*) and penicillin amidase (isolated from *Escherichia coli*)

PENICILLINS

penicillium chrysogenum: *see* penicillins

penicillium notatum: *see* penicillins

pentaacetyl glucose: *see* bleach activators

pentabromochlorocyclohexane: flame retardant used for polymers

pentachlorophenol: preservative for wood; hazardous chemical waste (EPA)

pentaerythritol: $C(CH_2OH)_4$; produced by threefold aldolization of acetaldehyde with formaldehyde in the presence of $Ca(OH)_2$ followed by reduction of the trimethylol acetaldehyde; use: for manufacture of explosives (*see* pentaerythritol tetranitrate), alkyd resins, drying oils, high melting waxes, varnishes, additives, emulsifiers, and chemicals

pentaerythritol tetranitrate: PETN; $C(CH_2ONO_2)_4$; high explosive; prepared by nitration of pentaerythritol with 94% HNO_3 at 50°C

pentanols: *see* amyl alcohols

pentazocine: analgesic drug that is less addictive than opioids

PENTAZOCINE

pentenes: *see* amylenes

pentolite: boosters for explosives made from

slurries of pentaerythritol tetranitrate and TNT

pentosans: polymers of pentoses; gums or resins derived from plant waste such as nut shells, oat hulls, corn cobs, wood; yield pentoses on hydrolysis; use: for production of furfural

pentose: pentaglucose; a mixture of sugars derived from pentosans; also any monosaccharide containing five carbon atoms such as arabinose, ribose, xylose

peptidizer: plasticizing agent; unlike other softening agents, peptidizers act chemically by increasing the oxidative decomposition of the polymer; use: reclaiming rubber, aid in mastication

peptize: to disperse a particulate solid in a liquid producing a sol

peracetic acid: CH_3COOOH; produced by air oxidation of acetaldehyde at -15 to $40°C$ and 25–40 bar or by oxidation of acetic acid with hydrogen peroxide and catalytic amounts of sulfuric acid at $40°C$; use: in chemical synthesis, caprolactam manufacture from cyclohexanone, and glycerol manufacture from allyl alcohol (Daicel process), as bleach and biocide

perbunans: *see* Buna

perchlorobenzene: *see* hexachlorobenzene

perchloroethane: *see* hexachloroethane

perchloroethylene: mixture of trichloroethylene and tetrachloroethylene; use: dry cleaning agent but its use is being limited because of potential health hazard

percolation: downward flow of a liquid through a porous body, e.g., water through rock or soil

performic acid: $HCOOOH$; produced by oxidation of formic acid with hydrogen peroxide and catalytic amounts of sulfuric acid; use: oxidizing agent, preferential agent for

the industrial epoxidation of unsaturated fatty acid esters

periclase: highly crystalline, hard-burned magnesia (MgO); use: in basic refractories

periodic kiln: kiln operated in batch mode as opposed to continuous mode

perlite: volcanic glass; low density filler; *see* soil conditioners

permanent press: durable press; *see* crosslinks

permanent set: in the polymer industry, a measure of inelasticity: the percent increase in length retained for a specified time after the stress is released

permeability: quality of a solid that permits gases or liquids to pass through

permethrin: pyrethroid insecticide

permittivity: specific inductive capacity; *see* dielectric constant

permselective membrane: film that exhibits different permeabilities toward different gases

peroxidase: enzyme that catalyses the reduction of hydrogen peroxide to water

peroxide value: indicator of oxidative rancidity of fats; peroxides formed during oxidation of fatty acids react with KI to liberate free I_2; peroxide value is expressed in milligrams of I_2 formed per kilogram of fat

perpropionic acid: CH_3CH_2COOOH; produced by oxidation of propionic acid with hydrogen peroxide; use: for liquid-phase production of propylene oxide; oxidizing agent

Persian Gulf oxide: Persian red; iron oxide pigment; use: primer, cheap paint for houses and ships

Persian orange: acid-azo pigment

Persian red: *see* Persian Gulf oxide

pesticides: lethal chemicals used against any undesired living organisms, e.g., insecticides, herbicides, rodenticides, fungicides; pesticides can be classified according to their chemistry such as carbamates, organophosphates, organophosphorothionates, halogenated hydrocarbons, methyl-, phenyl-, and sulfonyl-ureas, pyrethroids, triazines

petrochemicals: chemicals derived from petroleum or natural gas

petrol: gasoline (British)

petroleum: mixture of hydrocarbons found underground at various depths; can exist as a mixture of gas (natural gas), liquid (crude oil), and solid (paraffin wax); obtained by drilling and raising the petroleum through the hollow drill pipe; petroleum may be raised under its own pressure (primary recovery) or with the help of drilling fluids that are pumped down through a hollow drill (secondary recovery); drilling fluids range from compressed air or CO_2 to clear water to high-density muds and clays with various additives; finally petroleum may be raised using tertiary recovery or enhanced oil recovery (EOR); EOR technologies drive oil from rock cracks using surfactants, foams, or carbon dioxide; petroleum is used as fuel, lubricant, and chemical feedstock; *also see* naphtha

petroleum refining: processes that convert crude oil to gasoline, lubricants, and chemical feedstock: refining involves (1) the production of larger olefins and more aromatics by cracking; (2) isomerization of straight chain hydrocarbons to branched hydrocarbons by cracking and reforming; (3) catalytic alkylation of C3 and C4 hydrocarbons with olefins to produce branched hydrocarbons (*see* catalytic reforming); (4) removal of metals from the refinery stream by heating in the absence of air (*see* coking); and (5) conversion of sulfur and nitrogen to their volatile gases, H_2S and NH_3 (*see* hydrotreating). *see* HIGH CONVERSION PETROLEUM REFINERY

Petrotex process: manufacture of butadiene by oxidative dehydrogenation of butane using bromine or iodine as cocatalysts

peyote: *see* alkaloids

Pfund test: *see* hardness

phage: short for bacteriophage; viruses that invade bacterial cells

phase: any homogeneous volume of gas, liquid, or solid

phase transfer agents: *see* phase transfer catalysis

phase transfer catalysis: the use of a fat-soluble (lipophilic) salt (e.g. $[n\text{-}C_4H_9]_4$ $N^+HSO_3^-$) to transfer an ion (e.g., CN^-) from the aqueous to the organic phase where it can react with a water-insoluble substrate; the salt then moves back into the aqueous phase to pick up another ion; only small amounts of the ammonium salt are needed to facilitate the reaction as the salt continues to shuttle back and forth between the two phases; other phase transfer catalysts (phase transfer agents) include phosphonium salts, crown ethers, and enzymes; homogeneous phase transfer catalysts can be immobilized leading to better product separation

phenacetin: $p\text{-}C_2H_5OC_6H_4NHCOCH_3$; analgesic drug

phenelzine: $C_6H_5CH_2CH_2NHNH_2$; antidepressant, a monoamine oxidase (MAO) inhibitor

phenobarbital: luminal; anticonvulsant drug; sedative; *see* barbiturates

phenol: C_6H_5OH; carbolic acid; mostly produced by air oxidation of cumene (isopropyl benzene) to the hydroperoxide followed by cleavage to phenol and acetone (Hock process); the air oxidation can be carried out in an aqueous emulsion or catalytically with undiluted cumene; the exothermic acid cleavage of the peroxide leads to numerous by-products, but selectivity for phenol

phenol-formaldehyde resin

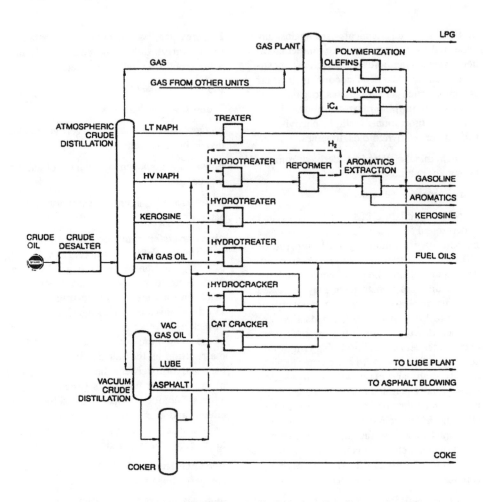

HIGH CONVERSION PETROLEUM REFINERY
[Copyright *Riegel's Handbook of Industrial Chemistry,* 8th ed., Van Nostrand-Reinhold Co., Inc.]

and acetone is ~91%; other but less important manufacturing processes are (1) liquid phase oxidation of toluene to benzoic acid (110–120°C, 2–3 bar, Co salt catalyst) followed by oxydecarboxylation at 220–250°C in the presence of steam/air (Cu salt catalyst); (2) formation of benzenesulfonic acid from benzene followed by hydrolysis (disadvantage: unavoidable formation of large amounts of sodium sulfite and sodium sulfate); (3) formation of chlorobenzene from benzene fol-

lowed by gas-phase hydrolysis (calcium phosphate/silica catalyst) (disadvantage: highly corrosive, energy intensive); (4) hydrogenation of lignin from sulfite waste liquor; phenol is also isolated from cracking units; major derivatives are phenol-formaldehyde resins, bisphenol A, and caprolactam; major end uses are adhesives and plastics

phenol-formaldehyde resin: Bakelite; thermosetting polymer; produced by condensation polymerization in several stages: stages

A and B produce preliminary condensation products that are not or are only slightly cross-linked, e.g., novolac; stage C achieves cross-linking and setting in the final form; heat- and acid-resistant material with good impact and tensile strength, good electrical properties, and slow burning rate; use: for molded and cast articles, decorative laminate backing, adhesive for particle board, coatings, insulation

phenolic resins: phenolics; thermosetting resins made from reaction of phenols with aldehydes; *see* phenol-formaldehyde resin

phenolics: *see* phenolic resin

phenolphthalein: produced by reacting phthalic anhydride with phenol in the presence of sulfuric acid; use: as a laxative, acid–base indicator, dye

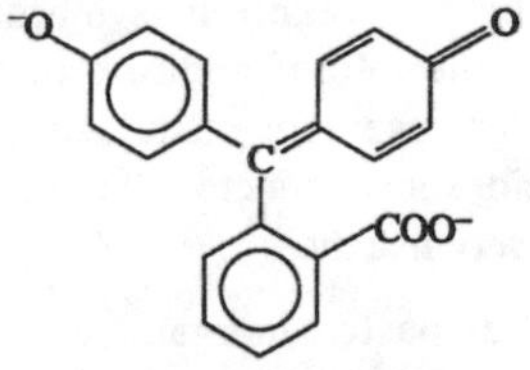

PHENOLPHTHALEIN

phenothiazines: PTZ; produced by catalytic cyclization of secondary amines using elemental sulfur; the parent compound phenothiazine is used as polymerization inhibitor for acrylic monomers, additive in lubricants, rubber, gasoline, waxes, and as anthelmintic drug; phenothiazine derivatives comprise classes of pharmaceuticals (e.g., the tranquilizer chlorpromazine and antiallergy drug promethiazine); phenothiazone dyes contain the characteristic three-ring structure in which one of the benzene rings is quinonoid (e.g., methylene blue)

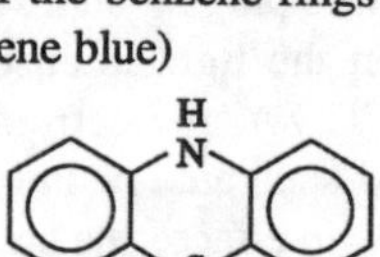

PHENOTHIAZINES

phenoxy resins: $-(OC_6H_4C(CH_3)_2C_6H_4O$

$CH_2CH(OH)CH_2)_n-$; thermoplastic resins made by reacting equivalent amounts of epichlorhydrin with bisphenol A in the presence of NaOH and $(CH_3)_2SO$

phenylbutazone: antiinflammatory agent, analgesic

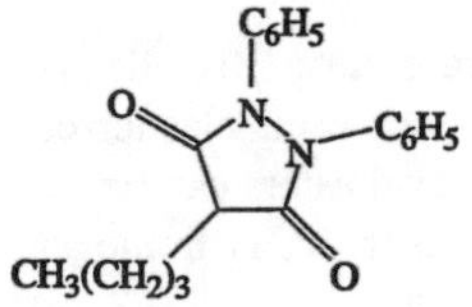

PHENYLBUTAZONE

m-phenylene diamine: $1,3-(NH_2)_2C_6H_4$; monomer for an aramid polymer

p-phenylene diamine: $1,4-(NH_2)_2C_6H_4$; raw material for dye manufacture and aramid polymer

m-phenylene-isophthalamide: *see* isophthalic acid

phenylethylene: *see* styrene

phenyl mercuric acetate: $C_6H_5HgOOCCH_3$; fungicide; banned in food products

phenylmethyl ether: *see* anisole

phenylpropanolamine: $C_6H_5CH(OH)CH (CH_3)NH_2$; nasal decongestant used in preparations to ease symptoms of colds

phenytoin: anticonvulsant drug used against epilepsy

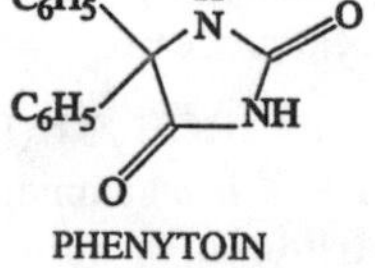

PHENYTOIN

pheromones: sex attractants; biological activity is intimately related to stereochemistry; use: for control of insect populations

phosgenation: *see* phosgene

phosgene: Cl_2CO; highly toxic, colorless gas; prepared from carbon monoxide and chlorine over active charcoal; a highly reactive intermediate; war gas in World War I; use: production of isocyanates by reaction with amines (e.g., toluene diamines and manufacture of urethane polymers), production of chloroformates by reaction with alcohols (chloroformates react with amines to give urethanes)

phosphate esters: tri-phenyl and tri-cresylphosphates; use: plasticizers, only meta- and para-cresol esters can be used, the ortho-cresol ester is toxic, a carcinogen

phosphate fertilizers: include superphosphate, triple superphosphate, ammonium phosphate, diammonium hydrogen phosphate, monoammonium phosphate, nitric phosphate, potassium phosphate

phosphatides: *see* phospholipids

phosphatidylcholine: lecithin; RCH_2CH $(R')CH_2OP(O)(OH)OCH_2CH_2N(CH_3)_3OH$; glyceride with fatty acids R and R'; phospholipid; use: antioxidant, emulsifier, and stabilizer in foods, viscosity regulator in casein paints and printing inks, curing accelerator in synthetic rubbers

phosphazene polymers: *see* polyphosphazenes

phosphazote: fertilizer consisting of urea and superphosphate

PHR: parts per hundred resin (or rubber)

phospho-gypsum: by-product during wet phosphoric acid processing

phospholipids: $RCH_2CH(R')CH_2OP(O)$ $(OH)OY$ where Y is an amine; phosphatides; widely distributed in nature; *see* phosphatidyl choline; lecithin

phosphorescence: emission of light some time after the molecules have been excited; phosphorescence increases the brightness of dyes

phosphoric acid: H_3PO_4; produced by (1) burning phosphorus in air and hydrating the resulting phosphoric oxide to yield furnace phosphoric acid; (2) acidulation of pulverized phosphate rock with sulfuric acid yielding wet process phosphoric acid; major derivatives are ammonium phosphates and triple superphosphate; major uses are in fertilizers, detergents, food and feed products, and for metal-treating processes

phosphorus: P; element; solid phosphorus occurs in several forms: white phosphorus is soft and waxy; it ignites spontaneously in air at 30°C and is highly toxic; it is produced by electrothermal reduction of apatite with coke in the presence of silica; used for manufacture of phosphoric acid, phosphorus sulfides, oxides, and halides; a more stable allotropic form is red phosphorus, which is produced by adding white phosphorus to a ball mill that is half-filled with iron shot; the temperature of the exothermic reaction is maintained at 350°C by the rate of addition of the white phosphorus; use: for manufacture of phosphoric acid, semiconductors, luminescent materials, safety matches

phosphorus paste: inorganic pesticide used against ants and roaches

photo-: prefix denoting light

photoactivation: activation with light; some biologically active materials (e.g., those having pyrroles in their structure), such as herbicides and anticancer agents, can be activated by light of specific wavelengths

photochromism: change of color due to exposure to light

photoelectrochemical cells: the electrons needed for the chemical reaction are produced by light impinging on a semiconductor anode; when the light is absorbed on the anode (e.g., TiO_2) an electron-hole pair is produced and separated at the semiconductor–electrolyte interface; the hole is driven to the anode surface to react with the anolyte; the electrons are conducted through an external circuit to the cathode

photolithography: *see* resist; photoresist

photopolymer: polymer capable of forming photographic images due to an incorporated photosensitive substance

photopolymerization: polymerization initiated by light

photoresist: in semiconductor chip manufacture, the light-sensitive polymeric material that covers the silicon wafer; to produce the circuit pattern a latent circuit image is delineated on the resist; if exposure to light degrades the polymer the circuit pattern is transferred to the underlying oxidized chip surface by etching away the degraded polymer (positive photoresist); if exposure to light causes the polymer resist to gel the circuit pattern is the area not covered by the gelled polymer and the pattern is transferred to the chip by etching away the unexposed polymer (negative resist); resist materials include poly(phenylene sulfide) (PPS), acrylonitrile-butadiene-styrene (ABS), and diazonaphthoquinone-novolac (DQN)

photovoltaic cells: *see* solar cells

phthalates: *see* phthalic acid esters

phthalic acid esters: phthalates; produced by esterification of phthalic anhydride; use: esters with C4 to C19 alcohols are plasticizers

phthalic anhydride: PA; produced by catalytic (V_2O_5/Al_2O_3) air oxidation of naphthalene or o-xylene; fixed bed and fluid bed processes are used for both oxidations; use: important intermediate for manufacture of chemicals, notably plasticizers, resins, phthalocyanine, and xanthene dyes

phthalocyanines: class of intensely colored metal benzoporphyrins; copper phthalocyanine is produced in a continuous process by heating a mixture of phthalic anhydride, urea, and cupric chloride with an ammonium molybdate catalyst in trichlorobenzene to 200°C; substituted derivatives may be obtained by substitution in the phthalocyanine

or by use of appropriate starting materials; the color of phthalocyanines is usually green or blue and depending on the metal, on the substituents on the isoindole rings, and on whether the pigment consists of one or more porphyrins linked together; use: colorfast paints and enamels; potential uses include electro- and photocatalysts, for thin-film photovoltaics, chemical sensors, and in batteries

PHTHALOCYANINES

physical testing: evaluation of the mechanical properties of metals, alloys, fibers, films, machined or molded test pieces as to their break strength, break elongation, modulus, ageing and environmental exposure stability, etc.; routine evaluation methods are expanded to every possible physical method to correlate structural and superstructural features of a material with its processing history and its physical and chemical properties; many of the plastics testing methods were originally developed for metallurgical evaluations

Physician's Desk Reference: PDR; book published by Medical Economics, listing drugs and their side effects

phytosterols: plant products characterized by the four-ring steroid carbon skeleton

pickling steel: removing oxides from steel by treatment with acid

picloram: herbicide; plant sterilant

PICLORAM

picric acid: 1,3,5-trinitrophenol; prepared from benzene by treatment with nitric acid in the presence of mercuric nitrate, or by nitration of phenol or chlorobenzene; use: explosive, yellow dye, organic reagent; hazard: explosive, toxic by skin absorption

Pieter's process: obsolete method of hardwood distillation

piezoelectricity: creation of electrical charges in a material upon exposure to stress such as pressure; piezoelectric materials expand along one axis upon exposure to an electric field; for example, quartz slices and polyvinylidenefluoride have piezoelectric properties

pigments: insoluble natural and synthetic coloring materials (including black and white); particle size and crystal structure are critical to the production of coloring effect and luster; in pigment dyeing or printing, dispersions of water-insoluble pigment are applied to the fabric surface and then bonded by heating; to obtain dull or matte yarns, small amounts of titanium dioxide (rutile or anatase) are added to the polymer solution or melt before extrusion; specific sets of colored pigments are metered into spinning solutions to produce the desired color in the spun yarn; use: for surface coatings, printing inks and coloration of plastics, rubber, and fibers

pile: in nuclear technology, the first reactors consisting of uranium fuel and carbon moderator

Pilkington's float glass process: *see* glass

pill: in textiles, small amounts of matted material rubbed off during wear; pilling occurs when short fibers pull out of the fabric and entangle with others still tied to the bulk of the fabric; this has been a problem with some fabric constructions with new and strong synthetic fibers; in the plastics industry a synonym for preform

pilot plant: facility in which laboratory reactions can be scaled up to develop economic processes, equipment, as well as evaluate raw materials, product yields, and by-products (useful or waste) before the design and construction of a commercial plant

pinenes: $C_{10}H_{16}$; terpene hydrocarbons made up of two isoprene units; obtained from wood turpentine; use: in perfumery, as solvent, lube oil additive, organic synthesis; *see* isoprene

pine oil: a mixture of secondary and tertiary terpene alcohols; obtained by steam distillation of pine wood; use: odorant in household cleaners and cosmetics, wetting agent and preservative

piperonyl butoxide: insecticide; acts as synergist for pyrethroids and rotenones

PIPERONYL BUTOXIDE

pitch: carbonaceous material obtained from distillation of coal, peat, petroleum, wood, or other organic material; use: adhesive, wood preservative, pipe insulation, raw material for manufacture of carbon fibers, pine pitch is used for marine applications

pivalic acid: trimethyl acetic acid; produced by reacting isobutene with CO at 20–80°C and 20–100 bar followed by addition of H_2O; catalyst is H_3PO_4/BF_3; use: chemical intermediate; *see* Koch carboxylic acid synthesis

pivalolactone: lactone of β-hydroxypivalic acid; its use as raw material for polypivaloactone (a high melting polyester) has been discontinued because the monomer is a carcinogen

planographic printing: *see* printing

plantation white: in the sugar industry the direct-consumption sugar

plant hormones: agents that regulate physiological processes in higher plants; for example, auxins (e.g., indoleacetic acid) promote cell elongation, cytokinins (e.g., zeatin) stimulate cell division, gibberellins (e.g., gibberellic acid) promote seed germination, ethylene promotes fruit ripening and abscission of leaves, and abscisic acid promotes seed germination and aging

plasma-enhanced chemical vapor deposition: PECVD; process with which amorphous films are deposited at low temperature over large areas using radicals, ions, or excited neutral particles that are produced in the plasma

plasma etching: pretreatment of plastic surfaces before etching using ionized gas in vacuo

plasmid: in biological cells, a circular DNA that is outside the nucleus; genetic engineers use plasmids to introduce foreign genes into the cell

plasticity: ability of a material to flow or deform without rupture when exposed to stress

plasticization of wood: treatment with anhydrous ammonia gives wood a temporary plasticity by entering into the cell walls and swelling both the cellulose and the lignin systems; after the wood has been given the desired shape and the ammonia has evaporated the wood loses plasticity and retains the new shape

plasticizers: additives that make polymers more flexible, workable and extensible; a plasticizer reduces the melt viscosity and the elastic modulus, lowers the glass transition temperature (T_g); plasticizers are nonvolatile organic liquids or low melting solids that reduce the intermolecular forces between polymer chains; different plasticizers are available for general or specific applications; primary plasticizers can be used alone but may be enhanced by secondary plasticizers; phthalate esters such as dioctyls account for

~60% of their use, followed by epoxy and phosphate esters; water serves as plasticizer for natural polymers

plastics: materials that contain as an essential ingredient one or more organic polymeric substances of large molecular weight, are solid in their finished state and at some stage in their manufacture or processing into finished articles that can be shaped by flow (ASTM [D 883-75a] definition); wider meaning of the term encompasses compounds that contain polymers, plasticizers, stabilizers, fillers, and other additives

plastigel: dispersion of finely divided resins in a plasticizer that contains enough gelling agent and/or filler to produce putty-like consistency; *see* plastisol

plastisol: dispersion of finely divided resins in a plasticizer; on heating the resin dissolves in the plasticizer and upon cooling it forms a plastic mass; used for molding thermoplastics; *see* organosol

plastomer: polymer that shows plasticity during processing at a certain temperature range

platelets: in crystals, minute flat particles such as tabular alumina crystal; in the blood, cells responsible for blood clotting

plate-out: in plastics technology, the objectionable extraction of additives such as dyes and lubricants from plastics

platforming of petroleum: reforming using platinum catalyst

plictran: tricyclotin hydroxide; acaricide; pesticide

plumbago: *see* graphite

plumber's cement: *see* litharge

plus sieve: portion of powder sample retained on a sieve

plustruder: twin-screw extruder used for solution polymerization; the extruder recovers the solvent and the unreacted monomer for recycle to the reactor

pneumatic: using a compressed gas to exert a force

poise: metric unit of viscosity, dyne-sec/cm^2 or g/cm-sec

polar molecules: molecules in which the electrons are not symmetrically distributed resulting in relatively positively and negatively charged regions; this polarity affects characteristics such as chemical reactivity and solubility; for example, water is a polar molecule in which polar substances dissolve and nonpolar materials do not; on the other hand, benzene is nonpolar and dissolves only nonpolar materials

polarography: analytical method based on the current-voltage curves that arise at the microelectrode when diffusion is the rate-determining step in an electrochemical reaction

pollutant: substance that makes the surroundings unclean

poly: generic for polymer

poly-: prefix denoting many

polyacetals: *see* poly(oxymethylene)

poly(acrylamide): $[-CH_2-CH(CONH_2)-]_n$; produced by (1) free radical polymerization from acrylamide in water solution; (2) copolymerization in mixed solvents (water, alcohol); or (3) dispersed phase polymerization in inert organic materials (xylene, paraffin oil); polyacrylamides provide strong hydrogen bonding with low ionic character; hydrolysis or copolymerization with acrylic acid gives anionic properties; copolymerization with basic monomers (vinyl amine, vinylpyridine) gives cationic properties; the polymers do not present a health hazard but the monomer is a neurotoxin and a cumulative toxicological hazard; use: as a flocculant in water solutions, for liquid/solid separations in mineral processing, water and waste water treatment, paper manufacture, oil recovery, as thickeners

polyacrylic esters: $[-CH_2-CH(COOR)-]_n$; emulsion and solution polymers and copolymers; emulsion polymerization with a water-soluble peroxide (e.g., potassium persulfate) yields emulsions with 30–60% (wt) of polyacrylate; use: coatings, finishes, binders for textiles, paints, paper and leather finishes, adhesives; solution polymerization gives clear solutions used for industrial coatings; *see* emulsion polymerization; solution polymerization

poly(acrylonitrile): PAN; $[-CH_2-CH(CN)-]_n$; polymeric fibers made from acrylonitrile, $CH_2=CHCN$, the monomer is water soluble (7.5 wt% at room temp.), thus, polymerization is in water with redox initiation; copolymers with ≤85 wt% acrylonitrile (generic name: acrylic fibers) use acrylamide, acrylic esters, vinyl acetate, vinyl chloride, vinylpyridine as comonomers; copolymers with 35–85% acrylonitrile (generic name: modacrylic fibers) usually have vinyl chloride or vinylidene chloride as comonomers to improve flame retardancy; original solvent for PAN was sodium thiocyanate in water, now many organic compounds have been developed, with dimethylformamide and dimethylacetamide the preferred spinning solvents; the polymers are dry or wet spun and oriented by drawing; they have good strength, toughness, abrasion resistance, flex life, are relatively insensitive to moisture, are stain resistant, have outstanding weatherability, good performance in carpets, wrinkle resistance, dyeability, soft, wool-like hand; use: staple and endless yarns; filament yarn is a major precursors for carbon and graphitic fibers

polyaddition: polymer formation by addition reaction; for example, formation of polyurethanes from isocyanates and alcohols, nylon 6 from caprolactam or polyoxyethylene from ethylene oxide

polyallomers: thermoplastic copolymers consisting of regularly alternating units of two or more homopolymers; made with Ziegler catalyst; best known: ethylene-propylene copolymer; *see* block polymer or copolymer

polyamides: $[-R-CO-NH-]_n$, $[-CO-R-CO-NH-R'-NH-]_n$; these polymers are called by a coined name: nylons; used primarily in the textile and plastics industries; they can be made by condensation of amino acids or by condensation of dibasic acids with diamines; nylon 66 is made from adipic acid and hexamethylene diamine, nylon 6 from omega-aminocaproic acid or caprolactam; nylons have good strength, elasticity, toughness and abrasion resistance; polyamide-imides have high temperature resistance; *see* nylon nomenclature

polyazo dyes: azo dyes containing more than one azo group

poly(benzimidazoles): reaction product of an aromatic tetramine (e.g., 3,3' diaminobenzidine) with dibasic aromatic acids (e.g., iso- or terephthalic acid), e.g., poly(benzimidazole) PBI; produced by condensation of 3,3'-diaminobenzidine and diphenylisophthalate under N_2 at $\leq 450°C$ and is solvent spun with dimethylformamide (DMF) to a high temperature-resistant fiber, that is nonflammable and with good moisture regain; used for protective clothing

POLY(BENZIMIDAZOLES)

polyblend: homogeneous mixture of two or more polymers; also known as alloy

polybutadiene: *see* elastomers

poly(butyleneterephthalate): PBT, poly-(tetramethylene terephtalate); produced from terephthalic acid and 1,4-butadiol; an engi-neering plastic used alone and in blends with other polyesters; *see* polyesters; poly(ethylene terephthalate)

polycarbonates: PC, $[-OC(O)-OR-]_n$; thermoplastic polymers; produced by reaction of diphenols with phosgene or by ester exchange of diphenols with diphenylcarbonate; best known is the polymer ($H[-OC_6H_4-C(CH_3)_2-C_6H_4-OC(O)-]_n$) made from bisphenol A; this polymer has good electrical properties, high impact, and temperature resistance

polycaprolactam: *see* nylon 6

polycarbosilane: $[-SiH(CH_3)-CH_2-]_n$; produced from polysilane by thermal decomposition and polymerization; the polycarbosilane oligomer (molecular weight 1000–2000) can be melt spun and cured in air to yield an infusible fiber; heat treatment of this fiber under nitrogen yields a continuous silicon carbide (SiC) fiber

polychlorinated biphenyls: PCB; produced by chlorination of biphenyls; used as plasticizers, fire retardants, insulating fluids in electrical transformers; they have high persistence in the environment; toxic, hazardous chemical waste (EPA)

poly(chloroparaffins): produced by chlorination of paraffins; use: plasticizers for PVC, raw materials for paint, impregnation aids for flame retardant and water-repellant finishes

poly(chloroprene): neoprene; $[-CH_2-C[Cl]=CH-CH_2-]_n$; polymer of 2-chloro-1,3-butadiene (chloroprene); rubber-like polymer or co-polymer; preparation by emulsion polymerization; can be vulcanized by heat with zinc oxide or magnesium oxide as vulcanizing agents; oil-resistant rubber suitable for general purpose uses; has high strength, is resistant to oxidation and degradation and is stable at elevated temperature; use: for wire and cable coatings, industrial hoses, belts, shoe heels

poly(chlorotrifluoroethylene): Kel-F; [–C(Cl)(F)–CF$_2$–]$_n$; thermoplastic polymer; produced by addition polymerization in emulsion or solvent; high resistance to chemicals; use: for packing, gaskets, coatings; *see* fluorocarbon polymers

polycondensation: *see* condensation polymerization

poly(dicyclopentadiene): a thermoset, highly cross-linked material; produced from dicyclopentadiene by reaction injection molding using WCl$_6$ catalyst; characterized by high stiffness and impact strength

polydispersity: nonuniformity of molecular weight of a substance, usually a polymer; *see* molecular weight; molecular weight distribution

polyelectrolyte: polymer containing ionic groups, examples: polyacrylic acid, sulfonated cross-linked polystyrene, copolymers with maleic acid, polyvinylamine, polyvinylpyridine

polyesters: [–O–C[O]–R–C[O]–O–R–]$_n$; thermoplastic and thermosetting polymers; produced by addition and condensation polymerization; unsaturated polyesters (i.e, containing fumaric, maleic, or other unsaturated units) can be reacted and cured with additional vinyl monomers; use: for fibers, laminates, moldings, and in composites; *see* poly(ethylene terephthalate); poly(butyleneterephthalate)

polyester fibers: *see* polyethylene terephthalate; aromatic copolyesters

polyether: *see* poly(phenylene oxide), ethers

poly(etheretherketone): PEEK; [–O–C$_6$H$_4$–C[O]–C$_6$H$_4$–O–C$_6$H$_4$–C[CH$_3$]$_2$–C$_6$H$_4$–]$_n$; high temperature-resistant copolymer of diphenyl ether and diphenyl ketone; can be melt-processed; use: for high-performance applications

polyethylene: [–CH$_2$CH$_2$–]$_n$; polythene; thermoplastic polymer; produced (1) under high pressure by free radical polymerization of ethylene; (2) at medium pressure by precipitation in the presence of metal oxide catalysts; (3) at low to medium pressure by precipitation in the presence of Ziegler catalysts; chemically resistant with good electrical properties and high impact strength; use: for films, coatings, flexible containers; *see* gel spinning; low and high density polyethylenes

polyethylene glycol: PEG; *see* poly(oxyethylene); nonionic surfactants

polyethylene glycol esters: RCOO(–CH$_2$–CH$_2$–O–)$_n$H; anionic detergents; prepared by reacting fatty acids such as lauric, palmitic, stearic, or oleic acids with ethylene oxide or by esterification of fatty acids with polyethylene glycols

poly(ethylene terephthalate): PET; polyester produced by melt condensation of dimethylterephthalate with ethylene glycol or from biethyleneglycol terephthalate; it is melt extruded to fibers or films; fibers are spun and drawn or spin oriented; used as endless yarns after finishing treatments (voluminizing, crimping, and heat setting) or cut into staple for blends with natural and synthetic staple fibers and fabrics; industrial PET fibers require several drawing steps for optimum strength properties; spin-oriented yarns of both POY (partially oriented yarns) and HOY (highly oriented yarns) types can be finished directly for textile applications; fibers are dyed with disperse dyes; films are extruded as wide sheets, are biaxially oriented parallel and perpendicular to the extrusion direction and heat stabilized; they have high tensile and impact strength, high stiffness and high flex life and toughness; use: in high-performance markets: magnetic recording tape, photographic film, microfilm, electrical insulation, also in blow molding of soft drink bottles; *see* polyesters

POLY(ETHYLENE TEREPHTHALATE): PET

poly(hexamethylene adipamide): *see* nylon 6,6

polyimides: PI; produced by condensation of an aromatic dianhydride with an aliphatic or aromatic diamine initially to give a soluble polyamic acid, the polyimide forms on further condensation; the imides are stable at 300°C and above, have high flame, radiation and oxidation resistance, and good electrical qualities; use: structural components in aircraft when reinforced with glass or boron, adhesives for metal surfaces, films, coatings, circuit boards

polyisobutylene: *see* isobutylene; elastomer

polyisoprene: *see* isoprene; elastomers

polymer: high molecular weight substance consisting of low molecular weight repeating units (monomers)

polymer block: portion of a polymer molecule comprising many constitutional units that has at least one constitutional or configurational feature not present in the adjacent portion (IUPAC definition); *see* block polymer; copolymer

polymer structure: chemical and physical structures of polymers; includes the molecular and steric structure as well as the arrangement of the chains in crystalline and amorphous areas; *see* morphology

polymeric methylene diphenylene isocyanate: PMDI; *see* urea resins

polymeric modifier: polymer blended with another polymer to modify its properties; *see* blends

polymerization: chemical reaction to link low molecular weight molecules (monomers) to give high molecular weight molecules (polymers); two major types of polymerization mechanisms can be distinguished: step or condensation polymerization and addition or chain polymerization with an active center (free radical); step polymerization is analogous to simple organic reactions like esterification, amide-, urethane-, or ether-formation, but polymer formation requires two or multifunctional molecules; chain polymerization needs a free radical to initiate the polymerization; *see* condensation polymerization, free radical polymerization, ionic polymerization, "living" polymers, stereoregular polymerization; initiators, catalysts, redox systems

polymerization methods: *see* addition polymerization, bulk polymerization, condensation polymerization, suspension polymerization, emulsion polymerization, interfacial polymerization, solution polymerization, thermal polymerization

polymerization number: average number of monomers in a polymer

polymers, inorganic: polymers without carbon backbone, e.g., ceramics, silica, silicates; silicone resins and elastomers (SI), polyphosphazenes; *see* fibers, inorganic; silicones

polymers, organic: polymers having a carbon backbone; *see* the individual names of polymers

polymethine dyes: coloring materials whose chromophores are conjugated $[-C=C-C=C-]_n$ or $[-C=N-C=]_n$ systems; conjugated carbon systems need at least 11 double bonds to absorb in the visible, whereas those containing heteroatoms such as nitrogen or oxygen absorb at longer wavelengths, therefore, shorter heteroatomic systems are colored; polymethine dyes include polyenes such as carotenes and carotenoids, di-and triarylmethine dyes such as phenolphthalein, azo analogs such as methylene blue, which has a phenothiazine structure, and metal phthalocyanine complexes

poly(methylmethacrylate): "plexiglass";

[$-CH_2-C(CH_3)(COOCH_3)-]_n$; a thermoplastic polymer; produced by (1) polymerization in suspension; (2) casting sheets in bulk between glass layers; (3) bulk polymerization and conversion to pellets for molding; transparent, rigid, with good impact strength; use: moldings, rods, tubes, heavy sheets, light fixtures, automobile lenses, furniture

polymorph: one of two or more geometric crystalline forms having the same composition

polynorbornene: prepared by catalytic ($RuCl_3$) ring opening polymerization of norbornene (CdF Chimie); use: vibration damping material

trans POLYNORBORNENE

polyoctenamer: [$-(CH_2)_6-CH=CH-]_n$; the *trans*-isomer is prepared by catalytic ring opening polymerization of cyclooctene (Hüls); use: plasticizer for rubber

polyolefins: thermoplastic hydrocarbon polymers produced from olefins in the presence of Ziegler–Natta type catalysts, e.g., polyethylene ([$-CH_2CH_2-]_n$), polypropylene ([$-CH(CH_3)-CH_2)-]_n$), poly(4-methylpentene) ([$-CH_2-CH(CH_2CH(CH_3)_2)-]_n$; these polymers are stereospecific and crystalline; *see* polyethylene; polypropylene; poly-4-methylpentene

polyols: substances that contain many hydroxyl groups, e.g., polyhydric alcohols, sugars

polyorganosiloxanes: *see* silicones

polyoxetane: thermoplastic polymer; produced by catalytic polymerization of 3,3-*bis*(chloromethyl) oxetane; the oxetane monomer is synthesized from pentaerythritol; use: for components where resistance to chemicals and high temperature is needed

poly(oxybutylene): *see* polytetramethylene ether glycol

poly(oxyethylene): $HO[-(CH_2)_2-O-(CH_2)_2-]_n$ OH; polyethylene glycol; PEG; produced by condensation polymerization of ethylene glycol or by addition polymerization of ethylene oxide; average molecular weight 200–6000; serves as the water-soluble terminal of nonionic surfactants; use: as chemical intermediate, in detergents, plasticizers, lubricants, cosmetics, solvents, feed additive

poly(oxymethylene): [$-CH_2-O-]_n$; POM; thermoplastic polymer produced by condensation of formaldehyde or trioxane in ionic medium; tough material with good abrasion and chemical resistance and good dimensional stability; the polymer has to be stabilized against thermal degradation by end-group substitution or by addition of small amounts of comonomer (ethylene oxide)

poly(oxypropylene): [$-CH(CH)_3CH_2-O-]_n$; used as the water-soluble terminal in nonionic surfactants; slightly less water soluble than polyoxyethylene

poly(*m*-phenylene-isophthalamide): MPD-I, Nomex; *see* aramids

poly(phenylene oxide): ([$CH_3]_2C_6H_3O)_n$; PPO; a polyether; produced by catalytic (copper-amine complex) polymerization of 2,6-dimethylphenol; the polymer has excellent dimensional stability at elevated temperature, good electrical properties, low moisture absorption, and is self-extinguishing in flame tests; use: an engineering thermoplastic for pumps, valves, and electric system parts where high temperature moisture resistance is required, for electronic systems; used with an added flame retardant for motor housings, appliances, business machines; also used in a blend with impact polystyrene

poly(phenylenesulfide): ($-C_6H_4-S-)_n$; PPS; produced from sodium sulfide and *p*-dichlorobenzene at elevated temperatures in water-free polar solvents; the polymer has a glass transition temperature of 85°C, is injection molded into electrical, electronic, and mechanical parts; curing at high temperature in the presence of oxygen causes a change of properties due to oxidation, increase of molecular weight and cross-linking; use: for coatings by a slurry coating process, as matrix material for composites

poly(*p*-phenylene terephthalamide): PPD-T; Kevlar (DuPont), Twaron (Akzo); *see* aramids

polyphosphates: salts of polyphosphoric acid; produced by heating orthophosphates to eliminate water; use: chelating agent for metal ions in water

polyphosphazenes: *see* inorganic polymers

polyphosphoric acid: polymers with a backbone of $(-P-O)_n$ having the general formula $H_{n+2}P_nO_{3n+1}$ where $n > 1$; derived by heating phosphoric acid to >250°C to expel water; use: chemical reagent

polypropylene: isotactic crystalline polypropylene is produced by Ziegler–Natta catalyst; fibers and films are produced by melt extrusion; unaffected by microorganisms; use: for indoor–outdoor carpets, outdoor furniture, nonwoven materials, microporous materials; *see* "hard elastic" materials

poly(propyleneglycol): *see* polyoxypropylene

polyquinoxalines: polymers with superior heat stability; *see* ladder polymers

POLYQUINOXALINES

polysilane: Si_nH_{2n+2} where $n = 1–6$; silicon hydrides; produced by acid hydrolysis of $MgSi_2$; they are spontaneously flammable in air; *see* silanes

polysiloxane: *see* silicones

polystyrene: PS; $(-CH[C_6H_5]CH_2)_n$; thermoplastic polymer; produced by bulk, suspension, or emulsion polymerization; transparent, rigid material with good electrical properties and molding speeds; used for moldings, records, and rigid foams, copolymers with butadiene, acrylonitrile; *see* acrylonitrile-butadiene-styrene copolymers; styrene polymers

poly(styrene-butadiene) polyblends: thermoplastic used for moldings and packaging

polysulfides: vulcanizing agents for rubber, e.g., alkylphenol disulfides, tetramethylthiuram disulfide, aliphatic disulfides

polysulfones: $(-CH[R]CH_2-S[O_2]-)_n$; thermoplastics; the aliphatic polymers are produced by copolymerization of olefinic compounds and sulfur dioxide; the sulfones with two carbon atoms between the sulfone groups are alkali sensitive and thermally unstable, this instability is used in computer chip manufacture; the aromatic polymers $(Ar-SO_2-Ar-O-)_n$ (Ar with *p*-substitution) are produced by condensation of aromatic sulfonyl chlorides using $AlCl_3$ as catalyst; the aromatic polymers have high heat distortion resistance and are self-extinguishing in flame tests; use: engineering plastics

poly(tetrafluoroethylene): $-(CF_2CF_2-)_n$, PTFE; Teflon; produced by emulsion polymerization; applied by first compressing at room temperature and then fusing under high pressure at 450–500°F; insoluble, high melting, chemically inert polymer; used for gaskets, packaging, tubing, corrosion resistant coatings; also *see* fluorocarbon polymers

poly(tetramethylene ether glycol): $HO([CH_2]_4O-)_nOH$, produced from tetrahydrofuran by catalytic ring opening polymerization; used for high quality polyurethane block copolymers, e.g., spandex elastic fiber

polythene: *see* polyethylene

polythiazoles: low molecular weight aromatic polymer consisting of alternating benzene and thiazole rings; heat stable

polyurethane: (–RNH–C[O]OR'–OC[O] NH–)$_n$; thermoplastic and thermosetting polymers; produced by condensation polymerization of diisocyanates with diols or polyols; use: rubbers with high abrasion resistance, foams, binders for paints to provide flexibility, biomaterials; also as block copolymers of diisocyanate glycol blocks with poly(oxybutylene) or aliphatic polyester blocks for elastic fibers (spandex); *see* urethane block copolymers

poly(vinyl acetals): tough flexible thermoplastic and thermosetting polymers with good adhesion; synthesized from aldehydes and polyvinyl alcohol; use: surface coatings for metals, wood; plasticized poly(vinylbutyral) sheets are the interlayer in safety glass laminates

poly(vinyl acetate): PVAc; (–CH$_2$CHOC[O] CH$_3$–)$_n$; vinyl acetate is primarily polymerized by free radical emulsion method; use: water-resistant paper adhesives, binders in coatings for paper and paperboard, latex paints are excellent as interior paints, used also in exterior paints; dry resins are used in solvent adhesives for technical applications such as brushing, roller coating, spraying; PVAc emulsions are widely applied as textile finishes for natural and synthetic fibers and nonwoven fabrics due to good adhesion and low cost

poly(vinyl alcohol): PVA; (–CH$_2$CHOH–)$_n$; produced from poly(vinyl acetate) by hydrolysis; used in manufacture of dispersions for paints and binders, raw material for paints, wet strength papers, PVA fibers after insolubilization by partial reaction with formaldehyde (acetal formation)

poly(vinyl butyral): thermoplastic polymer; good resistance to UV light; used as layer in safety glass; *see* poly(vinyl acetals)

$$-(CH_2CHCH_2CH)_n$$

POLY(VINYL BUTYRAL)

poly(vinyl chloride): PVC; –(CH$_2$CHCl)$_n$; thermoplastic polymer; produced by suspension, precipitation, or emulsion polymerization; rigid material with good electrical properties and flame and chemical resistance; can be made flexible with plasticizers and heat resistant with stabilizers; use: copolymers with vinylacetate, acrylonitrile, vinylidenechloride, with plasticizers and stabilizers in films, sheets, floor covering, rainwear, handbags, wire coatings, moldings

poly(vinyl formal): used as insulating material

poly(vinylidene chloride): thermoplastic polymer; produced by suspension or emulsion polymerization; chemically resistant material with low vapor transmission; self-extinguishing flame resistance; used as copolymer with vinyl chloride, as film and in pipe lining

pontianak: natural resin

popcorn: in polymer technology, hard, tough, insoluble polymer particles formed by continuing polymerization of dienes inside fractionation columns, storage containers, or other parts of the plant; addition of NO can prevent this undesireable side reaction

porcelain: hard, fine-grained, vitreous ceramic having a high kaolin content

porcelain clay: *see* kaolin

pore: small opening or channel in a consolidated solid

porogen: substance that promotes the production of pores within a polymer matrix; a blowing agent

poromeric: ability of a material to transmit some water vapor but not liquid water

porosity: percent volume of air or void in a consolidated solid

porphyrins: group of biologically active metal containing compounds; substances include chlorophyll (Mg) and the nonprotein components of hemoglobin (Fe) and cytochrome (Cu); *see* phthalocyanines

Portland cement: hydraulic cement made of impure and variable calcium aluminum silicate; use: for construction, as binder for paints; *see* cement

positive dyes: basic or cationic dyes

POT: publicly owned (waste) treatment works

potash: K_2CO_3, K_2O, KOH, KCl, or K_2SO_4; use: in fertilizers, glass, pottery, chemicals

potash-lime glass: *see* glass

potassium nitrate: KNO_3; niter, nitre, saltpeter produced by reaction of KCl with HNO_3 or by metathesis with other nitrates; use: in explosives, fireworks, fertilizers where chloride must be avoided e.g., growing tobacco, and steel manufacture, as oxidizer in rocket fuel

potassium persulfate: $K_2S_2O_8$; water-soluble initiator for emulsion polymerization and copolymerization; *see* emulsion polymerization

potassium phosphate: tetrapotassium diphosphate, $K_4P_2O_7$; use: in liquid cleaners due to its high water solubility; similar to sodium tripolyphosphate, use restricted in United States for ecological reasons

pot spinning: *see* package spinning; viscose

pour-point depressants: viscosity index improvers, VI improvers; motor oil additives are needed because paraffin wax present in motor oil has undesirably high pour points and interferes with oil flow in cold weather;

the more highly branched the paraffin the higher the pour point; VI improvers modify the crystal structure of paraffins to improve flow; they are polymers that control viscosity across a wide temperature range; examples: polymethacrylates, ethylene-propylene copolymer (EP), ethylene-propylene-diene terpolymer (EPDM), and hydrogenated styrene-diene copolymers

power density: in batteries, the ratio of power available to the weight (watt/kg) or volume (watt/L) of the battery

power factor: in an AC electric circuit the ratio of actual power to apparently used power; *see* dissipation factor

POY: partially oriented yarn; *see* poly(ethylene terephthalate)

prazosine: blood pressure lowering drug that acts by dilating blood vessels; *see* antihypertensive agents

PRAZOSINE

precipitation polymerization: solvent–nonsolvent polymerization; polymerization of neat monomer results in polymer precipitate; used for polymers that are not soluble in their monomers, e.g., acrylonitrile, chlorotrifluoroethylene, ethylene, vinyl chloride, and vinylidene chloride; solvents are sometimes added to modify polymer solubility; this process results in high molecular weight polymers

prednisolone: *see* prednisone

prednisone: corticoid antiinflammatory agent; steroid derivative; used against certain forms of leukemia; the related drug, prednisolone, has an OH rather than a keto group on position 11; *see* steroids

preform: a compressed shaped plastic material prepared before molding

pregel: unintentional layer of cured resin on part of the surface of a reinforced plastic

prepolymer: short chain polymer or oligomer, capable of further reaction

prepreg: in composite manufacture an intermediate product of fine fiber bundles (e.g., fiberglass, carbon fiber, boron nitride, aramids or other reinforcing materia) embedded in semiprocessed plastic sheets or ribbons; these sheets are subsequently arranged and molded to give the desired composite structure; brittle fine fibers (e.g., carbon or graphitic fibers, fiberglass) are often transported as prepregs to prevent breakage and loss of mechanical properties

press-cake: paste containing 20–60% solids obtained from filter press

pressure-jet dyeing: *see* jet dyeing

pressure-sensitive adhesives: *see* adhesives

pressurized water reactor: PWR; nuclear electricity-generating plant that uses pressurized water as coolant, e.g., the 1957 Shippingport plant in Pennsylvania

pretreatment: in waste treatment any reduction in pollution load before waste stream enters the sewer system or before it is delivered to the treatment plant

prill: small round aggregates formed by forcing melted material through a nozzle and cooling to form granules; fertilizers are often applied in the forms of prills

primary batteries: *see* batteries

primary creep: in textile technology, the recoverable portion of creep

primary explosives: detonators; compounds that develop detonation waves in extremely short periods of time after ignition, e.g., lead azide

primary oil recovery: *see* petroleum

primary plasticizer: *see* plasticizers

primary treatment: in waste water treatment the first stage in which floating and settleable solids are removed by sedimentation or screening

printing: there are four distinct impact printing processes (1) from raised type (relief, letterpress, typographic, flexographic); (2) from planar surfaces (planographic, lithographic, off-set); (3) from recessed or engraved type (intaglio, rotogravure, gravure); and (4) through a stencil (porous, silk or wire screen); each process places special requirements on the printing ink; for modern methods of non-impact printing: *see* electrography; ink jet printing; laser xerography; thermal transfer printing; xerography

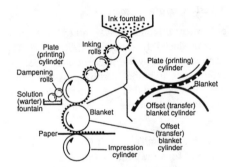

OFFSET PLANOGRAPHIC
(LITHOGRAPHIC) PRINTING
[Copyright *Riegel's Handbook of Industrial Chemistry*,
8th ed., Van Nostrand-Reinhold Co., Inc.]

printing inks: fluids or pastes consisting of pigment or dye that is dispersed in a vehicle; these formulations can be printed on a solid substrate such as paper, metal, plastic, or ceramic; printing inks are classified according to usage or mode of application; *see* printing

process development unit: *see* pilot plant

processing oils: in plastics technology, mineral oils used to improve processing conditions by facilitating dispersion of fillers, softening the compounded polymers, and acting as diluents

prochloraz: *see* fungicides

producer gas: low Btu gas; mixture of CO and H_2 obtained by reacting coal with steam and air; use: for steam and power generation

progesterone: progestin; female sex hormone that suppresses ovulation; *see* steroids

Progil process: manufacture of allyl alcohol by isomerization of propylene oxide using Li_3PO_4 at 280°C

prolamines: vegetable proteins such as zein (corn), gliadin (wheat)

proline: *see* amino acids

PROLINE

promazine: tranquilizer; *see* phenothiazines

promethazine: phenothiazine antihistamine

prometrone: substituted triazine; preemergence and postemergence nonspecific herbicide for noncrop land

promoter: weak catalyst; substance that increases activity of catalyst (e.g., the decomposition of peroxide in thermosetting plastics); activator for blowing agents

propagation: term used in polymerization kinetics to describe the mechanism of the chain growing steps.

propane: C_3H_8; gaseous hydrocarbon obtained from natural gas; use: fuel (bottled gas, cooking stoves), chemical synthesis; highly flammable

propane diol: *see* propylene glycol

propanil: *see* herbicides

propanoic acid: *see* propionic acid

propargyl alcohol: $HC\equiv CCH_2OH$; pro-

duced as intermediate product in the manufacture of 1,4-butane diol from acetylene and formaldehyde; use: as chemical raw material, to prevent embrittlement and corrosion in steel manufacture

propazine: substituted triazine; herbicide against broadleaf and grassy weeds

propellant explosive: low explosive; deflagrating explosive; explosive with low burning rate and relatively low peak pressure, e.g., black powder

propene: *see* propylene

propene nitrile: *see* acrylonitrile

propenol: *see* allyl alcohol

propham: carbamate herbicide

propiolactone: $C_3H_4O_2$; chemical intermediate; *see* ketene

propionic acid: CH_3CH_2COOH; propanoic acid; produced (1) by hydrocarboxylation from ethylene (*see* Reppe reaction) at 270–320°C and 200–240 bar using nickel propionate catalyst; (2) as by-product of butane oxidation; (3) by fermentation of glucose; use: food preservative; propionate esters are used as solvents

Propionobacter shermanii: microorganism used for the production of propionic acid from glucose

propoxyphene: $(CH_3)_2NCH_2CH(CH_3)C$ $(OOCCH_2CH_3)(C_6H_5)CH_2C_6H_5$; the *d*-isomer is an analgesic with less addiction potential than morphine but abuse leads to addiction; the *l*-isomer (levopropoxyphene) is an anticough drug

proppant: in petroleum recovery, hard, high-strength beads that are forced into the cracks and fissures of oil and gas wells to keep the fractures from collapsing, thus increasing the flow of oil and gas

propranolol: a β-blocker; use: blood pressure lowering agent, antiangina medication

237

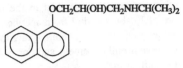

OCH$_2$CH(OH)CH$_2$NHCH(CH$_3$)$_2$

PROPRANOLOL

propylamide ammonium betaine: amphoteric surfactant; *see* betaine

propylene: propene; CH$_3$CH=CH$_2$; produced by thermal cracking of petroleum stock (natural gas, refinery gas, or crude oil); the cracked C3 fraction also contains methylacetylene and allene that are hydrogenated into propene and propane; use: raw material for manufacture of acrolein, acrylic acid and acrylonitrile, cumene, polypropylene, propylene oxide, isopropanol, hydroformylation products, acetone, allylic compounds; propene is not used as fuel because it tends to polymerize and form gums

propylene chlorohydrin: CH$_3$CH(OH)CH$_2$Cl; produced by reaction of propylene and HOCl; intermediate in the manufacture of propylene oxide

propylene glycol: CH$_3$CHOHCH$_2$OH; propanediol; 1,2-propylene glycol; produced by hydration of propylene oxide; use: solvent, antioxidant, food preservative, antifreeze, deicing agent; intermediate for polymer synthesis (polypropylene glycol, polyurethanes, cellophane)

propylene oxide: PO; produced by treating propylene chlorohydrin with milk of lime [Ca(OH)$_2$] at room temperature; by direct liquid-phase oxidation of propylene with hydroperoxide complex (*see* oxirane process); use: solvent; important intermediate for manufacture of propylene glycol, dipropylene glycol, polyethers, polypropylene glycol, polyurethane, surfactants, acetone, allyl compounds

CH$_3$CHCH$_2$

PROPYLENE OXIDE

prostaglandins: group of C20 fatty acids that contain a five-membered ring and modulate hormone activity in the body; the pain-reducing effect of aspirin is due to its interference of prostaglandin synthesis from arachidonic acid

protamines: low molecular weight, sulfur-free proteins

proteases: proteolytic enzymes; enzymes that hydrolyze the peptide bond (–CONH–) in proteins; produced from microorganisms by fermentation; use: in detergents to remove protein stains, in the tanning industry for removal of wool, as meat tenderizer, for the preparation of protein hydrolyzates, milk coagulation, and for cheese manufacture

protective colloids: high molecular weight water-soluble substances that locate themselves at the oil–water interface; examples: gelatin, polysaccharides, polyacrylic acid, poly(methylacrylic acid), poly(vinyl alcohol), poly(vinyl pyrrolidone), and sulfonated polystyrene

protein isolates: in the food industry, products from soybean or other sources containing ≥85% protein

protein plastics: produced from animal and plant proteins; casein plastics are tough and can be made in pastel shades; peanut fiber (made from peanut meal by forcing the dissolved protein through spinnerettes into a bath) is soft, resilient, moisture absorbent and dyes easily; fiber produced from zein (wheat) is light yellow, soft, tough, and mildew resistant

proteins: natural polymers of amino acids; use: nutrients, nutritional supplements (e.g., soybean protein, wheat gluten, whey granules, and mushroom powder), source of amino acids and amino acid derivatives (e.g., sodium monoglutamate); *see* single cell protein (SCP)

proteolytic enzymes: *see* proteases

protozoa: single-celled microorganisms

Prussian blue: Chinese blue; $Fe_4[Fe(CN)_6]_3$; blue pigment; produced by reacting ferric chloride with potassium ferrocyanide; *see* ferrocyanate pigments

prussic acid: *see* hydrocyanic acid

pseudocumene: 3,4-dimethyltoluene; produced during crude oil cracking and reforming; oxidation leads to trimellitic anhydride

Pseudomonas aeruginosa: microorganism; use: for the industrial production of pyruvic acid from glucose and for the production of salicylic acid from naphthalene

Pseudomonas fluorescens: microorganism: use: for the industrial production of 2-keto-gluconic acid from glucose

pseudoplastic liquids: liquids whose viscosity decreases suddenly with increased stirring

psi: pounds per square inch; pressure measurement

psia: pounds per square inch absolute

psig: pounds per square inch gauge

pugmill: mixer consisting of one or more rotating shafts bearing arms or blades within a drum

pullulanase: enzyme obtained from *Aerobacter aerogenes* and *Klebsiella aerogenes*; use: to increase fermentability of starch syrups and for structure determination of polysaccharides

pulp: fibrous products; natural pulp is derived from cellulosic material and is used to manufacture paper, hardboard, and fiberboard; synthetic pulp is almost always made from high-density polyethylene (HDPE) or polypropylene (PP); such synpulp is thermoplastic, chemically inert, and has good dielectric properties; in addition to the usual pulp applications synpulp is used as sealant, adhesive, spray cement, as filler in textured paints and vinyl tile

pulping: in the paper industry processes to separate wood fibers mechanically and/or chemically; in semichemical pulping wood chips are first soaked in caustic or sodium sulfite solutions and then ground in disc refiners to produce groundwood-type pulp; chemical pulping involves the solubilization and removal of the lignin portion of wood while leaving the cellulosic fibers intact; the two main chemical processes are the alkaline (sulfate or kraft) process and the sulfite (soda) process; the soda process uses NaOH, the kraft process uses NaOH and Na_2S to remove the lignin; addition of anthraquinone as pulping catalyst improves the kraft process

pultrusion: in fiber technology a process in which continuous filaments are coated with liquid resin before being extruded and cured; reinforcement materials include glass, alumina, aramid, and graphite; the product has high unidirectional strength and can be used for structural materials such as I-beams, shafts, ladders, tent frames, rails and rods

pumice: porous igneous rock; use: abrasive, filler for plastics

Purex process: in nuclear technology, solvent–extraction of spent nuclear reactor fuel with nitric acid; uranium, plutonium, and residual radioactive materials are separated into three streams, the recovered U-235 and Pu are recycled and the waste is dehydrated and stored or used for isotope recovery

Purisol process: in petroleum technology, removal of H_2S from synthesis gas using N-methylpyrrolidone

PWB: in computer technology, printing wiring boards

pycnometer: device to measure the density of liquids

pyranose ring: six-membered cyclic form of sugars

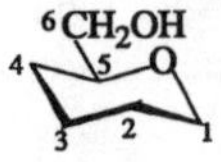

PYRANOSE RING

pyrazophos: *see* fungicides

pyrethrin: natural insecticide isolated from the pyrethrum flower; use: knock-down agent for indoor use

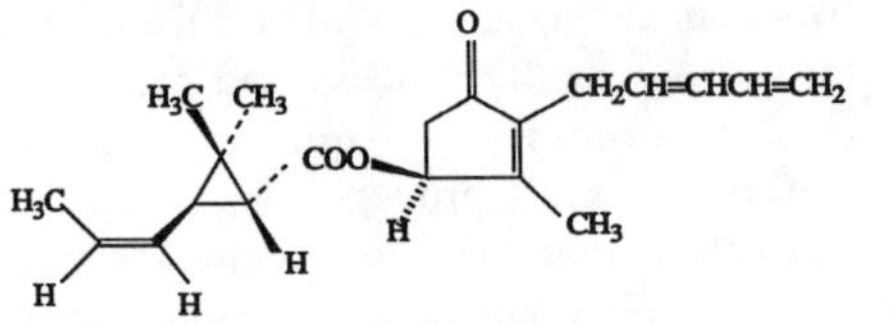

PYRETHRIN

pyrethroids: synthetic insecticides that are analogs of pyrethrin

pyridine: isolated from coal tar; use: in chemical synthesis

PYRIDINE

pyrite: *see* iron pyrite

pyrogallol: 1,2,3-trihydroxybenzene; *see* tannins

pyrogen: fever-inducing agent

pyrolysis: destructive distillation; reaction induced by heating in the absence of oxygen; *see* cracking

pyrolytic graphite: *see* graphite

pyromellitic anhydride: *see* 1,2,4,5 benzenetetracarboxylic acid

pyrometer: device for measuring heat

pyrophosphoric acid: $H_4P_2O_7$; diphosphoric acid; manufactured by heating phosphoric acid to 250–260°C; at higher temperatures trimers and tetramers of phosphoric acid are obtained; use: catalyst, raw material for manufacture of phosphates

Pyrotol process: catalytic hydrodealkylation of aromatic hydrocarbons

pyroxylin: collodion cotton; soluble guncotton; nitrocellulose rayon; mixture of cellulose nitrate; flammable, but less explosive than guncotton; use: for coating fabrics and for the manufacture of explosives, artificial leather and lacquers; *see* dope

2-pyrrolidone: produced by reaction of γ-butyrolactone and ammonia; use: manufacture of N-vinylpyrrolidone for special polymers

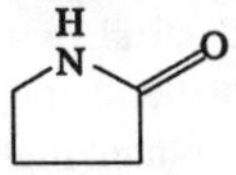

2-PYRROLIDONE

Q

quad: unit of energy, 10^{15} Btu

quality assurance: *see* quality control

quality control: independent section within a manufacturing unit that ensures the required standards of safety, purity and effectiveness of the products

quartz: one of several crystalline forms of silica (SiO_2)

quaternary ammonium compounds: "quat"; *see* cationic surfactants

quaterpolymer: polymer derived from four monomers (IUPAC)

quebracho: South American tree that is the main source of tannins

quench: cool rapidly to retain some high-temperature properties

quicklime: *see* calcium oxide

quick-setting inks: inks that dry by absorption and coagulation of the solvent (vehicle); after printing the special resin-oil combination separates into a solid dry field that remains on the surface and an oil that penetrates the paper

quinine: antimalaria drug; *see* alkaloids

quinoline: *see* antiozonants

quinolones: family of pharmaceuticals active against gram-negative microorganisms

QUINOLONES

quinones: in polymer technology used as inhibitors and chain terminators

R

R: symbol for alkyl group

R acid: 3,6-di-SO_3H-β-naphthol; raw material for dye manufacture

rad: unit of energy absorbed by ionizing radiation; 1 rad = 100 erg/g

radial tires: tires in which plies of textiles are placed at right angles to the direction of motion; *see* tires

radiation curing: ionizing radiation of α, β or γ rays, X-rays, high energy electrons, or UV radiation can produce cross-linking reactions in polymers; as chemical bonds are broken by radiation of ~100 eV or less the energy is adequate to form free radicals; visible light is effective when photosensitive materials are present; the free radical causes degradation or cross-linking; polymers which have no tertiary hydrogens or with halogens degrade, e.g., methacrylates, polyisobutylene, poly(vinyl chloride), poly(vinylidene chloride), poly(tetrafluoroethylene); polyethylene, other olefin polymers, acrylates, and

rubber cross-link; curing yields mechanically and thermally improved polyethylene

radiation polymerization: polymerization initiated by radiation rather than by chemical means; *see* radiation curing

radical: *see* free radical

radio frequency: electromagnetic waves with a frequency between 1.4×10^4 and 10^{11} cycles/sec; use: supply of heat for welding thermoplastics (high-frequency welding); modulates the external magnetic field in NMR spectroscopy

radioimmunoassay: RIA; *see* immunoassays

radiolysis: decomposition caused by high energy radiation; for example, γ radiation of water produces hydroxyl radicals and hydrogen peroxide; in chosing materials and sites for nuclear waste disposal, radiolysis of containment materials must be considered

raffinate: during extractive oil purification, the undissolved oil fraction

raffinose: trisaccharide consisting of fructose, glucose and galactose; isolated from sugar beets and cottonseed

Raleigh scattering: *see* Raman spectroscopy

ram: in molding technology, the press member that enters the cavity block to exert pressure on the molding compound

Raman spectroscopy: methodology used for structure analysis; monochromatic radiation of a laser beam in the visible range brings molecules into excited states; the molecules re-emit the same frequency (Raleigh scattering) except for a few whose emission is modified by the vibrational or rotational transitions frequencies of the excited molecules; the differences between the laser frequency and the emitted frequencies are the Raman spectrum of that molecule; the information is similar to that gained by infrared (I.R.) and the two spectroscopies are complementary; in contrast to I.R., lasers run in the visible light range and the method has no problem with water solutions; resonance Raman spectroscopy permits the study of vibrations of specific areas in complex molecules; fluorescence is a problem with some samples; the Raman Fourier transform method is being developed

ram mixer: machine with a closed mixing chamber in which ingredients are mixed by pressing down in the mixing chamber with a ram, e.g., Banbury internal mixer

ramie: natural fiber; *see* fibers, organic

random copolymer: *see* copolymerization

Raney nickel: hydrogenation catalyst; produced from nickel–aluminum alloy by leaching the aluminum with strong base

ranitidine: H_2 blocker; antiulcer drug

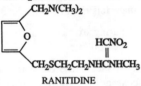

RANITIDINE

Rankine: R; absolute temperature scale based on °F; to convert add 460° to °F

rapeseed oil: vegetable oil derived from the rape plant which grows in northern climates; new hybrids have been bred to produce oil that contains large amounts of oleic acid and lesser amounts of the C22 monounsaturated erucic acid that leads to heart lesions; use: salad oil, margarine manufacture, soft soaps

Raschig rings: packing material for distilling towers

rauwolfia: *see* reserpine

rayon: artificial silk; wood silk; regenerated cellulose; *see* viscose

reaction injection molding: RIM; liquid injection molding (LIM), liquid reaction molding (LRM), high-pressure injection molding

(HPIM); used for production of thermosetting polymers; for example, when a mixture of isocyanate and polyol is injected into a closed mold with a high-pressure impingement mixer the monomers copolymerize to form polyurethane; RIM is used for production of plastic parts such as automotive body panels and bumpers

reaction spinning: chemical spinning; chemical wet spinning; process in which a prepolymer is extruded into a bath where reaction occurs, e.g., the prepolymer poly(tetramethyleneoxide) is tipped at its ends with diisocyanates and is extruded into a bath containing diamines or triamines; a rapid reaction leads to the immediate formation of a linear or cross-linked block elastomer of the spandex type

reactive dyes: dyes that form covalent bonds with the hydroxyl or amino groups of fibers; for example, in triazine dyes the cyanuric chloride moiety is attached to chromophoric groups and that chlorine reacts with the OH groups of cellulose when the dye is applied in the presence of base; basic or acidic dye stuffs react with synthetic fibers containing acidic or basic end groups

ream: in paper manufacture, 500 sheets; in plastics technology, layers of unhomogeneous material that are parallel to the surface of a transparent or translucent plastic

Rebuttable Presumption Against Registration: RPAR; process by which the EPA cancels registration of hazardous substances; this process has been used against some chlorinated pesticides such as chlordane and heptachlor

receptor: in biotechnology, the molecular site in the body to which active materials must bind to exert their effect; *see* antagonists

reciprocating screw injection molding: plastic material is advanced by means of the screw as usual; but unlike in conventional extrusion the screw rotation is stopped when an injection shot is made and the material is pushed into the mold using a hydraulic cylinder; *see* injection molding, thermoplastics

recombinant DNA: piece of DNA that contains sequences from more than one source; produced by splicing a DNA molecule with an enzyme (which is called restriction enzyme or endonuclease) and inserting another DNA sequence

Rectisol process: purification of synthesis gas by washing with pressurized methanol

recycle fermentor: continuous single-stage fermentor in which the biomass is recycled to increase unit productivity

recycling solid waste: the limited availability of landfill space has increased interest in postconsumer recycling of metals, plastics and paper; steps involved are collection, sortation, and reclamation; metals are separated from waste streams electrostatically; paper is separated at the source and thus does not enter the waste stream; most "as is" recycled paper is of poor quality, high quality recycled paper is produced by treating waste paper with NaOH and steam followed by bleach; recycled mixed plastics are used for products that do not depend on color, clarity, or strength, e.g., strapping or carpet backing; higher value recycled products require sorting the plastics in the waste stream using differences such as density, physical properties, solubility, or light sensitivity

red lead: Pb_2O_3; made by calcining litharge in air at ~640°F; use: pigment, paints

red liquor: *see* aluminum acetate

red oil: impure oleic acid recovered as by-product during stearic acid manufacture

redox: abbreviation for oxidation–reduction; metal oxides often serve as redox catalysts or redox reagents; in polymer technology redox systems may serve as initiators, e.g. hydrogen peroxide with ferrous ions (Fe^{++}) or

potassium persulfate with thiosulfate; *see* emulsion polymerization

redrawn inviscid fiber spinning: RIMS; technique used to boost modulus of ceramic fibers; *see* inviscid melt spinning

reduced crude: *see* residuals

reducing sugar: sugars that reduce copper or silver salts in alkaline solution, e.g., dextrose, sucrose

refiner's blackstrap: molasses; in sugar refining the residual syrup left after crystalline sugar from concentrated affinity syrup is returned to refinery; *see* sugar refining

refinery blending stock: butane fraction obtained in natural gas processing plants

refinery sour gas: gas containing large amounts of sulfur; use: source of elemental sulfur

refining: purification of products by various methods of separation; frequently both the desired product and the removed materials are of value so that the separation leads to further purifications; *see* petroleum; sugar refining; vegetable oils

reflectivity: light reflected from a surface/total incident light; coefficient of reflection; reflectivity varies with wavelength and angle of incidence; mirrors are surfaces that have perfect reflection, they are used in spectroscopes (e.g., IR, UV, visible, lasers) to focus light beams; *see* gloss

reformate: product of petroleum reforming process; *see* reforming

reforming: process for upgrading gasoline by catalytic cracking; the product contains increased amounts of polyolefins and aromatics; typical catalysts are Pt or Re on alumina; process is the major source of BTX (benzene, toluene, and xylene)

refractive index: measure of the ratio of light velocity in vacuum to light velocity in a transparent medium; expressed as the ratio of the sine of the angle of incidence/sine of the angle of refraction; refractive index varies with wavelength, temperature, and pressure

refractories: materials with low thermal conductivities; that can withstand high temperatures (1650–2200°C); examples of refractory materials are aluminum silicates (fire clay), alumina, carbon, dolomite (CaO-MgO), zirconia and silicon carbide; primarily used as lining in furnaces and ovens

refrigerant 12: *see* dichlorodifluoromethane; chlorofluorocarbons

regenerated cellulose: polymer made by treating native cellulose with NaOH and carbon disulfide to make the xanthate, dissolving it in dilute NaOH to make viscose and then regenerating the cellulose with acid; *see* viscose; cellophane

regenerated protein: proteins from milk, soy bean, corn, or peanuts that have been denatured by strong alkali, regenerated with acid and spun; the products are low strength, soft fibers that are easily dyed

regular polymer: polymer made up of one monomer in one configurational arrangement; a linear polymer

reinforced plastic: RP; *see* reinforcements

reinforced thermoplastic: RTP; *see* reinforcements

reinforcements: in plastics technology, fillers that improve tensile and flexural strength; reinforcements are strong, inert materials such as fibers of asbestos, boron, carbon, ceramics, glass, graphite, textile fibers, chopped paper, flock, jute, macerated fabrics, sisal and synthetics; pigments such as carbon black; reinforcements must form strong adhesive bonds with the resin so that coupling agents that promote adhesion are often added to the fibers to form a strong interface

relaxation: time required for the release of stress from an applied strain; the time varies widely: relaxation for electronic polarization is ~10^{-15} sec while recovery from deformation of solids may be hours

release agents: materials that free cured products from metal molds; *see* abherents

relief printing: *see* printing

rennet: enzyme that precipitates casein from milk; obtained from stomach secretions and from microorganisms such as *Bacillus polymyxia, B. subtilis, Endothia parasitica, Mucor miehei, M. pusillus*; used in cheese manufacture

replacement drugs: *see* agonists

replicate: to make a copy, e.g., viruses replicate within cells

Reppe processes: catalyzed reaction of acetylene or olefins with CO and a nucleophile (e.g., water): (1) carbonylation: hydrocarboxylation; used industrially for the liquid-phase production of propionic acid from ethylene, CO and H_2O in the presence of nickel propionate at 270–320°C and 200–240 bar; under these conditions the catalyst, $Ni(CO)_4$ is formed in situ; compare with Koch acids; (2) hydrocarbonylation: production of butanol from propylene, CO and water in the presence of $Fe(CO)_5$ and *N*-butylpyrrolidine; hydrocarbonylation occurs via a Fe-CO-H complex; (3) carbonylation of acetylene: production of hydroquinone from acetylene, CO and H_2 in the presence of Rh and $Ru(CO)_4$ catalysts at 100–300°C and 100–350 bar; (4) carbonylation of acetylene: production of acrylic acid from acetylene, CO and water (or alcohol) using a stoichiometric amount of $Ni(CO)_4$; (5) vinylation of acetylene: liquid phase production of vinyl ethers in the presence of alkali hydroxide

research octane number: RON; measure of automotive fuel ability to prevent engine knocking based on the octane number 100 for

2,2,4-trimethylpentane; RON is measured at an engine speed of 600 rpm with an air inlet temperature of 125°F and a fixed spark advance; the motor octane number, MON, is measured at 900 rpm and 300°F and uses variable spark timing; RON runs about 10 octane numbers higher than MON

reserpine: blood pressure reducing agent; alkaloid isolated from the roots of the rauwolfia plant

RESERPINE

residual monomer: unpolymerized monomer remaining within the polymer after polymerization

residuals: reduced crude; in petroleum technology, crude oil fraction boiling above 650°F; used as industrial fuel oil

resilience: energy returned upon recovery from stress–work input to produce deformation

resin-filled paper: produced by impregnating the paper sheet or the pulp with resin; *see* beater-addition process

resin-impregnated paper: *see* wet strength resins

resins: natural resins are water-insoluble exudates from plants that on exposure to air harden initially to soft products and after long periods polymerize to solids or semisolids; use: in varnishes and lacquers; examples of natural resins: accroides, amber, Canada balsam, Congo copal, dammar, elemi, and sandarac; synthetic resins are defined by ASTM (D 883-75a) as "solid or pseudosolid materials often of high molecular weight, which exhibit a tendency to flow

under stress and usually have a softening or melting range"; in common usage the word resin is also used for uncured fluid thermosetting materials and often as a synonym for plastics and polymers

resin transfer molding: RTM; in composite manufacture, a fibrous mat is inserted into a mold and resin is injected under low pressure

resist: in computer chip technology, layers covering the silicon base; to produce the required pattern either the resist covering the pattern is removed (negative resist) or the resist covering the background is removed (positive resist); the most common method to produce the circuit pattern is photolithography: a latent image is formed by exposing the resist with UV light through a mask that contains outlines of the circuit pattern; the next steps are development, etching, and stripping; common resist materials are diazonaphthaquinone-sensitized novolac (DQN) systems that can resolve images <0.5 µm

resistance: in biotechnology the ability of microorganisms to overcome toxic effects of pesticides or drugs; *see* drug resistance

resist printing: in textile technology, the production of a pattern by protecting certain areas before placing the fabric into a dye bath; after dyeing, the protective material is washed away

resite: fully cured thermosetting resin (C-stage); *see* phenol formaldehyde resins

resole: uncured thermosetting resin before heating (A-stage); *see* phenol formaldehyde resins

resolite: partially cured thermosetting resin (B-stage); *see* phenol-formaldehyde resins

resorcinol: 1,3-dihydroxybenzene; produced from benzene via benzenedisulfonic acid; use: intermediate for pharmaceuticals and dye manufacture, UV stabilizer for polyolefins, in adhesives for steel and rubber in the form of resorcinol/formaldehyde condensates, water-stable adhesives for wood veneer in the form of polycondensates with formaldehyde

restriction enzyme: endonuclease; enzymes that cut DNA molecules at specific sites; used for genetic engineering; *see* recombinant DNA

retardants: *see* inhibitor

retinol: *see* vitamin A

retrovirus: infective agent whose single-chain RNA is synthesized into a double-chain DNA within the host cell; this DNA is then integrated into the host DNA to form a provirus so that the host cell now generates the virus

reverse electrodialysis: *see* inverse electrodialysis

reverse impact test: test in which one side of a sheet is struck and reverse side is inspected for damage

reverse osmosis: RO; a membrane process in which pressure is applied across a semipermeable membrane to force solvent to move from the more concentrated to the more dilute solution; use: for water purification in industrial processes, desalination of seawater, production of ultrapure water, color removal of effluents in the pulp and paper industry; *see* TYPICAL REVERSE OSMOSIS SYSTEM

reverse phase chromatography: *see* high performance liquid chromatography

Reynolds number: dimensionless function used in fluid flow calculations; $N_{re} = DV\rho/\mu$ where D is the inside pipe diameter, U is the average velocity of flow, ρ is the density of the fluid, and μ is the viscosity of the fluid; the critical Reynolds number corresponds to the transition from turbulent to laminar flow; for a circular pipe the Reynolds number ranges from 2000–3000

rhenania: $CaNaPO_4$; produced by calcining apatite, silica, and sodium carbonate in a rotary kiln <2300°F; use: granulated fertilizer

rheniform process: production of reformate gasoline using bimetallic platinum/rhenium catalyst; *see* reforming

rheology: study of flow

rheometer: instrument to determine flow properties; viscometer

rheopecticity: rheopexy, increase of viscosity under constant stress, viscosity decreases upon removal of stress

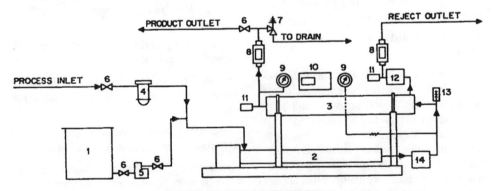

TYPICAL REVERSE OSMOSIS SYSTEM
(1) chemical feeder (day tank); (2) pump; (3) permeator; (4) micron filter; (5) pump; (6) isolating valves; (7) relief valve; (8) flow indicator; (9) pressure gage; (10) conductivity monitor; (11) conductivity cell; (12) flow regulator; (13) thermometer; (14) pressure regulator. [Courtesy D. M. Considine, P.E.]

rheopexy: *see* rheopecticity

Rhizopus delemar: microorganism used to manufacture fumaric acid from glucose

rhodamine B: xanthene dye; basic violet 10; brilliant red dye used for paper, wool, and silk; biological stain

RHODAMINE B

rhodium complexes: catalysts for numerous reactions including Fischer–Tropsch synthesis, water gas shift reaction, isomerization of alkenes, hydroformylation and hydrosylilation; examples of complexes: chlorotris(triphenylphosphine)rhodium(I) and chlorodicarbonylrhodium (I) dimer

ribbed smoke sheets: RSS; field latex that has been coagulated with acid and dried in wood smoke; the resultant brown rubber resists deterioration by mold and bacteria

riboflavin: vitamin B_2; produced from *Ashbya gossypii* by fermentation; *see* vitamin B complex

ribonucleic acid: RNA; a single stranded biological polymer consisting of a ribose-phosphate backbone and four bases (adenine, uracil, guanine, and cytosine); in the cell three different RNAs collaborate to translate the DNA genetic code to manufacture proteins from the amino acids present in the protoplasm: messenger RNAs (mRNA) transfer the information from the DNA to the protein-forming system (ribosome) in the cell, transfer RNA (tRNA) carries the amino acid to the ribosome, ribosomal RNA (rRNA) synthesizes the protein; in some viruses, i.e, retroviruses, the viral RNA carries the code (template) for manufacture of viral RNA to the host cell; there it forms viral DNA, is incorporated into the host DNA, that then manufactures new viruses; HIV is the retrovirus that causes acquired immune deficiency syndrome (AIDS)

ribosome: cell organelle in which protein synthesis takes place

ricinoleic acid: a monohydroxy, monounsaturated C16 fatty acid; produced by saponification of castor oil; use: manufacture of undecenoic acid (precursor for nylon 11 synthesis), soap, textile finishes, and dyes

Riga hydrolysis method: hydrolysis of wood using small amounts of concentrated sulfuric acid while feeding the wood through a press

RIM: reaction injection molding; *see* inviscid melt spinning

ring spinning: process step in yarn spinning from staple that continuously drafts the roving (a loose assembly of oriented and slightly twisted staple filaments), gives it a high twist (~20–30 turns/in.) and winds it on a bobbin, in the staple yarn spinning process the twist is essential to give the yarn strength and hold it together; in cellulose acetate spinning the same process is used to put a slight twist (<1 turn/in.) onto the as-extruded yarn

rinse agent: a nonionic surfactant that is used in the last dishwasher cycle to overcome spotting due to hard water

riser cracking: petroleum cracking with shortened contact time; yields gasoline with higher octane numbers (2–3 numbers higher)

RNA: *see* ribonucleic acid

mRNA: messenger RNA; *see* ribonucleic acid

tRNA: transfer RNA; *see* ribonucleic acid

roasting: heating in the presence of air, used in metals

Rockwell hardness: hardness of material measured by the difference in the depths of indentation caused by the impact of a sphere 0.25 in diameter and loads of 10–60 kg

roentgen: the quantity of X- or γ-rays that will produce 1 esu of electricity in 1 mL of dry air at 0°C and standard pressure

Roga test: see coal rank

rosin: resin obtained from pine trees; contains abietic acid; use: rosin oil, varnishes, paint driers, paper sizing, thermoplastics, linoleum, printing inks; see naval stores; abietic acid

rosin acids: acids contained in rosin, mainly abietic acid; see rosin

rosin oil: produced by dry distillation of rosin; use: plasticizer for rubber

Ross–Peakes tester: instrument for measuring the temperature at which a given amount of molding powder will flow through a standard orifice during a given time and pressure

rotary molding: process using several mold cavities mounted on a rotating table

rotational casting: process for forming hollow articles from fluids by rotating a mold so that fluid coats the inner mold walls where it is hardened by curing

rotational injection molding: modified injection molding process in which the mold is rotated during the molding cycle until the material has hardened; used for making cups and beakers

rotational molding: variation of rotational casting in which sinterable powder rather than fluid is used; the powder is sintered and fused against the mold walls

rotenone: insecticide; extracted from derris root

ROTENONE

rotogravure: impact printing using recessed or engraved type; requires inks with resin-solvent vehicles that dry by evaporation of solvent

rotomolding: see rotational casting; rotational molding

rouge: iron oxide pigment grade used for polishing glass and lenses

roving: in fiber technology, a loose, slightly twisted bundle of staple fibers, an intermediate product in yarn spinning; in plastics technology, bundles of glass fibers imbedded in composites or laminates

RP: reinforced plastic

RTM: see resin transfer molding

RTP: reinforced thermoplastics

RTV: room temperature vulcanizing

rubber-base paints: pigments with rubber-base binders that are solvent-thinned (as differentiated from latex binders); rubbers used are chlorinated rubber, styrene-acrylic, styrene-butadiene, and vinyltoluene-butadiene; use: outdoor applications

rubber hydrochloride: $(C_5H_8 \cdot HCl)_n$; nonflammable thermoplastic obtained by treating dissolved latex with anhydrous HCl; use: packaging, water-resistant textiles, filters

rubber hydrofluoride: $(C_5H_8 \cdot HF)_n$; tough thermoplastic obtained by treating rubber

248

with 50–85% HF; use: adhesive for metal/rubber bonding

rubber, natural: (India rubber, caoutchouc), *cis*-1,4-polyisoprene obtained as latex from *Hevea brasiliensis*

rubber, synthetic: *see* elastomers

rubber transition: *see* glass transition

run-of-mine coal: untreated, unsized coal as it comes from the mine

rust removers: reducing agents such as sodium hydrogen sulfite, added to laundry soap formulations

rutile: crystalline form of TiO_2; use: white pigment, raw material for production of titanium tetrachloride

S

saccharase: *see* invertase

saccharides: carbohydrates; used primarily as food and fodder; cellulose in food serves as bulk; digestible saccharides important in human nutrition are classified as single sugars (monosaccharides—glucose, fructose, galactose), double sugars (disaccharides—sucrose, maltose, lactose) and complex sugars (polysaccharides—starches, dextrin, glycogen); hemicelluloses (agar, algin, pectin) are polysaccharides made up mostly of hexoses (glucose, galactose, arabinose) and derivative acids (glucuronic acid, O-methyl-glucuronic acid, and galacturonic acid); hemicelluloses swell in water and are used as gelling agents; cellulose is used as raw material for fiber production, paper, explosives

saccharification: hydrolysis of wood resulting in the conversion of cellulose to glucose and of hemicellulose to mannose and xylose

saccharomyces: yeast strain family; *Saccharomyces cerevisiae*; use: for fermentation of glucose for manufacture of ethanol and vinegar; *S. ellipsoidens*; use: for manufacture of vinegar

safety glass: shatterproof glass, laminated glass, bulletproof glass; produced by placing a sheet of transparent resin between two sheets of plate glass and molding the three layers together under heat and pressure; resins commonly used are acrylics, polyvinyl acetate, polyvinyl butyral, and silicones

safflower oil: vegetable oil; the common variety contains 75–80% linolenic acid and 13% oleic acid; in the new variety these percentages are reversed; use: in soft margarines, salad oils, and imitation ice cream

Saflex process: production of colored thermoplastic interlayers in safety glass; it involves coextrusion of the thermoplastic and a predyed thermoplastic dye layer; extrusion temperatures depend on the plastic, e.g., for polyvinyl butyral processing temperatures are 190–205°C

safranine: azine dye; used in color photography

sagger: ceramic container that holds ware during firing in the kiln

salad oils: blends of unhydrogenated or lightly hydrogenated, winterized oils that have been refined and deodorized; an exception is virgin olive oil which is prized for its flavor

salbutamol: albuterol; antiasthma drug; acts by dilating lung passages (bronchi); unlike epinephrine it does not increase the heart rate (a β_2 agonist)

salicylic acid: *see* acetylsalicylic acid

salt: *see* sodium chloride

salt cake: impure sodium sulfate

salt index: *see* Hottenroth number

salting out: reducing the solubility of compounds by addition of salt

saltpeter: KNO_3 ore; niter; *see* potassium nitrate

samarium: Sm; rare earth element; important industrial uses for samarium or its compounds: permanent magnets ($SmCo_5$); ingredient of optical glass fibers (Sm_2O_3); for coded inks (Sm)

SAN: *see* styrene-acrylonitrile copolymer

sandarac: natural resin

sandwich: term used for laminates consisting of three layers such as in safety glass; in chemical science the term sandwich compound refers to a metallocene

sanitation: control of all environmental factors that may have deleterious effects on human health

saponification: in the soap industry, the reaction of a fat with caustic soda yielding fatty acid sodium salts and glycerine; in organic synthesis the alkaline hydrolysis of an ester

saponification value: weight in milligrams of KOH needed to completely saponify 1 g of fat; a rough indication of the molecular weight of the fat

saponified acetate yarn: steam stretched cellulose acetate regenerated to cellulose by NaOH; strong yarn (6–7 g/denier) with low elongation (~6.5%)

sapphire: gem quality corundum (Al_2O_3) with a hardness of 9 Mohs; synthetic sapphires are produced by flame-fusing pure alumina

saprophytes: plants such as fungi that live on dead organic material

sarin: isopropyl methylphosphonofluoridate; nerve gas; banned warfare agent under chemical arms accord

SASOL process: South African fixed-bed coal gasification process; *see* coal gasification

SBR: *see* styrene-butadiene rubber

scale: solid layer formed on working surfaces of reaction vessels, boilers, and pipes; caused by precipitation of material from circulating solutions (e.g., calcium and magnesium carbonate from hard water)

scale-up: increase of process dimensions, i.e, going from laboratory to pilot plant scale or from pilot plant to manufacture; in this task one must consider such effects as heat dissipation, stirring, and the relative increase in volume and surface of the reaction vessel

scanning electron microscope: SEM; high-power microscope that utilizes a high-energy electron beam and magnetic lenses for magnification; because of its high energy the electron beam does not focus on the object for a long time but sweeps (scans) across the surface permitting a 5-nm image resolution

scanning tunneling microscope: STM; technique with a resolution of about 1 Å that allows the viewing and manipulation of individual atoms; STM utilizes the ability of electrons to tunnel through energy barriers; electrons flow between the fine tip of a needle and projections of an adjacent surface; the needle is supported by piezoelectric crystals; the tip is moved by applying precise voltages to the crystals, changing their dimensions slightly to permit the atomically sharp tip to go across the energy gap between the outer electrons of the surface atoms; if the voltage is kept low the tip-to-projection distance controls the current; what is recorded is the tip motion needed to keep the tunneling current constant as the tip moves across a 10-nm square of the surface; the result is the image of the surface

scavengers: agents that remove unwanted chemicals or materials from reactions or processes, e.g., free radical scavengers in polymerizations (short-stops) and in biological systems, formaldehyde scavengers during curing of formaldehyde-derived resins, removers of contaminants from waste streams

Schaeffer's acid: $6\text{-}HO_3S\text{-}\beta\text{-naphthol}$; starting material in dye manufacture

schapping: in textile technology, removal of gums from silk waste by fermentation

Schweizer solution: in rayon manufacture, a solution of cuprammonium cellulose of known viscosity that is used to determine the average molecular weight of cellulose

scintillometer: device to detect radiation by recording the visible excitation of inorganic or organic compounds on impact by α-, β-, or γ-particles

scleroscope: *see* hardness

scorch: in rubber technology, the premature start of vulcanization of rubber latex during mixing or shaping

scratch hardness: *see* hardness; Mohs scale

screen: mesh for separating and/or grading particles of different diameter

screening: removal of fines from powdered mixtures, of solids from the waste stream before it enters the sewer treatment plant, a process during beneficiation

screen process printing: silk screen printing; *see* printing

screw-injection molding: *see* injection molding

screw melting: melt spinning process in which polymer chips are fed into an extrusion-type screw inside a heated tube

scrubber: in air pollution control, device that cleanses an air stream by spraying a liquid that absorbs or reacts chemically with the pollutant

sealant: soft substance that hardens after application to form a stable bond with the substrate; unlike adhesives sealants are used to provide load-bearing and elasticity in joints;

they are also used to reduce noise and vibration; examples of sealants include linseed oil (putty), asphalt, silicones, rubbers, urethanes, and waxes

sebacic acid: $HOOC(CH_2)_8COOH$; produced from castor oil by alkaline cleavage or from naphthalene via naphthalene hydroperoxide; use: raw material for nylon 6,10, high-quality plasticizers, engineering plastics, paints, and hydraulic fluids

sebacyl chloride: *see* interfacial polymerization

secant modulus: in fiber technology, the slope of a line drawn between the origin and any point on the stress–strain curve; measured in units of force/unit area; *see* modulus

secondary batteries: *see* batteries

secondary ion mass spectroscopy: SIMS; a microanalytical method: bombardment with low-energy ions yields elemental and molecular information of the outermost atomic layer of a solid; lateral resolution is 50 nm

secondary oil recovery: use of flooding agents to extract petroleum remaining after the oil has been drawn from the well using gas pressure (primary recovery); *see* petroleum

secondary plasticizer: extender plasticizer used together with primary plasticizer to reduce cost and improve properties of the polymeric material

secondary waste water treatment: biochemical removal of pollutants using a trickling filter or activated sludge process; the final step of this treatment is chlorination

second order transition point: in the plastics industry, the softening of a polymer; *see* glass transition temperature

sedigraph: device for measuring the particle size distribution by the relation between particle size and sedimentation rate

sedimentation: removal of solids from a liquid stream by gravitation; sedimentation is used to separate solids from liquids or solids from solids based on particle size; in waste water treatment, sedimentation tanks are used during primary purification; tanks are also used for clarification in industries such as ceramics manufacture

Seebeck effect: temperature difference that occurs at the junction of two metals when an electric current flows through a wire; this effect is the basis of thermocouples

seeds: small particles or crystals introduced into a melt or solution to serve as nuclei for crystallization; also used on clouds to initiate rain

selective catalytic reduction: in power production technology, removal of NO_x by mixing ammonia with flue gas, separating the mixture from the scrubber vessel and reducing it catalytically to nitrogen and water

selenium: Se; element; nutritional factor; because of its toxicity the metal is used in ship-hull paint and as pesticide against plant feeding mites; when amorphous Se is heated it becomes crystalline and a better conductor of electricity; this property is exploited by its use in electric eyes in which light energy causes a change in conductivity

Selexol process: purification of synthesis gas by a solvent mixture of polyglycol and dimethyl ether

semi-anthracite: lowest rank of anthracite coal

semichemical pulping: in wood processing, a mild chemical pretreatment before mechanical defibering

semiconductors: substances having an electric conductivity between that of conductors and nonconductors; semiconducting elements include the group III element boron, group IV elements silicon and germanium, group V element selenium, and compounds such as CdS, GaAs, CdTe, and $CuInSe_2$; to

modify semiconductivity impurities are added to the pure materials; the process is called doping, e.g., silicon doped with the group III elements boron or aluminum has electron-deficient areas (holes) and is called p-Si, silicon doped with the group V elements phosphorus or arsenic has electron-rich areas and is called n-Si; in a device containing doped silicon electrons can flow from n-Si to p-Si; use: in electronic devices, photovoltaic cells

semigelatin dynamite: soft gelatinous explosive consisting of nitroglycerin and nitrocellulose; *see* gelatin dynamite

sencor: *see* metribuzin;

senna: laxative derived from the leaflets of *Cassia acutifolia*

sensitivity: in the explosives industry, sensitivity differentiates between fast, shattering, and slow propellant substances

sensors: highly specific and sensitive measuring devices such as enzyme electrodes

separation processes: unit operations for the physical separation of mixtures of solids, liquids, and gases; most solid–solid separations are based on differences in particle size or density (e.g., centrifugation, jigging, screening, tabling); others take advantage of magnetic or electrostatic properties, adherence to gas bubbles (flotation), sublimation and freeze drying; solid–liquid separations include centrifugation, clarification, drying, evaporation, expressing, filtration, and leaching; solid–gas separations include bag filtering, cyclone separations, electrical precipitation and gravity settling; liquid–liquid separations include distillation (high energy cost), extractions and membrane processes; liquid– gas separations include cyclone separation, gravity settling and foam breaking; gas–gas separations include absorption, adsorption, chromatography, condensation, diffusion, and low-temperature distillation

sequestering agents: *see* chelates

sericin: gummy substance that must be removed from raw silk before dyeing

Serpek process: manufacture of aluminum nitride by passing hot nitrogen over aluminum

serum: liquid remaining after a colloid solution has been coagulated and the solid removed; in rubber technology, the liquid left after the coagulation and filtering of latex; in biomedical applications the liquid filtered off after coagulation of blood proteins

sesquimustard Q: 1,2-*bis*(2-chloroethyl-thio)ethane; warfare agent banned under chemical arms accord

set: in materials technology, the conversion of a material from a fluid or plastic to a hard state; the term is used in industries such as ceramics, cement, plastics, and rubber; setting time is the time required for transforming a plastic material to its required hardness

sethoxydim: cyclohexadione herbicide

set-off: *see* off-set printing

settleable solids: in waste disposal, the solids that are usually removed during primary treatment

settled soap: soap resulting from full-boiled kettle process

settling tank: *see* sedimentation

sevoflurane: *see* anesthetics

sewage: the total organic waste and waste water generated by domestic, commercial, and industrial activity

sewerage: system of sewage collection, treatment, and disposal

S-glass: magnesia-alumina-silica glass with high tensile strength

shaft kiln: a direct-fired vertical cylindrical kiln capable of firing temperatures >2000°C;

the charge is fed continuously from above and discharged from below

shale: fine-grained clay rock

shale oil: black, viscous oil contained in some shale formations (oil shales); the oil has a high sulfur content

shape memory materials: metal alloys and to a lesser extent rubbers, plastics, and some natural fibers that can return to their original shape after distortion; this return is achieved by heating to their "shape restoring temperature"; use: medical implants, circuit breakers, heat setting of fabrics

Sharples process: a four-stage continuous soap manufacture using high-speed centrifuges to separate lyes, nigre, and glycerol

shatterproof glass: *see* safety glass

Shaw Intermix: *see* internal mixers

shear: action or stress due to forces that result when parallel planes in a material slide relative to each other; in polymer technology, deformation when polymer chains in a material are displaced relative to one another

shear modulus: ratio of shearing stress to shearing strain; *see* modulus

shear strength: maximum force required to separate a moving portion from a stationary one

sheet: in the plastics industry, web or film with a thickness of ≥10 mils

sheet molding compound: SMC; mat of chopped fiberglass saturated with a mixture of thermosetting polyester, filler reactive thickener (such as MgO), and additives (such as initiators, internal mold release agents, colorants); SMC is first aged to complete the thickening effect, then cut to size and molded under heat and pressure

shellac: resin secreted by the south Asian insect, *Tachardia lacca* and deposited on twigs to protect egg masses; shellac for industrial use is prepared by melting or extracting these deposits with solvents; shellac substitutes include rosin-modified maleic resin and a polymer made from phthalic anhydride and polyol; use: protective films

shell flour: ground nut shells used as filler and extenders for plastics

shift conversion: the adjustment of the CO/H_2 ratio during coal gasification; *see* carbon monoxide shift

shift reaction: *see* carbon monoxide shift

shock loading: in biological manufacturing systems, conditions under which the concentration of substrate becomes inhibitory

SHOP process: Shell Higher Olefin Process; production of terminal C10 to C18 linear α-olefins by metathesis of appropriate olefins with ethylene:

$$CH_3(CH_2)_mCH=CH(CH_2)_nCH_3 +$$
$$CH_2=CH_2 \rightarrow CH_3(CH_2)_mCH=CH_2$$
$$+ CH_3(CH_2)_nCH=CH_2$$

The C11 to C14 fraction is converted to alcohols by hydroformylation; these alcohols are used for the manufacture of nonionic detergents

Shore hardness: *see* hardness

short: in the plastics industry, surface imperfection caused by absence of a surface film in reinforced plastics

short-oil varnish: rapidly drying varnish that contains <20 gallons of oil per 100 lb of resin; hard material that is highly resistant to water, alcohols, bases, and acids; too brittle for use on floors; use: varnish for furniture

short shot: in injection molding, failure to fill the mold completely

short-stop: *see* scavengers

short-term exposure limit: STEL; maximum allowable exposure of humans to an agent for any 15 min period, e.g., for benzene the STEL is 25 ppm

short ton: 2000 lb

shot: in molding, one complete cycle

shot capacity: in injection molding, the maximum weight of material that the machine can deliver with one stroke

shrink film: prestretched film used for shrink packaging

shrink packaging: sealed packaging with oriented film that shrinks to its preoriented size upon heating

SI: silicone resins

SI: Systéme Internationale for metric units; SI units are replacing units from other systems

sialon: family of engineering ceramics whose main ingredients are silicon, aluminum, oxygen, and nitrogen; produced from powder mixtures by sintering; sialon is lightweight, has outstanding strength and hardness, and excellent abrasion and heat resistance

SIC: Standard Industrial Classification (United States); the chemical and allied products industry forms section 28, e.g., plastic materials and synthetics (SIC 282); plastics materials and resins (SIC 2821); synthetic elastomers (SIC 2822)

siemens: SI unit of conductance

siennas: group of yellow, brown, and red iron oxide pigments; the presence of $\leq 1\%$ MnO_2 gives a brown-yellow hue that turns red-brown on sintering; use: in paints

sieve: standard wire mesh or screen used to determine mesh size of particulates

silanes: class of compounds with the generic formula, Si_nH_{2n+2} where $n = 1$–6; produced by acid hydrolysis of Mg_2Si or by reaction of lithium aluminum hydride on $SiCl_4$; various organic silanes can be made by addition of alkenes or alkynes to silanes in the presence of platinum; industrially important (chloro)silanes are manufactured by reacting silicon powder with alkyl or phenyl chloride in the presence of CuO and ZnO catalyst in a fluidized bed under pressure at ~300°C; use: as coupling agents, starting material for silicones

silica: SiO_2; the mineral constitutes ~60 wt% of the earth crust in the form of sand and quartz; silica products include glass, ceramics, silica gel; use: raw material for silicon production, adsorbent, catalyst, drying agent, filler, refractory material; *see* cement, ceramics, glass, silicate ceramics, silica gel, zeolites

silica gel: colloidal or dry hydrated silica; made from sodium silicate solutions by treatment with mineral acids to produce hydrated silica; the latter yields silica gel upon drying; use: drying agent, adsorbent in column chromatography, as catalyst substrate

silica glass: *see* glass

silicate ceramics: ceramics in which SiO_2 is the main ingredient, e.g., clay ceramics are composed of different mixtures of clay, feldspar, and quartz; clay ceramics are used to produce bricks, clay pipes, structural ceramics, earthenware, stoneware, and porcelain; other silica ceramics are used as refractories, cast bricks, and insulators

silicate cotton: *see* slag wool

silicates: *see* alkali silicates; cement; glass; hexafluorosilicates; fibers; inorganic; optical fibers; zeolites

siliceous earth: *see* diatomaceous earth

silicic acid esters: $Si(OR)_4$; produced by reaction of $SiCl_4$ with ROH; sol-gel processing of $Si(OC_2H_5)_4$ (tetraethoxysilane, tetraethyl orthosilicate, TEOS) is used for coating glass and making binders for ceramic pastes

silicides: metal silicides are produced by melting metal–silicon mixtures in an arc furnace; these hard, brittle materials are applied by chemical vapor deposition (CVD) to form protective layers on high melting metal surfaces

silicon: Si; metallurgical grade (MG) silicon is produced by reduction of quartz with coke in an arc reduction furnace; electronic grade (EG) silicon is made by reacting MG Si with HCl in a fluidized bed at 300°C to produce $SiHCl_3$ that is then again reduced with hydrogen at 1000°C; MG Si is used as deoxidizing agent in steel production, as component in aluminum alloys and for the production of silicones; crystalline and amorphous EG Si is used for semiconductor devices, photoelectric cells; n-Si contains small amounts of impurities such as As or P, p-Si contains small amounts of impurities such as B or Al; these doped silicons are used in photoelectric cells

silicon carbide: SiC; silicon carbide ceramic is produced from high purity quartz sand and petroleum coke or anthracite in an electric resistance furnace at 2200–2400°C; used for high-temperature applications that require hard, high strength materials; silicon carbide fibers are produced by chemical vapor deposition (CVD) of CH_3SiCl_3 at 1200°C; the fibers and fibers coated with silicon carbide can be used as reinforcement for various matrices; *see also* polycarbosilane

silicon chlorides: the most important industrial silicon chlorides are silicon tetrachloride, $SiCl_4$ and trichlorosilane, $SiHCl_3$; manufactured by reaction of Si with Cl_2 >300°C or by reaction of SiO_2 with burning carbon and Cl_2 at ~1400°C; chlorides are used in manufacture of electron grade silicon for semiconductors, for siliconizations of metals

silicon dioxide: *see* silica

silicones: siloxanes, polyorganosiloxanes; polymers with a [–Si–O–Si–O] backbone and organic side chains bonded to Si; produced by hydrolysis of alkylchlorosilanes (the latter are made by heating silicon with alkylchloride) to cyclic siloxanes or low molecular weight silanols; catalyzed anionic polymerization leads to gums and nonsilanol-terminated fluids; silanol-terminated polymers may be produced by hydrolysis of chlorosilanes or from cyclic siloxanes by equilibration with low molecular weight silanol fluids; silica is used as filler for silicone polymers to improve their stiffness, tensile and tear strength, and abrasion resistance; block and graft copolymers of silicones with organic polymers (urethane, polycarbonate, methacrylate, methylstyrene, and polypeptides) are made to achieve specific properties; blocks are connected by C-Si or C–O–Si bridges; side chain liquid crystalline materials with polysilicone main chains have liquid crystalline transitions at or close to room temperature; use: lubricants, heat transfer agents, break fluids, defoaming agents, transformer oils, adhesion promoters, cosmetics, deaerating agents, water repellant and heat-resistant coatings, varnishes, elastomers, antifoaming agents, heat-resistant paints, thermoplastics, medical soft-part prostheses, encapsulation of electrical components, liquid crystals; *see* polymers, inorganic

silicon nitride: Si_3N_4; manufactured by reacting Si and N_2 at 1200–1400°C; use: for high-temperature applications of hard, high-strength materials; incorporation of oxides such as alumina (sialon) leads to as yet experimental ceramics; *see* sialon

silk: natural protein filament fiber

silk screen printing: *see* printing

siloxanes: *see* silicones

silver: Ag; element; obtained together with gold and copper by mining; use of silver and its compounds: photography (silver halide), electrical contacts, batteries, jewelry, catalysts, medicinal ($AgNO_3$), silver solder (AgCu), photochromic glass (silver chloride and silver molybdate)

simazine: triazine herbicide for control of broadleaf and grassy weeds

single cell protein: SCP; cells of organisms such as algae, bacteria, fungi, or yeast are grown for their protein content as nutrient supplement; cells have been grown on carbon dioxide, methane and on petroleum; SCP is not commercially successful because of cost and consumer resistance; *see* SCHEMATIC REPRESENTATION OF A CONTINUOUS, SINGLE-CELL, PROTEIN-PRODUCING PROCESS FROM PETROLEUM

single stage biox system: in waste disposal a one step biological oxidation of sewage

sintering: agglomeration of particles by heating just below their melting point; use: for fusing, coating, molding

sisal: natural cellulosic fiber

SITC: Standard Industrial Trade Classification (United Nations); the chemical and allied products industry forms section 5

sitosterol: plant sterol; constituent of tall oil that is produced as by-product of sulfite pulping process; use: for production of cortisone; *see* steroids

size: *see* sizing

size enlargement: *see* agglomeration, flocculation

size reduction: achieved by fracturing, crushing, and grinding; fine particles are produced in mills

sizing: size; additives used in paper and textile production to fill pores, coat fibers, and add body; *see* textile sizing

skim rubber: rubber produced from the serum left after the natural latex has been centrifuged; exhibits considerable variability, gives hard vulcanizates

slag: during metal manufacture, a layer of metal oxides floating on top of the molten metal

slag wool: silicate cotton; fibers into which

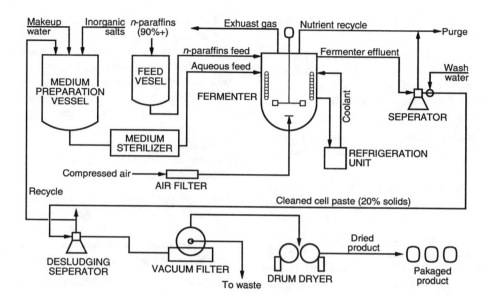

SCHEMATIC REPRESENTATION OF A CONTINUOUS, SINGLE-CELL,
PROTEIN-PRODUCING PROCESS FROM PETROLEUM
[Courtesy *Chemical Engineering,* August 26, 1968]

molten slag has been blown; used as felt for packing and for insulation

slake: reaction with water of anhydrous materials such as lime; a slaker is a device in which water is added to lime

slaked lime: $CaO \cdot H_2O$; hydrated lime

slate: metamorphic rock; use: for roofing, blackboards, and as powdered filler for polymers

slip: in ceramics manufacture a water suspension of fine clay having the consistency of cream; use: for cementing and glazing, for slip casting (a process in which slip is poured into a mold, is removed after drying and fired)

slip agent: in plastics processing, an internal lubricant that rises to the surface during processing

slow release: term refers to the controlled release of medicinals or agriculturals through a membrane; the method permits the maintainance of a constant dose over time

sludge: residue in centrifuges and storage containers; soft mud; in the rubber industry, the dirty precipitates from latex on sieves and in tanks; in waste disposal, raw sludge is the mass collected in sedimentation tanks from screened waste water; sewage sludge is considered a potential source of agricultural nutrients because of its high N and P content

sludge digestion: in waste disposal, biological step using aerobic or anaerobic bacteria

sludge nitrification: biological step in waste disposal, oxidation of ammonium compounds to nitrites and nitrates by bacteria

slurry: concentrated suspension of solid particles in a carrier liquid, e.g., paints, ink, drill muds, and cements; coal and ore slurries can

257

be transported over long distances through pipes

smectic crystal: liquid crystal phase in which rod-like molecules are arranged in two-dimensional order, i.e, molecules are parallel and their ends are in line rather than random; *see* liquid crystals

smelting: during metal production, a process of fusing or melting metals to separate them from ores

smoke point: temperature at which a fat will produce continuous wisps of smoke; free fatty acids lower the smoke point of the fat

soaps: salts of long-chained fatty acids (C12 to C18); most soaps are produced from animal fat, tallow (containing C16 to C18 fatty acids) and vegetable oils such as palm kernel and coconut oils that contain lauric acid (C12) and myristic acid (C14); the fats are heated with lye in batch and continuous processes; batch production of kettle soap takes 4 days, whereas continuous processes can be completed in a few hours; the various commercial continuous processes differ in the design of the washing towers, the temperature of the reactants, and the direction of flow of fat and lye; *see* full-boiled kettle process; *see* SOAP PRODUCT FLOW CHART

Society of Dyers and Colourists: British organization that has established a system of dye nomenclature; the comparable United States organization is the American Association of Textile Chemists and Colorists

soda ash: Na_2CO_3; *see* sodium carbonate

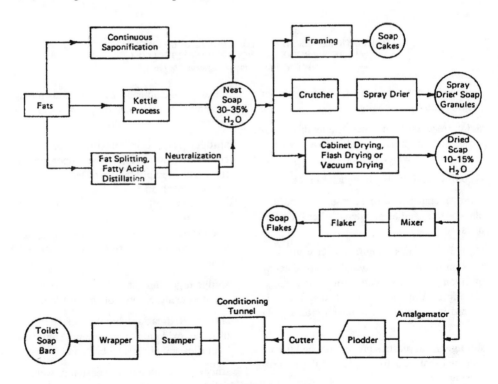

SOAP PRODUCT FLOW CHART
[Copyright *Riegel's Handbook of Industrial Chemistry*, 8th ed., Van Nostrand-Reinhold Co., Inc.]

soda ash-sulfur process: old process for manufacture of sodium thiosulfate by pumping a hot solution of soda ash into SO_2-absorbing towers and heating the resulting Na HSO_3 with powdered sulfur

soda cellulose: alkali cellulose; in rayon manufacture the product obtained by steeping α cellulose (cotton linters, woodpulp) in 17–20% NaOH

soda niter: *see* sodium nitrate

soda process: old pulping process using NaOH; no longer in much use because it yields inferior pulp

sodium: Na; element; manufactured by electrolysis of molten NaCl (Downs process) using graphite electrodes; use: in manufacture of tetraethyl lead (antiknock agent), as starting material for chemicals including $NaBH_4$, Na_2O_2, sodium azide, catalyst, coolant for nuclear reactors, in sodium–sulfur batteries

sodium aluminum hexafluoride: Na_3AlF_6; synthetic cryolite; produced by reaction of ammonium fluoride with sodium aluminate; use: electrolyte during production of aluminum by electrolysis of bauxite (Hall–Heroult process)

sodium bicarbonate: $NaHCO_3$; sodium hydrogen carbonate; obtained by reaction of CO_2 with aqueous sodium carbonate; byproduct of the Solvay process for soda ash (sodium carbonate) production; major use: as baking powder; other use: in the chemical, pharmaceutical, surfactant, leather, paper and textile industries; also used in fire extinguishers

sodium bichromate: *see* sodium dichromate

sodium bisulfate: $NaHSO_4$; sodium hydrogen sulfate; produced (1) by heating NaCl and H_2SO_4 in cast iron retorts; (2) by aqueous reaction of Na_2SO_4 with H_2SO_4; use: cleaning agent, flux

sodium bisulfite: $Na_2S_2O_5$; sodium disulfite; manufactured by treatment of sodium carbonate with SO_2, or by reacting NaOH and SO_2 in a saturated solution of $NaHSO_3$; use: in the dye, leather, paper, photography, and textile industries, also for manufacture of sodium hydrosulfite

sodium borate: $Na_2B_4O_7 \cdot nH_2O$; borax; produced from minerals and brines; use: for manufacture of glass, ceramics, enamels, detergents, fertilizers, flame retardants, corrosion inhibitors, fluxes

sodium carbonate: Na_2CO_3; soda ash; produced (1) from naturally occurring trona ore, a sodium sesquicarbonate, $Na_2CO_3 \cdot NaHCO_3 \cdot 2H_2O$ found in the United States (mainly in Wyoming and California); (2) by the Solvay process and its modification by Asahi; and (3) from natural brines; use: in the manufacture of glass, chemicals, alkaline cleaners, and in the pulp and paper industry; also used as water softener and as builder in detergent formulations

sodium carboxymethylcellulose: use: detergent builder and antiredeposition agent; *see* carboxymethylcellulose

sodium cellulose xanthate: *see* cellulose xanthate

sodium chlorate: $NaClO_3$; produced by electrolysis of NaCl in diaphragmless cells using activated Ti anodes and steel cathodes; oxidizing agent; use: in the pulp and paper, leather, dye, and explosives industries; starting material for manufacture of perchlorates

sodium chloride: NaCl; produced from salt mines, brines, and ocean water; use: for manufacture of chlorine and NaOH (chlor-alkali process), sodium carbonate, HCl, metallic sodium; also for salting out solutions, regeneration of ion exchange resins, as filler in detergent formulations, in food preparation, in pharmaceuticals

sodium chromate: Na_2CrO_4; produced by

heating a ground mixture of chrome iron ore and sodium carbonate in air at 1000–1100°C; use: pigments, inks, anticorrosion agent, wood preservative; toxic

sodium citrate: chelating agent; use: builder in nonphosphate detergent formulations

sodium cooled breeder reactor: liquid metal fast breeder reactor, LMFBRs; in nuclear technology, breeders of plutonium that use metallic sodium as coolant; fuel doubling time is 7–10 years; the reactor core has a five times greater power density than light water reactor (LWR) cores

sodium dichromate: $Na_2Cr_2O_7$; prepared from sodium chromate by heating with concentrated sulfuric acid; use: oxidizing agent, catalyst, corrosion inhibitor, mordant, pigment, preservative; for manufacture of chromium(VI) oxide, pigments, chrome tanning agents, and oil drilling

sodium dihydrogen diphosphate: *see* sodium pyrophosphate

sodium dimethyl dithiocarbamate: SDD; use: short stop for polymerizations; *see* scavangers

sodium disulfite: *see* sodium bisulfite

sodium dithionite: $Na_2S_2O_4$; sodium hydrosulfite; produced by reacting zinc dust with SO_2 and then treating an aqueous solution of the resulting ZnS_2O_4 with NaOH; also manufactured by reacting SO_2 under pressure with an alkaline methanolic solution of sodium formate; use: reducing agent in textile and paper industry

sodium formaldehyde sulfoxylate: SFS; accelerator used in SBR cold rubber formulation

sodium hexametaphosphate: *see* sodium metaphosphate

sodium hydrogen carbonate: *see* sodium bicarbonate

sodium hydrogen sulfate: *see* sodium bisulfate

sodium hydrogen sulfide: NaHS; produced by reacting H_2S with NaOH or Na_2S; use: mainly in paper manufacture, also in the dye stuff, rayon, and leather industries; *see* sodium sulfide

sodium hydrosulfite: *see* sodium dithionite

sodium hydroxide: NaOH; produced by electrolysis of aqueous NaCl (chlor-alkali process); use: as strong base (lye) in the production of organic and inorganic chemicals, in the paper and pulp, detergent and soap, oil, and textile industries, in aluminum manufacture; *see* chlor-alkali process

sodium hydroxymethanesulfinate: $HOCH_2SO_2Na$; produced by reacting dithionite, formaldehyde, and NaOH; used in direct printing

sodium hypochlorite: NaOCl; produced by (1) reacting chlorine with aqueous NaOH; (2) electrolysis of seawater; use: bleach and disinfectant

sodium metaphosphate: $(NaPO_3)_n$; sodium cyclophosphate; sodium hexametaphosphate is a cyclic compound consisting of six tetrahedral PO_3 units linked by shared oxygens; produced by heating sodium pyrophosphate; chelating agent; use: water treatment

sodium metasilicate: Na_2SiO_3; produced by melting sand with sodium carbonate in a 1:1 molar ratio; highly alkaline ingredient of some detergent formulations

sodium nitrate: $NaNO_3$; soda niter, Chile saltpeter; mineral; use: for making nitric acid, explosives; fertilizer, flux in welding

sodium perborate: $NaBO_3 \cdot 4H_2O$; produced by heating sodium borate (borax) and aqueous NaOH and then oxidizing the resulting metaborate, $NaBO_2$, with hydrogen peroxide; use: bleach for textiles, laundry and paper, for water purification and for production of chemicals

sodium percarbonate: $Na_2CO_3 \cdot 1.5H_2O_2$; sodium carbonate perhydrate; produced by spraying sodium carbonate and hydrogen peroxide onto a fluidized bed of sodium percarbonate and collecting the large particles while the fines are recycled; use: bleach and oxidizing agent

sodium perchlorate: $NaClO_4$; produced by electrochemical oxidation of sodium chlorate; use: oxidizing agent

sodium peroxide: Na_2O_2; produced in two steps by air oxidation of Na to Na_2O at 200–700°C followed by air oxidation at 350°C to the peroxide; use: oxidizing agent

sodium phosphates: monophosphates are produced by reaction of sodium carbonate or sodium hydroxide with phosphoric acid (NaH_2PO_4 [sodium dihydrogen phosphate], Na_2HPO_4 [sodium monohydrogen phosphate], Na_3PO_4 [sodium phosphate]); diphosphates, also called pyrophosphates ($Na_2H_2P_2O_7$, $Na_4P_2O_7$), are produced by heating sodium hydrogen phosphates to 245–900°C; sodium polyphosphates have the general formula $Na_{n+2}P_nO_{3n+1}$; they are produced by dehydrating sodium hydrogen phosphate and sodium dihydrogen phosphate in appropriate stoichiometric amounts; sodium phosphates are primarily used in detergents and cleaning formulations, for metal cleaning, boiler water treatment, for food preparations and for animal feed; disodium dihydrogen diphosphate is used as baking powder; sodium tripolyphosphate (STPP) was formerly widely used in detergents but use has been reduced because of ecological considerations

sodium pyrophosphate: $Na_2H_2P_2O_7$; disodium dihydrogen diphosphate; produced from NaH_2PO_4 by heating at 245°C; use: baking powder, chelating agent in detergent formulations; *see* sodium phosphates

sodium sesquicarbonate: $Na_2CO_3 \cdot NaHCO_3 \cdot$ $2H_2O$; isolated from trona ore during production of sodium carbonate (soda ash); use: in hard surface cleaners for pH control

sodium silicate: $Na_2O(SiO_2)_n$; waterglass; water-soluble silicate; made by fusing Na_2CO_3 with silica in a batch process; marketed as dry powder and as aqueous solutions (22°–69°Bé); use: includes gels, surfactants, pigments, boxboard adhesives, water treatment and support for catalysts

sodium sulfate: Na_2SO_4; salt cake; produced (1) from natural brines; (2) during manufacture of HCl from NaCl and H_2SO_4 (Mannheim process); (3) as by-product during manufacture of viscose rayon and other chemical processes; use: in the pulp and paper (kraft), detergent, glass and chemical industries

sodium sulfide: Na_2S; produced (1) mainly by controlled scrubbing of H_2S with aqueous NaOH to give NaHS that is converted to Na_2S by further treatment with caustic; (2) as by-product in the manufacture of $BaCO_3$ from barite ore, $BaSO_4$; (3) an earlier process consisted of the reduction of salt cake (Na_2SO_4) with coal; use: in numerous industries including metal industry for ore flotation, leather industry for hair removal, and textile industry for dyeing with sulfur dyes

sodium sulfite: Na_2SO_3; thionite; produced by reaction of SO_2 and NaOH in a saturated Na_2SO_3 solution; use: reducing agent, fire retardant, textile bleach, food preservative, antioxidant, photographic developer

sodium tetraborate: *see* borax

sodium tetraborohydride: $NaBH_4$; water-soluble reducing agent used for organic synthesis

sodium thiosulfate: $Na_2S_2O_3$; produced (1) as by-product from the manufacture of sulfur dyes; (2) by reduction of SO_2 with Na_2S in the presence of Na_2CO_3; (3) by heating Na_2SO_3 suspensions in the presence of sulfur;

use: fixer (hypo) in photography, bleach in laundry detergent formulations

sodium trimetaphosphate: *see* sodium tripolyphosphate

sodium tripolyphosphate: STP; $Na_5P_3O_{10}$; sodium trimetaphosphate, pentasodium tripolyphosphate; produced by (1) reacting a hot solution of sodium carbonate with phosphoric acid in a mole ratio of 1.67:1; (2) heating a 2:1 stoichiometric mixture of disodium hydrogen phosphate and sodium hydrogen phosphate to 300–550°C; use: fertilizer, chelating agent for Ca^{+2} and Mg^{+2}, detergent builder

soft alkyl benzenes: group of biodegradable detergents

softeners: in the polymer and rubber industries, plasticizers that promote softening of the resin; in water purification, agents (e.g., ion-exchange resins) that remove the slightly soluble $MgCO_3$ and $CaCO_3$

soil conditioners: materials that improve the texture and chemical composition of soil, e.g., gypsum serves as a source of calcium, helps to retain organic nitrogen without affecting the pH; lime serves to stabilize soil against temperature and moisture changes; vulcanic material (perlite) aids soil aeration; partially heat-decomposed peat (charred peat) serves as fertilizer; pesticides are added to sterilize soils

soil suspending agent: *see* antiredeposition agent

sol: dispersion of a solid in a liquid (e.g., colloidal gold in water) or a solid in a solid (e.g., gold in glass: ruby glass)

solar cells: photovoltaic cells; solar voltaic cells; devices that convert sunlight into electricity; they consist of semiconductors (e.g., crystalline or amorphous Si (a-Si), CdS, GaAs, CdTe, and $CuInSe_2$) that can absorb light in the visible range to promote electrons from the valence band into the conduction band; unlike crystalline Si, a-Si can be made into thin films (0.5 μm) leading to cost reduction and production of large area solar modules

solar ponds: bodies of saline water in which the salt concentration increases with depth reaching saturation at ~2 m. Because the density gradient inhibits free convection solar radiation reaching the bottom is trapped and the heat can be withdrawn into a heat exchanger for useful energy

solar voltaic cells: *see* solar cells

solar year: in energy technology, amount of solar energy reaching the earth each year

solder: brazing alloys; mostly alloys of lead and tin that melt between 70 and 183°C

solder glass: *see* glass

sol-gel process: formation of glass, ceramics, or composites by conversion of a sol into a gel, which upon aging and drying results in a dense, crack-free gel and finally a solid; process is carried out below 1000°C using drying control agents and catalysts; for composite manufacture, powder or whisker reinforcements are added to the sol before processing; use: for manufacture of high-grade coatings and monolithic ceramics, glasses, and composites

solid fat index: SFI; indicates percentage of solids in a fat at a given temperature

solid phase peptide synthesis: SPPS; automated method for the synthesis of polypeptides; the carboxyl-terminus is bound to a polymer and is reacted with protected amino acids; after each step the bound peptide can be separated by filtration and purified; *see* Merrifield resin

solid state polymerization: the conversion of a crystalline monomer to polymer with radiation or other intiators; monomers include methacrylate, styrene, acrylonitrile, trioxane (BF_3 initiator)

solubility parameters: values that relate

molecular structure to solubility; substances with similar parameters are likely to be compatible with each other, i.e, be soluble in each other; parameters can be calculated from such measurements as the energy of vaporization

soluble guncotton: *see* pyroxylin

solution dyeing: dope dyeing; process in which dyes are introduced into polymer solutions before they are extruded through a spinnerette

solution polymerization: polymerization process in which the solvent for the monomer also serves as solvent for the resulting polymer; advantage of the process is the easy temperature control; direct use for film casting, surface coatings

solvation: in the polymer industry, sorption of solvent by molecules or colloid particles; solvation assists solubility and dispersion of colloids; in organic chemistry the action of a solvent on a solute (e.g., water is a stronger acid than alcohol leading to ROH_2^+ species in solution)

Solvay process: ammonia-soda process; production of soda ash (sodium carbonate, Na_2CO_3) from limestone ($CaCO_3$) and ammoniated brine (NaCl); the reaction is based on the liberation of CO_2 from a coke-fired lime kiln and the precipitation of sodium bicarbonate from the ammoniacal solution resulting from the reaction of CO_2 and NaCl; the $NaHCO_3$ is then calcined to Na_2CO_3; ammonia is recycled; the process is giving way to a modification, the Asahi process, that has lower energy requirements and improved $CaCl_2$ recovery

solvent naphtha: *see* naphtha

solvent nonsolvent polymerization: *see* precipitation polymerization

solvent polishing: in the plastics industry, immersion in (or spraying with) a solvent to dissolve surface irregularities on a molded plastic article

Solvent Refined Coal process: SRC I, SRC II; process developed by Pittsburgh & Midway Coal Co. to produce ash and sulfur-free solid (SRC I) and liquid (SRC II) fuel from high-ash, high-sulfur coal; the pulverized coal is mixed with a process-derived hydrocarbon solvent and hydrogenated; the operation is carried out at 2000 psig and 840–870°F and the solvent is recycled

somaclonal variation: regeneration of plants from cultured somatic cells leading to genetic variation; *see* somatic cells

soman: $(CH_3)_3C(CH_3)CHOP(CH_3)(F)=O$; nerve gas; banned war gas under chemical arms accord

somatic cells: plant or animal cells that take part in the generation of organs; any cell other than germ cells (egg or sperm cells)

sonication: use of sound waves (audible or ultrasound) to induce chemical reactions

sonochemistry: *see* ultrasound

sorbate: *see* sorption

sorbent: *see* sorption

sorbic acid: *trans,trans*–$CH_3CH=CH–CH=CHCOOH$; *trans,trans*-2,4-hexanedienoic acid; produced by (1) catalyzed (Zn or Cd) addition of ketene to crotonaldehyde at 30–60°C followed by acid catalyzed hydrolysis or thermolysis; (2) catalyzed (Ag) oxidation of sorbic aldehyde (*trans,trans*-2,4- hexadienal) that is made by aldol condensation of acetaldehyde; use: food preservative

sorbitol: $CH_2OH(CHOH)_4CH_2OH$; produced by catalytic hydrogenation of glucose; used in the food industry as sweetener, anticaking agent, emulsifier, anticrystallizing agent, softener, and thickener; as moisture retainer in cosmetic creams and as filler in the pharmaceutical, paper, and textile industries

sorption: binding of a substance (sorbate)

on a solid material (sorbent); term includes processes such as adsorption, absorption, ion exchange, desorption, and chemisorption

sour gas: petroleum gas with high sulfur content

sour oil: petroleum with high sulfur content

soybean meal: residue remaining from soybeans after oil extraction; contains >40% proteins; use: animal feed

soybean oil: highly unsaturated oil; use: for salad oil after winterizing and deodorizing, to manufacture margarine and shortenings, as binder in paint

spandex: generic term for fibers based on elastomeric urethane polymers; extensibility with good recovery is present in both rubber and spandex fibers, however, rubber is a cross-linked material whereas urethane polymers are thermoplastics that can be solution or melt spun; properties of urethanes polymers are based on hard polyurethane blocks interspersed by flexible polyether or polyester units; due to better aging properties, durability, and higher stiffness, spandex has replaced rubber in swimwear, foundation, and other critical performance fabrics; *see* urethane block copolymers

Spanish oxide: iron oxide pigment; use: paint

sparger: porous pipe to introduce fine streams of steam, air, or water to be sprayed into a liquid

spar varnish: long-oil varnish; contains >30 gallons of oil per 100 lb resin; drying oils in the formulation provide elasticity and resistance needed for exterior surfaces

specialty chemicals: low-volume, high-value-added chemicals

specific: term used to denote a measurement expressed in relation to some reference, i.e, specific gravity, specific heat

specific gravity: density of a substance related to the density of water, i.e, the density of water is 1; specific gravity has no units; also *see* Baumé

spectrophotometer: instrument measuring the brightness of spectral lines

spectroscopy: study of absorption or emission of electromagnetic waves by materials; use: determination of atomic and molecular structure; the various techniques are listed separately

specular reflection: gloss; degree to which a surface has the light reflecting property of a mirror (angle of reflection equals the angle of incidence); the gloss of any surface increases as the angle of incidence departs from the perpendicular to the surface; gloss is measured by a goniophotometer: light from a prefocused incandescent source falls on the sample surface at a given angle; light specularly reflected at an equal but opposite angle falls on a photocell and its intensity is measured; the goniophotometer measures the reflected light as function of the angle of incidence

spent fuel processing: used in nuclear technology; *see* Purex process

spermicides: birth control agents that kill the sperm

sperm whale oil: *see* jojoba oil

spherodizers: producers of small spherical particles, used for fertilizers

spherulite: spherically arranged aggregate of crystallites, recognizable by their characteristic appearance in the polarizing microscope; they show a dark Maltese cross (a specific birefringence effect) due to the orderly change in orientation of the crystallites in each spherulite; electron microscopy of crystalline polymers has shown that their crystalline unit is the lamella, the spherulite being formed by many specially arranged lamellae

spiegeleisen: iron–manganese alloy; produced from franklinite; used to make manganese steel

spin dyeing: dope dyeing; *see* solution dyeing

spinel: mineral ($MgAl_2O_4$) that has an isometric crystal structure and is subject to twinning; use: as a refractory

spinneret: extrusion die for spinning polymers into fibers from a melt or a solution

spinning: for synthetic fibers, processes to convert polymers into filaments and yarns, primarily by melt or solution extrusion; the latter is used when the polymer degrades at its extrusion temperature or when specific fiber properties are essential [*see* gel spinning; ultra-high molecular weight HDPE); fibers may be oriented during (*see* poly(ethyleneterephthalate); nylon-6] or after spinning (*see* fiber drawing; orientation) to obtain tensile properties for textile or industrial applications; for spinning of staple; *see* yarn spinning

spiro polymers: noncross-linked polymers in which two molecular strands are bonded to each other, i.e, a polymer based on pentaerythritol

spore: in biotechnology, a single or multicell plant reproductive body capable of giving rise to a new organism

spray drying: rapid evaporation process in which the material is injected under high pressure into a rotary drier

spun-bonded materials: nonwoven fabrics prepared by a single step process from thermoplastic polymers (i.e, polyamides, polyesters, polypropylene, polyethylene) by extrusion, drawing, web formation, and bonding; bonding of the webs is achieved by mechanical means (i.e, needle punching, treatment of a fiber web with barbed needles that entangle the web or by thermal or chemical bonding (area or point bonding); use: carpets, furniture, disposables

squib: small size pyrotechnic or explosive device

SRC: *see* Solvent Refined Coal process

stabilization pond: lagoon, oxidation pond; in waste disposal, 3–5-ft deep pond in which reactive organic matter in sludge is converted into an inert, harmless material by the interaction of sunlight, algae, and oxygen

stabilizers: agents used to maintain the properties of phases or substances, such as emulsions and plastics; stabilizers protect against harmful effects of light, oxygen, solvents, and chemicals

standard moisture regain: commercial moisture regain; in textile technology, arbitrary standard to compare the percent moisture absorbed by dry material (measurement at 65% relative humidity at 70°F); moisture regain is one indicator of comfort level of materials; comfort level is low when moisture regain is low; moisture regain of cotton is 7.4%, silk 9.4%, wool 13%, flax 12%, nylon 66 4.2%, acrylics 1.8%, polyester 0.4%

Staphylococcus: disease-causing bacterium; historically, the inability of *S. aureus* to grow on a nutrient when contaminated by the mold *Penicillium notatum* resulted in the discovery of penicillin

staple fiber processing: staple fibers have a length of 0.5–5 in.; natural fibers such as cotton, flax, ramie, and wool are staples; synthetics are cut into staples from endless as-spun yarns; staple fibers are spun into yarns either alone or in blends to combine desirable properties of cotton or wool with strength and ease of care of the synthetics

starch: polysaccharide derived from plants; consists of a linear (amylose) and/or branched (amylopectin) polymer of glucose connected via α acetal linkages; in the textile and laundry industries the term covers products that provide body or stiffness to fabrics; examples of synthetic starches are polyvinyl acetate and carboxymethylcellulose; use: filler for paper industry

staypak: product made by heating wood to 165–175°C at 100–140 kg/cm^2; staypak has improved resistance to moisture, swelling, and shrinking, and has improved strength

steam coal: coal used in power production; has lower Btu content than metallurgical (met) coal

steam cracking: *see* steam reforming

steam reforming: steam cracking; cracking in the presence of water vapor; steam cracking of oil or natural gas leads to the production of synthesis gas (CO + H$_2$) (1) a hydrocarbon vapor and steam mixture is passed over nickel catalyst at 1200–1800 °F and 600 psig leading to the reaction: CH$_4$ + 2H$_2$O → 4H$_2$ + CO$_2$, and CH$_4$ + H$_2$O → 3H$_2$ + CO; or (2) steam can be reacted with partially oxidized hydrocarbons at 2300–2700 °F and 600 psig yielding hydrogen and oxides of carbon; the gases are separated by fractionation and by removal of impurities such as ammonia, water, and hydrogen sulfide with sorption; steam cracking of naphtha yields olefins and aromatics

steam spraying: *see* paints

stearic acid: straight chain saturated C18 fatty acid

stearin: glyceryl tristearate; fat; use: food, for soap manufacture

Steffan's process: in sugar production, separation of sucrose from beet molasses by precipitation of calcium saccharate with pulverized lime; the calcium cake is filtered off, added to carbonated water and separated into sucrose and calcium carbonate

step polymerization: *see* condensation polymerization

stereograft polymer: polymer consisting of chains of an atactic polymer grafted to chains of an isotactic polymer

stereolithography: several technologies that allow computer-controlled production of solids directly from computer aided designs; the solid is manufactured (1) by curing a photosensitive monomer point by point with a laser moving across the surface; after the first layer has solidified the laser proceeds to cure the next layer; acrylates are used as monomers; (2) by feeding a thermoplastic into a heated extrusion head where it melts to a temperature just above solidification; extrusion is computer controlled and the solidified plastic is laid down as successive laminates; (3) by passing a colloidal binder over a thin layer of powder using a computer-controlled single-nozzle printhead and repeating the process to build up layers to form the solid

stereoregular polymerization: reaction using catalysts such as the Ziegler–Natta type that control the stereochemistry of the forming polymer chains; these catalysts are complexes of aluminum alkyl and titanium halides; *see* stereoregular polymers; tacticity; poly-1,4-isoprene; natural rubber; guttapercha; Ziegler–Natta catalysts

stereoregular polymers: polymers consisting of units all of which have the same configuration; examples: linear polyethylene, isotactic or syndiotactic polypropylene, pure polydienes: *cis*- and *trans*-1,4 polybutadiene, the isotactic and syndiotactic 1,2 polymers; because of the regularity of their structures, these materials may crystallize and they have become important commercial products; *see* tacticity

stereospecific: referring to a specific three-dimensional molecular structure; particularly important in pharmaceuticals because biological activity (beneficial as well as toxicity) generally depends on the steric configuration of the molecule; *see* stereoregular polymers

stereospecific polymers: *see* stereoregular polymers

sterilant: substance that is toxic to all microorganisms and plant life; because they are not selective, sterilants exhibit no useful differential activity in agricultural practice

sterilization: in biotechnology, process to kill all interfering organisms before experimentation; sterilization can be done by heat, oxygen, air, or radiation

steroids: biologically active compounds characterized by a fused four-ring carbon skeleton; synthetic derivatives include fertility and contraceptive drugs, antiarthritis drugs, antiinflammatories; *see* sterols

STEROIDS

sterols: alcohols of the steroid family, e.g., cholesterol, sex hormones, adrenal hormones, bile acids

Stex process: styrene extraction from pyrolysis gasoline that contains 6–8% styrene

stiffness: ability of a material to resist deformation under stress; *see* modulus

stigmasterol: plant sterol; isolated from plant galls; used in manufacture of cortisone

stilbene: $C_6H_5CH=CHC_6H_5$; chemical intermediate; *see* alternating copolymers

stoichiometry: weight relations in chemical formulas and equations; these relations are translated into atomic ratios within chemical formulas and molecular ratios in chemical equations

stoke: the cgs unit of kinematic viscosity

stoneware: vitreous or semivitreous ceramic made from moderate-firing clays

storage batteries: devices to convert chemical energy to electrical energy; these batteries are recharged by reversing the chemical

reaction; for example, in the car battery electrons flow through a wire from the Pb anode to the PbO_2 cathode giving solid $PbSO_4$ as product; to recharge the Pb and PbO_2 are regenerated from $PbSO_4$; *see* batteries

Stormer viscometer: instrument described in ASTM D 562; *see* Krebs unit

straight gelatins: dynamites based on sodium nitrate dopes; *see* gelatin dynamites

straight run gasoline: *see* light naphtha

strain: under stress, the change in length of a material (fiber, film) per unit of original length

streptomycin: antibacterial agent and tuberculostatic; produced by fermentation from the soil bacterium *Streptomyces griseus*; derivatives include streptomycin B, streptomycin pantothenate

stress: force per unit area that produces a deformation in a material; measured in pascal (SI system)

stress relaxation: decay of stress under constant strain

stretch fabric: *see* spandex

stretching: orienting; during fiber spinning an operation that provides desired properties to fibers by producing oriented polymer chain structures

stretch recovery: in fiber technology, the rate and degree of recovery of a yarn or fabric from extension (1) immediate recovery on removal of stress; (2) delayed recovery when it is slower than decrease of stress; (3) the nonrecoverable part of the extension (irreversible extension); *see* physical testing

stripping: removal of unwanted material, such as color, from textiles, volatile components from gasoline distillation, unreacted monomer during polymerization

structure–activity relation: SAR; in the pharmaceutical industry a method to aid the design of drugs by relating structural features such as electronic and solvent effects to activity; extension of the technique leads to quantitative SAR (QSAR) and computer-aided design (CAD)

strychnine: alkaloid; extracted from seeds of *Strychnos* plant species; use: rodent baits; extremely toxic

styrene: $C_6H_5CH=CH_2$; produced by catalytic dehydrogenation of ethylbenzene preferably with steam and oxides of iron, chromium, and potassium; another route is peroxidation, followed by reaction with propylene to give α-phenylethanol and propyleneoxide; the phenylethanol is dehydrated using tungsten, molybdenum, or vanadium compounds; use: monomer for thermoplastics, elastomers, thermosetting resins, and packaging foams; *see* oxirane process

styrene–acrylonitrile copolymer: SAN; manufactured by emulsion, suspension, or bulk polymerization; the latter process is initated thermally or by catalyst, is carried out at 100–200°C with solvents to low conversion levels (40–70%); residual monomer is removed and the molten product extruded and pelletized; use: most SAN is incorporated into acrylobutadiene styrene (ABS) resins that are used in housewares, appliances, industrial and automotive parts, camera parts, telephone, and electric outlets

styrene–butadiene-rubber: SBR; manufactured by emulsion polymerization in the presence of emulsifiers, initiators, chain transfer agents (thiols), antioxidants, antiozonants; use: passenger car tires, adhesives, binder, chewing gum; *see* compounding

styrene copolymers: *see* styrene-acrylonitrile copolymer; styrene-butadiene rubber

styrene plastics: polystyrene is a thermoplastic, amorphous, clear, stiff material manufactured by bulk, emulsion dispersion polymerization; it is converted by injection molding to finished products, by extrusion to pipes, sheets, by blow molding into hollow objects; a major use is in foamed beads for packaging materials; it is blended with flame retardants and antistatic compounds; high impact polymer (HIPS) is tough due to dispersion of small discrete rubber particles in its matrix; glass reinforcement of polystyrene or SAN improves mechanical properties (strength, stiffness, toughness, creep resistance); copolymers lead to tough and elastomeric materials; copolymerization with acrylonitrile produces SAN which with polybutadiene gives acrylobutadiene styrene (ABS) resin; SBR is a styrene–butadiene elastomeric copolymer in which butadiene is the major component

subbituminous coal: the third rank of coal; has a heating value of 8300–11,500 Btu/lb; located in the United States mainly in Alaska, Colorado, Montana, New Mexico, Utah, Washington, and Wyoming; *see* coal rank

suberic acid: $HOOC(CH_2)_6COOH$; produced by oxidation of cyclooctane; use: for manufacture of nylon 6,8

sublimation: direct vaporization of a solid; process used for purification

subsieve size: particle size definition: <37 μm diameter

substantive dyes: *see* direct dyes

substitute natural gas: SNG; produced by gasification of coal using air (oxygen) and steam: $2C + O_2 \leftrightarrows 2CO$, $C + H_2O \leftrightarrows CO + H_2$; additional reactions with CO_2 produce more CO and H_2: $C + CO_2 \leftrightarrows 2CO$, $CO + H_2O \leftrightarrows CO_2 + H_2$ (water gas shift reaction); final reactions lead to methane: $C + 2H_2 \leftrightarrows CH_4$, $CO + 3H_2 \leftrightarrows CH_4 + H_2O$; the many different gasification processes require high temperature and the heat may be supplied from external sources or from energy released by the reaction of coal with oxygen; process technology also varies as to type of coal (hard or brown), reactor (fixed-bed, flu-idized-bed, entrained-bed, reaction with par-allel or counter flow), gasification agent (water and oxygen or air, hydrogen and CO_2), and process conditions (pressure and temperature)

succinic anhydride: produced by hydro-genation of maleic anhydride as intermediate for 1,4-butanediol production; use: for man-ufacture of organic chemicals

SUCCINIC ANHYDRIDE

succinonitrile: produced by reaction of acrylonitrile with HCN; use: reduction to 1,4-diaminobutane which is the raw material for nylon 4,6

sucralfate: aluminum-containing antiulcer drug

suds stabilizer: *see* alkanol amides

suds suppressor: long-chain soaps used in laundry and dishwasher detergents

sugar refining: series of processes starting with affination, i.e, the removal of the mo-lasses film that contains most of the nonsugar solid impurities of raw sugar: the raw sugar is mixed with saturated sugar syrup, cen-trifuged, the washed crystals are dissolved in hot water and the impurities (e.g., sand and fiber) are filtered off; in the next step, the sugar liquor is clarified by heating to 60–85°C in the presence of lime (sometimes with additional phosphoric acid, CO_2 or cationic surfactant) to a final pH 7–9; the treated liquor is filtered using diatomaceous earth; the filtered dark brown liquor is decol-orized with carbon; final crystallization to granulated sugar is initiated by seeding

sulfa drugs: sulfonamides; bacteriostatic agents; derivatives or analogs of sulfanil-amide (p-$NH_2C_6H_4SO_2NH_2$) such as sul-

fabenzamide, sulfadiazine, sulfaguanidine, sulfamerizine

sulfallate: CDEC; $(C_2H_5)_2NC(S)SCH_2C(Cl)=CH_2$; preemergence herbicide; mild skin irritant

sulfamic acid: H_2NSO_3H; produced by reac-tion of urea with SO_3 and H_2SO_4; use: a sul-fating agent, for chemical cleaning, for manufacture of dyes; sulfamates are flame re-tardants, ammonium sulfamate is a herbicide

sulfanilamide: *see* sulfa drugs

sulfate naval stores: by-product from kraft pulping of pine wood; a mixture of pinenes and turpentine

sulfate process: *see* kraft process

Sulfinol process: *see* sulfolane

sulfite process: pulp production from soft wood, mainly fir and spruce using HSO_3^-, which is derived from burning sulfur to SO_2; this old process has largely given way to kraft pulping; by-products include vanillin, ethanol, and torula yeast

sulfochlorination: production of *sec*-alkyl-sulfonates by treating *n*-paraffins with SO_2/Cl_2 at 20–35°C under UV light

sulfolane: produced by 1,4-addition of SO_2 to butadiene followed by hydrogenation of the sulfolene; used in syngas purification for removal of acidic gases (Sulfinol process) and as aid for the extractive distillation of aromatics

SULFOLANE

sulfonamides: *see* sulfa drugs

sulfonation: introduction of $-SO_3H$ group into an aliphatic or aromatic compound; di-rect sulfonation of aromatics is achieved by treatment with strong sulfuric acid, oleum, SO_3 dissolved in an organic solvent, or chlorosulfonic acid; this is an important process in manufacture of dyes and detergents; *see* alkylaryl sulfonate

sulfosuccinates: produced by peroxide catalyzed addition of $NaHSO_3$ to maleic acid or its esters; use: surfactants in emulsion polymerization, to accelerate drying time during coal purification, as wetting agents during textile dyeing

sulfoxidation: manufacture of *sec*-alkylsulfonates by treating *n*-paraffins with SO_2/O_2 at 25–30°C under UV light

sulfur: S; element; produced from elemental deposits, coal purification, natural gas and ores (pyrite); use: for manufacture of sulfuric acid, sulfur oxides, sulfides, rubber vulcanization agents, sulfur dyes, sulfur concrete

sulfur dichloride: SCl_2; produced by catalyzed reaction of S_2Cl_2 with chlorine; use: for sulfidizing and chlorination reactions, for manufacture of thionyl chloride; chemical to be monitored under chemical arms accord

sulfur dioxide: SO_2; obtained by roasting sulfide ores and by burning of sulfur; 100% SO_2 is produced by absorption of SO_2 in various solvents and followed by expulsion of the gas; SO_2 is considered an environmental pollutant that is emitted during the burning of coal; use: in manufacture of sulfuric acid, sulfites, thiosulfates, alkylsulfonates, as antioxidant in food preservation, in wood pulping, for cellulose production, as disinfectant

sulfur dyes: water insoluble dyes; when reduced with sodium sulfide they are soluble and have an affinity to cellulose; they are applied in the reduced state and then oxidized on the fabric to regenerate the insoluble dye within the fibers

sulfur hexafluoride: SF_6; produced by reaction of sulfur with elemental fluorine; this nontoxic inert gas has a high dielectric constant and is used as insulating gas for high voltage equipment

sulfuric acid: H_2SO_4; marketed as chamber acid (50°Bé), tower acid (60°Bé), oil of vitriol (66°Bé), and oleum or fuming acid; manufactured by catalytic (V_2O_5, $VOSO_4$) oxidation of SO_2 to SO_3 in a fixed-bed process (contact process): SO_2 is produced by burning sulfur at 1000°C, the gas is passed through four chambers containing granulated catalyst; it enters the first chamber at 600°C and the temperature is then lowered to 420°C to shift the reaction equilibrium in favor of SO_3; from the third chamber the gas mixture is fed to an absorption tower where SO_3 is hydrated to H_2SO_4; the remaining gas mixture is then cycled through a fourth chamber and passed into a final absorption tower; sulfuric acid is the most widely used of all industrial chemicals; major users are agriculture for manufacture of fertilizers and phosphoric acid, and the chemical industry, petroleum refining, iron, steel and other metal industries, paint and pigment manufacturing, and rayon and cellulose manufacture

sulfur monochloride: S_2Cl_2; produced by passing Cl_2 over molten sulfur; toxic by inhalation and ingestion; military poison gas, monitored under chemical arms accord

sulfur trioxide: SO_3; produced by distillation of oleum with low SO_3 content in evaporators; use: for sulfonation of organic compounds, manufacture of H_2SO_4, sulfonic acids (e.g., chlorosulfonic acid) and thionyl chloride

sulfuryl chloride: SO_2Cl_2; produced by catalyzed reaction of SO_2 and chlorine on activated carbon; use: as chlorination and sulfiding agent

sulindac: nonsteroid antiinflammatory drug

SULINDAC

sunflower oil: highly unsaturated vegetable oil; northern oil contains about 70% linolenic acid and 14% oleic acid, southern oil contains about half that much linolenic and twice as much oleic acid

superconducting ceramics: *see* superconductors

superconductivity: ability of materials to conduct electric current without resistance at low temperatures; according to theory conduction electrons are highly organized in the superconducting state; they exist as loose pairs having opposite momentum and spin; these "Cooper pairs" are not scattered by the usual lattice vibrations or defects; at high temperature the binding between Cooper pair electrons is broken; *see* superconductors

superconductors: materials that conduct electricity without resistance; the superconducting transition temperature (T_c) depends on the material: mercury becomes superconducting near 0 K, Nb_3Ge at 23.2 K, $La_{2-x}Ba_x CuO_{4-y}$ at ~40 K, and $YBa_2Cu_3O_x$ (known as 123) and $YBa_2Cu_4O_8$ (124) at 90 K; critical to the structure of the 123 and 124 "high-temperature superconductors" is the absence of oxygen in one corner of the crystal unit cell; to prepare 123 a ground mixture of $Y_2 O_3$, $BaCO_3$, and CuO in atomic ratios of 1:2:3 is heated at 900°C for 10–12 h, reground, pressed into pellets and reheated overnight at 900–950°C; yttrium can be replaced by other transition metals, barium can be partially replaced by strontium and calcium; organic materials also can be formed into superconductors: salts of tetramethyltetraselenafulvalene [$(TMTSF)_2X$] have a T_c of 12 K; work on potential applications for high-temperature superconductors is in progress

supercritical fluid: SCF; fluid above the critical values of temperature and pressure; SCFs have the density and solvent power of liquids, and transport properties and compressibility of gases; these properties make SCFs excellent media for extraction of natural products, processing heavy hydrocarbons and polymers, chemical separations, tertiary oil recovery, chromatography and critical point drying; in polymer processing, SCFs can separate fractions with narrow molecular weight distributions and remove monomer from polymer leading to polymers with high purity; rapid expansion of supercritical solutions leads to the production of uniform fine particles, films, and fibers

super-fatted soaps: bar soaps with extra fat for increased emollient properties; the added fats include coconut oil fatty acid, cocoa butter, cold cream, and lanolin

superhardboard: *see* densified hardboard

superphosphate: mixture of monocalcium phosphate and calcium sulfate; used as fertilizer; *see* ordinary superphosphate (OSP); triple superphosphate (TSP)

superphosphoric acid: *see* polyphosphoric acid

surface active agent: substances that reduce the surface tension of a liquid or the interfacial tension between two immiscible liquids or between a liquid and a solid; use: detergents, emulsifiers, soaps, and wetting agents; *see* surfactants

surface methods of analysis: nondestructive

techniques used to examine the first few monolayers of a surface; some of these involve low-energy beams such as secondary ion mass spectrometry (SIMS), Auger electron spectroscopy (AES), and electron spectroscopy for chemical analysis (ESCA); higher energy techniques such as Rutherford backscattering spectrometry (RBS) and elastic recoil detection analysis (ERDA) offer information at greater depth

surface retorting: in the oil shale industry, the processing of shale above ground

surface tension: the force per unit area that maintains a surface due to intermolecular attractions

surfactants: surface-active agents that when added to a liquid, change the properties of the liquid's surface; surfactants serve as detergents, and wetting, foaming, emulsifying, dispersing, and penetrating agents; surfactants may be anionic, cationic, amphoteric, and nonionic; anionic surfactants are derived from aliphatic hydrocarbons and have high sudsing characteristics; most common are the sodium salts of linear alkylate sulfonates; unlike soaps they do not form insoluble salts with Ca^{+2} or Mg^{+2} and can thus be used in washing machines; most cationic surfactants are quaternary ammonium salts and are used as fabric softeners; amphoteric surfactants are used in personal care products where mildness is important; and nonionic surfactants include ethoxylated alcohols and alkyl amine oxides; they are low sudsing detergents used in washing machines and dishwashers; surfactants find broad use in industries such as agriculture, concrete and cement, cosmetics, metal processing, dry cleaning, food, oil field operations, paint, pulp and paper, polymer, textile; *see* detergents

surfation: contraction for surface modification; term is used in the plastics industry to indicate alterations of filler, colorant, or reinforcement that change the surface appearance

suspended solids: SS; in waste disposal, small particles that pass through the screening process and are removed by sedimentation; they sink to the bottom of sedimentation tanks to form sludge

suspension agents: *see* dispersants; antiredeposition agent

suspension fertilizers: fertilizers in which solids are held in suspension by addition of 2% gelling clay; these fertilizers are about twice as concentrated as liquid fertilizers and can carry herbicides in their formulations

suspension gasification: entrained-flow gasification; *see* coal gasification

suspension polymerization: pearl polymerization; dispersion polymerization process in which monomer is dispersed in a liquid (usually water) in which the resulting polymer is also insoluble while catalysts and additives are soluble; polymer and monomer are easily separated after polymerization; easy temperature control

Sward rocker: *see* hardness

sweating: in plastics, exudation of additives to the surface of an object

sweeteners: nonnutritive food additives that have greater sweetening power than sugar; saccharin is made from toluene, cyclamate (banned by the United States FDA) is made from cyclohexylamine, aspartame is made from L-phenylalanine and L-aspartic acid

sweet oil: crude oil containing 0.5 wt% sulfur or less

swelling agents: materials that absorb liquid or gas to form a gel

synapse: region of contact between two nerve fibers; chemicals called neurotransmitters are secreted at the synapse; by passing across the cell membranes they transmit nerve signals; the synapse is a target for drug and pesticide design

syncrude: crude oil derived from processing

carbonaceous material such as coal or oil shale

syndets: synthetic detergents

syndiotactic polymer: polymer in which the polymer side chains alternate regularly above and below a linear C-chain backbone model

syneresis: in a gel, the spontaneous contraction of colloid particles leading to agglomeration and the separation of the liquid

synergy: synergism; the mutual reinforcement of properties of two substances so that the total effect is greater than the sum of the two individual ones

synerizing: heating plastics near the softening point

synfuel: *see* synthetic fuels

syngas: *see* synthesis gas

syntactic foam: composite formed of microballoons bound with a polymer

Synthane process: *see* substitute natural gas

synthesis gas: mixture of CO and H_2 obtained from coal or petroleum; *see* petroleum; substitute natural gas; *see* PROCESS FOR PRODUCING SYNTHESIS GAS

synthetic cryolite: *see* sodium aluminum hexafluoride

synthetic detergents: surface-active agents to act as soap substitutes; detergents do not form solid precipitates in hard water

synthetic fuel: any liquid fuel made from coal, unconventional oil sources or biomass

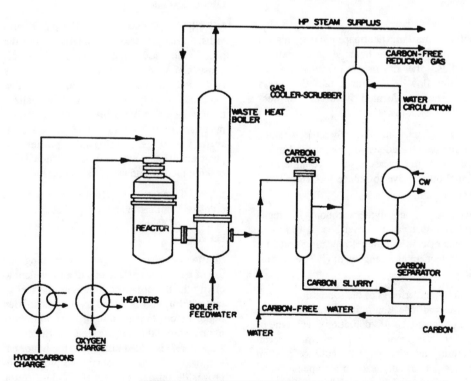

PROCESS FOR PRODUCING SYNTHESIS GAS
H_2 plus CO, or H_2 plus N_2. [*Shell Gasification Process (SGP), Shell Development Company.* Courtesy D. M. Considine, P.E.]

synthetic natural gas: *see* substitute natural gas

synthetic natural rubber: *cis*-polyisoprene; prepared by polymerization of isoprene over a triethyaluminum catalyst (Ziegler catalyst)

synthetic rubber: elastomers; polymers with rubber-like properties as differentiated from natural rubber, *cis*-1,4-polyisoprene; *see* elastomers

systemic: description of pharmaceuticals and pesticides that act by entering the organism rather than remaining in the area of application (topical agents), e.g., benomyl is a systemic fungicide

T

tablet formation: pharmaceutical agents are often compounded as homogeneous mixtures of solids; after dry or wet granulation these mixtures flow freely into a tableting press where they are formed by direct compression; *see* agglomeration

tackifiers: substances, mostly terpene resins, that increase the adhesive properties of materials; use: for pressure-sensitive adhesives, as fuel oil additive to reduce loss from bearings

tactic block: in polymer technology, a regular block that can be described by only one species of configurational repeating unit in a single sequential arrangement (IUPAC definition)

tacticity: orderliness of the succession of configurational repeating units in the main chain of a polymer (IUPAC definition); any type of regular arrangement in polymer structure can cause crystallinity while irregular or random arrangement usually leads to amorphous materials; polymers of the type $-(CH_2-CHR)_n$ may exist in two regular arrangements: "isotactic"

in which all R substituents are in the same direction along the chain and "syndiotactic" in which R alternates regularly between the two available directions along the chain; when R changes randomly along the chain; the polymer is said to be "atactic"; tacticity affects the polymer properties, e.g., isotactic polypropylene is highly crystalline and melts at 165°C, while atactic propylene is rubbery at room temperature; the discovery of stereospecific polymerization methods has lead to the development of a broad field of new polymers of vast commercial importance; *see* Ziegler–Natta catalysts; stereoregular polymers

talc: talcite; $Mg_3Al_4O_{10}(OH)_2$, the softest mineral on the Mohs hardness scale; use: lubricant and filler

talcite: *see* talc

tall oil: by-product from sulfite paper–pulp mills; a nonglyceride oil obtained from the pine wood used in paper making; tall oil consists of 35–40% rosin acids, 50–60% fatty acids, and varying amounts of unsaponifiable material; it may also contain ≤10% sitosterol, which is used to make cortisone; use: manufacture of scouring soaps, paper sizing, surface coatings, printing inks, putties, asphalt emulsions, cutting oils, insecticides, in making factice, and in paints and plastics

tallow: fat containing C16 to C18 fatty acids; obtained from cattle and sheep; use: in manufacture of soap, stearic and oleic acids, abherents

talofloc process: in sugar cane refining, the use of a cationic surfactant and a flocculant to clarify the liquor before filtration; the surfactant ("talofloc") is added to the unclarified liquor before the addition of lime and phosphoric acid; the polyacrylamide flocculant ("taloflote") is added just before the liquor enters the clarifying chamber; the process improves decolorization as well as clarification

taloflote: *see* talofloc process

tamoxifen: p–$(CH_3)_2NCH_2CH_2OC_6H_4C(C_6H_5)$=$C(C_6H_5)CH_2CH_3$; anticancer drug; antiestrogen that forms a stable complex with receptors for female sex hormones but does not stimulate protein synthesis; the citrate is used in breast cancer therapy

tangent modulus: on a stress–strain curve, the slope of the line at any point

tannic acid: *see* tannins

tanning: preservation of hides by stabilization of collagen fibers against bacterial attack and temperature while retaining their flexibility; achieved by soaking hides in solutions of tanning agents such as natural tannins, formaldehyde, melamine, or salts; in the United States chromic acid derivatives (chrome tanning) are used for 95% of the leather produced; other steps in tanning include initial cleaning of the hides, hair removal, oil finishing, and dyeing

tannins: complex, nonuniform mixtures of polyhydroxy phenols; commercial tannic acid is usually given the empirical formula $C_{76}H_{52}O_{46}$; natural tannins are obtained by extraction with warm water and/or alcohol from plants; they fall into two classes: hydrolyzable tannins (esters of a polyhydroxybenzoic acid and a sugar [usually glucose]) that yield pyrogallol on dry distillation and condensed tannins (derivatives of flavanol) that yield catechol on dry distillation; the most important source of tannins is the South American quebracho tree, other sources include hemlock, larch, oak, willow, sumac, chestnut, and gallnuts; use: tanning leather, mordant in dyeing, coagulant in rubber manufacture, manufacture of gallic acid and pyrogallol, clarifying beer and wine, astringent, heavy metal antidote, protein precipitant

tapping: incisions made into trees to release resins such as rubber latex

tar sands: deposits of finely divided clays or siliceous materials coated with bitumen and water

tartaric acid: $HOOCCH(OH)CH(OH)COOH$; produced by reaction of maleic acid or maleic anhydride with hydrogen peroxide using Mo or W catalyst; use: in pharmaceuticals, effervescent beverages, as mordant in dyeing

tartrazine: acid-azo dye

TARTRAZINE

Tatoray process: production of xylene by disproportionation of toluene at 350–530 °C and 10–50 bar

T cells: white blood cells (lymphocytes) that mediate the cellular immune response; T cells have receptors on their surface that permit them to recognize foreign molecules; one type of T cell (killer T cells) destroys foreign cells, whereas another type (helper T cells) modulates immune responses of other blood cells (B cells); destruction of T cells by the HIV virus causes AIDS syndrome

Technically Specified Rubbers: TSR; grade of natural rubber specified by the Rubber Manufacturers' Association

Techni-Chem process: manufacture of caprolactam from cyclohexanone via 2-nitrocyclohexanone followed by ring opening, hydrogenation and hydrolysis

telomer: low molecular addition polymer in which the molecular growth is terminated by a chain transfer reagent that supplies a free radical; for example, reaction of ethylene with CCl_4 using light or a free radical initiator leads to $Cl(CH_2CH_2)_nCCl_3$ where n is

1–7; term is also used for polymer with 2–10 repeating units; *see* oligomer

temazepam: benzodiazepine derivative; anxiety reducing drug, sedative; *see* diazepam

temik: *see* aldicarb

tenacity: in the textile industry, tenacity is measured in grams of breaking strength of a fiber or yarn per denier, gpd

tensile modulus: *see* modulus

tensile strength: resistance to pulling stresses; maximum load or stress sustained by a solid during a tension test (ASTM D 638-72); expressed in lb/in.^{2}, kg/cm^2, or gpd

tentering: process of stretching a material using pins on a board to maintain shape; orientation of films

teratogens: substances such as alcohol (ethanol) that cause birth defects by acting on the fetus; they are particularly damaging in the early stages of development

terephthalic acid: TPA; p-HOOCC$_6$H$_4$COOH; produced by liquid-phase catalytic oxidation of p-xylene; in most processes the initial product, p-toluic acid (CH$_3$C$_6$H$_4$COOH) is esterified with methanol before oxidation of the second methyl group; the intermediate after the second oxidation consists of mixtures that include dimethylterephthalate (DMT); final hydrolysis yields TPA; also produced by isomerization of the potassium salt of o-phthalic acid in the presence Zn-Cd catalyst; used for manufacture of polyethylene terephthalate (PET),–[–COC$_6$H$_4$COOCH$_2$CH$_2$O–]$_n$–

terfenadine: antihistamine that does not cause drowsiness

termination: in polymer technology, the stoppage of the polymerization reaction

terpenes: unsaturated hydrocarbons that can be considered dimers, trimers, or higher oligomers of isoprene, C$_5$H$_8$; derived from turpentine and natural oils; use: in perfumery, as antioxidants, as tackifiers; *see* naval stores

terpolymer: polymer consisting of three different monomers

tertiary butyl hydroquinone: antioxidant

tertiary oil recovery: *see* enhanced oil recovery

tertiary waste water treatment: advanced waste treatment; includes chlorination, chemical precipitation, reverse osmosis, activated carbon absorption, oxidation ponds, and electrodialysis; the result is a high-quality effluent

testosterone: male sex hormone; steroid; use: body-building after major surgery or to counter debilitation; use by athletes to improve performance, is forbidden by athletic associations

tetrabromobisphenol A: flame retardant for plastics, paper, and textiles

TETRABROMOBISPHENOL A

tetrabromophthalic anhydride: *see* fire retardant

tetrachloroethylene: Cl$_2$C=CCl$_2$; produced (1) together with trichloroethylene from 1,2-dichloroethane by combined dehydrochlorination and chlorination at 350–450°C using CuCl$_2$, AlCl$_3$, or FeCl$_3$ as catalyst; (2) by cracking propane-propene mixtures at 450–550°C in the presence of chlorine (chlorinolysis); a mixture of trichloroethylene and tetrachloroethylene is called perchloroethylene; use: dry cleaning agent; limited because of health problems

tetracyclins: group of broad-spectrum antibiotics characterized by a linear fused four-ring system; derived from *Streptomyces* species by fermentation

tetraethyl lead: $(C_2H_5)_4Pb$; TEL; anti-knock gasoline additive; phased out in the United States because of pollution hazard

tetraethylene glycol dimethyl ether: tetraglyme; $CH_3O(CH_2CH_2O)_4CH_3$; *see* ethylene glycol ethers

tetraglyme: tetraethylene glycol dimethyl ether; *see* ethylene glycol ethers

tetrahydrofuran: THF; diethylene oxide; C_4H_8O; prepared by dehydration of 1,4-butanediol in the presence of phosphoric or sulfuric acid at 110–120°C; use: solvent, especially for polymers, precursor for polyurethane and for polytetramethylene ether glycol; THF is an irritant to mucous membranes

TETRAHYDROFURAN

tetramer: oligomer of four monomer units

1,2,4,5-tetramethyl benzene: durene; derived from higher boiling petroleum fraction or from coal gasification; use: manufacture of 1,2,4,5 benzenetetracarboxylic acid and its anhydride, pyromellitic anhydride ("pyro"); both carboxylic compounds are important chemical synthons, pyro is used for manufacture of heat-resistant polyimides

tetramethylthiuram disulfide: $(CH_3)_2NC(S)SSC(S)N(CH_3)_2$; produced by oxidation of the dithiocarbamate, $(CH_3)_2NC(S)SH$; use: accelerator for vulcanization of rubber, fungicide

tetrapotassium pyrophosphate: TKPP; $K_4P_2O_7$; detergent builder; use: for liquid detergents because of its water solubility

tetrapropene: "iso-$C_{12}H_{24}$"; obtained from refinery gases; formerly used as olefin feedstock for starting materials for alkylarylsul-fonates; these have been phased out because they are not biodegradable

tetrasodium pyrophosphate: TSPP; $Na_4P_2O_7$; builder for heavy-duty laundry soaps

tetryl: $[NO_2)_3C_6H_2N(CH_3)NO_2$; nitramine; produced by nitration of the benzene ring of dimethylaniline followed by removal of one CH_3 on the nitrogen by oxidation, decarboxylation, and final nitration; a military explosive, used as booster; when cast after slurrying with molten TNT it is used as a bursting charge, called Tetritol

Tetritol: *see* tetryl

tetrole: *see* furan

textile sizing: generic term for coating yarns or fabrics to protect them from abrasion, yarn and filament breakage during textile operations; sizes are hydrophilic materials such as starch, soluble gums, natural and synthetic hydrophilic polymers, antistatic agents

texturing: in textile technology the conversion of as-spun yarns into a wavy, crimped form; air jet texturing uses hot air or steam, overfeeding yarn into a tube or box and setting it in the compacted (crimped) form with heat or steam; the textured yarn is cooled, unraveled and wound up

T_g: glass transition temperature; in polymer technology, T_g defines the transition from a glassy (stiff) structure to a more flexible, mobile plastic system; deformation and crystallization occurs above this temperature

thallium: Tl; metallic element; use: bait for insects and as rat poison, amalgam is used for switches that must operate <0°F; toxic: causes nausea, weakness in arms and legs, may lead to coma and death

thermal analysis methods: methods characterize the thermal behavior of compounds and materials (1) DTA, differential thermal analysis and DSC, differential scanning calorimetry observe the crystallization and melting temperatures (transition temperatures) and the energies evolved or absorbed in these transitions; (2) TGA, thermogravimetric analysis observes the weight changes in a material and in conjunction with mass spectrometry identifies the volatile products in this process; (3) TMA, thermomechanical analysis provides information on mechanical behavior of oriented or unoriented systems under a specific load, i.e, extension, shrinkage, transition temperatures, expansion coefficients; (4) DMA, dynamic mechanical analysis measures the resonance frequency and mechanical damping of a material as a function of temperature; DMA data correlate with the elastic modulus and energy dissipation of the sample

thermal black: carbon black produced by the thermal decomposition of natural gas

thermal cracking: thermocracking; heating hydrocarbons under pressure to 400–900°C; residence times decrease with higher temperature; use: production of olefins from petroleum

thermal decomposition: *see* destructive distillation

thermal fluids: liquids used as working fluids in heat transfer equipment; examples: oils for heat storage tanks, ammonia for converting the heat in ocean surface layers into usable energy (ocean thermal temperature conversion, OTEC)

thermal–mechanical pulping: in wood technology, the use of steam or hot water in conjunction with defibering; continuous version of the process operates by using steam at 170–190°C and 100–165 psi

thermal pollution: damage to aquatic life due to discharge of heated effluents from industrial plants and energy generators; control is achieved with cooling towers

thermal polymerization: polymerization at elevated temperature in the presence or ab-

sence of free radical initiators; use: production of acrylates, methacrylates, styrene

thermal printing: nonimpact printing method in which ink from an ink donor film (IDF) is transferred to the underlying paper; the waxy ink on the IDF is melted by a thermal print head that passes over the film and resolidifies on the receiving sheet; *see* printing

thermal transfer printing: *see* thermal printing

thermionic emission: emission of electrons or ions as a function of temperature

thermocouple: device consisting of a wire having a junction between two metals such as bismuth and antimony; emits a small voltage proportional to the temperature of the junction

thermocracking: *see* thermal cracking

thermogravimetric analysis: TGA; measurement of weight change of a sample with increasing temperature; samples usually lose weight due to decomposition or dehydration; thus the resulting thermograms give information concerning thermal stability of materials; *see* thermal analysis methods

thermolysis: decomposition of a substance by heating; *see* ketene

thermoplastic elastomers: these elastomers do not need vulcanization; example: spandex fibers that have physical cross-links but melt at elevated temperatures

thermoplastic polyesters: unfilled or glass-filled semicrystalline polyesters (poly[butylene terephthalate] and poly[ethylene terephthalate]); they have good physical and electrical properties, toughness, low moisture absorption and thermal stability, good chemical and hydrolytical stability; glass reinforcement produces a considerable increase in flexural modulus and flexural stability; thermoplastics have better processing and recycle options for scrap materials than thermoset resins; *see* thermoplastics

thermoplastics: materials that can be softened repeatedly by heating and hardened by cooling; for example, cellulose esters and ethers, polyolefins (e.g., ethylene and isotactic propylene), polystyrene, poly(methylmethacrylate), polyesters (e.g., polyethylene- and poly[butylene terephthalates]), nylons, linear polyurethanes, and polyureas

thermosets: *see* thermosetting plastics

thermosetting plastics: thermosets; plastics that cannot be softened after curing; usually obtained by cross-linking; examples: phenol-formaldehyde resins, polyesters made from acids and/or alcohols that have more than two reactive sites (e.g., glycerol, tribasic acids) or from unsaturated alcohols and/or acids, urea- and melamine-formaldehyde resins, polyurethane foams, and diallyl phthalates

thermotropic: compound that shows liquid crystalline phases as a function of temperature; *see* liquid crystals

thickeners: thickening agents; thixotropic agents; substances that increase the viscosity of fluids, e.g., starches, gums, gelatin, carboxymethyl-cellulose, polyvinyl alcohol, colloidal silica; use: in foods, detergents, coatings and paints

thinners: in the paint industry, solvents dilute the solution; in medicine, drugs that reduce the tendency of blood to clot

thiofuran: *see* thiophene

thiokol: organic polysulfide rubber, introduced in the United States in 1930

thionite: *see* sodium sulfite

thiophene: thiofuran; C_4H_4S; made by (1) dehydrogenation of butane in the presence of sulfur; (2) passing ethylene or acetylene into boiling sulfur; (3) passing acetylene and H_2S over hot bauxite; or (4) reaction of *n*-butanol and CS_2; use: solvent, for manufacture of

resins, conducting polymers, agrochemicals, dyes and pharmaceuticals

THIOPHENE

thixotropy: decrease in viscosity with stirring; when stirring ceases the original viscosity returns; materials providing thixotropic effects in polymeric materials are fine-sized compounds such as fumed silica, precipitated silica, organophilic clays, and some organics; these fine powders seem to have hydrophilic surface structures forming interparticle hydrogen bonds thus increasing the viscosity of the resin; fast flow reduces the viscosity by destroying these bonds; use: with unsaturated polyester resins, epoxies, acrylics, urethanes; *see* thickeners

threshold limit value: TLV; in public health, the allowable exposure to an agent for an average 8-hr day or a 40-hr week

thymine: one of the four bases in DNA

THYMINE

tires: there are three types of tires (1) the bias or cross ply tire in which 2–4 plies of rubber-coated reinforcing cords extend diagonally to circles of high-tensile brass-coated steel wires ("beads") that are attached to the rim of the wheel; the successive plies run at right angles to each other; (2) the bias/belted tire also consists of bias plies but has in addition an outer stabilizing ply ("belt"); (3) in the radial tire the plies are at a right angle to the travel direction of the tire and a restricting belt runs along the circumference of the tire; rayon that has previously been used for plies and belts has been replaced by nylon and polyester with increasing use of wire; lining the plies is a thin layer of rubber ("inner layer") that holds in the compressed air; topping the plies and belt is the tread that is usually made of a blend of styrene–butadiene rubber (SBR) and polybutadiene; the sidewall rubber sometimes contains white pigment and is specially formulated for high flexibility and impact resistance; design characteristics of the upper portion of the sidewalls just below the tread ("shoulder") affect heat behavior of the tire

Tishchenko reaction: formation of esters by catalytic disproportionation of aldehydes in alkaline solution; industrially used for manufacture of ethylacetate from acetaldehyde

titania: *see* titanium dioxide

titanium: Ti; metallic element; important minerals are anatase, ilmenite, and rutile; used in hard alloys; *see* titanium dioxide

titanium dioxide: TiO_2; titania; manufactured by oxidation of titanium tetrachloride; use: white pigment (in paints, inks, plastics, fibers, paper, rubber), opacifying agent (smoke screens), semiconductor (as single crystals), gem stone, optical lenses, refractory and electrical insulator

titanium tetrachloride: $TiCl_4$; made by chlorination of the ore rutile in the presence of carbon; use: raw material for titanium dioxide manufacture

tolbutamide: $p–CH_3–C_6H_4SO_2NHCONH$ $(CH_2)_3CH_3$; made from butylisocyanate and 4-toluenesulfonamide; oral blood pressure lowering agent

tolerance: in government regulation of toxic substances such as pesticides, the residue level that is presumed safe

toluene: $C_6H_5CH_3$; obtained by catalytic reforming of petroleum and by fractional distillation of coal-tar oil ; use: fuel, solvent, raw material for organic chemicals; hazardous chemical waste (EPA)

toluene-2,4 diamine: TDA; toluene di-amine; $CH_3C_6H_3(NH_2)_2$; raw material for manufacture of tolylene diisocyanate

toluene diisocyanate: TDI; $CH_3C_6H_3$ $(NCO)_2$; *see* tolylene diisocyanate; urethane block copolymers

toluic acids: $CH_3C_6H_4COOH$; by-product during oxidation of xylenes to phthalic acids; use: intermediate for manufacture of polyester resin, chemical intermediate, bacteriostat (*o*-toluic acid), broad-spectrum insect repellant (*m*-toluic acid), animal feed supplement (*p*-toluic acid)

toluidine: $CH_3C_6H_4NH_2$; *o*-,*m*-, and *p*-aminotoluene; produced by nitration of toluene to nitrotoluene followed by hydrogenation; *m*-toluidine is made by reaction of *m*-cresol with ammonia; use: raw material for dyes, chemical intermediate

tolylene diisocyanate: TDI, toluene diisocyanate; the 2,4 and 2,6 isomers are industrially important; manufactured by hydrogenation of dinitrotoluene to toluene diamine followed by phosgenation; use: starting material for manufacture of polyurethanes

TOLYLENE DIISOCYANATE

toner pigments: organic coloring materials that are insoluble in pure form

Topham box: *see* package spinning

torula yeast: produced as by-product during wood pulping; use: source of single cell protein (SCP), animal feed

tower fermentor: column fermentor with a perforated plate; permits recycling of cells that are retained above a perforated plate while the product is removed from the bottom; aeration occurs through the plate; use: for manufacture of vinegar and ethanol

toxaphene: chlorinated camphene, a complex but reproducible mixture of polychloro compounds; use: insecticide; toxic: can be absorbed through skin, causes central nervous system stimulation that may lead to tremors, convulsions, death

trace analysis: see microanalytical methods

tranquilizers: antianxiety drugs (minor tranquilizers) include diazepam and meprobamate; antipsychotic drugs (major tranquilizers) include chlorpromazine, thioridazine, and haloperidol

transalkylation: catalyzed transfer of alkylgroups from polyalkylated aromatics to less alkylated aromatics

Transcat process: production of vinyl chloride by chlorination and dehydrochlorination of ethane in a copper chloride/alkali chloride melt at 450–500°C

transcription: in genetic engineering, the formation of an RNA copy from the DNA template

transdermal drug delivery systems: skin patches in which drugs are encapsulated or imbedded in a nonactive matrix and attached to the patient's skin with an adhesive; a semipermeable membrane permits the constant, slow release of the drug that then passes through the skin into the body; the system provides constant drug dose levels and bypasses the digestive hormones that might inactivate the dugs

transducer: device for transforming one form of energy into another; device activated by a form of energy that is then converted into a suitable signal

transduction: in genetic engineering, the process of incorporating a foreign DNA into the genetic apparatus of the host cell

transesterification: exchange of the alcohol moiety of two esters; use: for the production of higher dialkyl and diphenyl carbonates and higher esters of methacrylic acid

transfer molding: process for molding thermosetting plastics: a preheated softened material is pushed through an orifice into a heated mold cavity and is held there to polymerize and cross-link to the final rigid product

transfer RNA: tRNA; transfer ribonucleic acid; tRNAs receive the code for amino acids from messenger RNA (mRNA); they collect the specific amino acid from the cell interior and carry it in an activated state to the ribosome where protein synthesis takes place; at least one tRNA exists for each of the 20 amino acids; see ribonucleic acid

transformation: in genetic engineering, a cell's acquisition of new heritable properties upon uptake of foreign DNA

transformation toughening: reduction of ceramic brittleness achieved by solid phase transformation, e.g., change from tetragonal zirconia to the less dense, more stable monoclinic form

translation: in genetic engineering, the process by which the genetic code carried in the messenger ribonucleic acid (mRNA) directs the synthesis of proteins

translocation: in genetic engineering, during biological protein synthesis the process by which one tRNA displaces its predecessor as it moves through the ribosome

translucent soap: soap with a high glycerol content but not as high as transparent soap

transmission electron microscopy: TEM; microanalytical technique that yields surface area images of a few hundred nanometers with atomic resolution; incoming beam consists of high-energy electrons that pass through materials thinner than 0.5 μm; best image resolution is ~0.15 nm and requires samples thinner than 0.05 nm

transparent soap: soap with higher glycerol content than tranlucent soap

tread: *see* tires

treflan: herbicide; *see* trifluralin

triacetate: *see* cellulose acetate

triacetin: *see* glycerol

2,4,6-triamino-s-triazine: *see* melamine

triamterene: diuretic

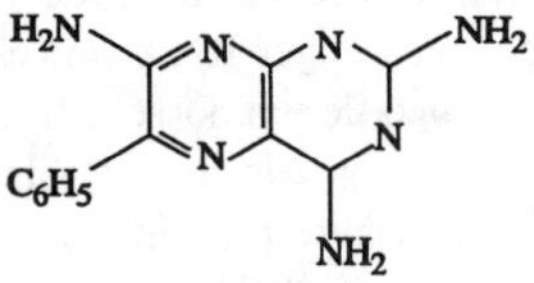

TRIAMTERENE

triarylmethanes: *see* dyes

s-triazine derivatives: these compounds include resins, dyes, and pesticides; triazines react with H_2S and not with CO_2 and are thus used for sweetening gas

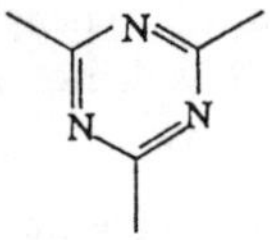

S-TRIAZINE DERIVATIVES

triazine dyes: dyes for cotton and other cellulosic fibers; produced by reaction of a dye that contains a –OH or -NH_2 group with cyanuric chloride; compound then reacts with cellulose to form a dye–cyanuric acid–fiber compound

triazine herbicides: group of heterocyclic nitrogen herbicides, e.g., atrazine

tricalcium phosphate: $Ca_3(PO_4)_2$; use: anticaking and suspension agent, polishing agent in toothpaste; *see* bone phosphate of lime

trichlorocarbanilide: $C_6H_3Cl_2NHCONH$ C_6H_4Cl; bacteriostat used in plastics, detergents and soaps; acts as deodorant in soaps by preventing growth of skin bacteria

trichloroethylene: 1,1,2-trichloroethylene; $Cl_2C=CHCl$; use: solvent and extractant for fats, oils resins and waxes; *see* tetrachloroethylene

trichlorohydroxy diphenyl ether: *see* triclosan

2,4,6-trichloro-1,3,5-triazine: cyanuric chloride; produced by gas-phase trimerization of ClCN; use: precursor for 1,3,5-triazine pesticides and dyes; *see* s-triazine derivatives

trickling filters: in waste water technology, sewage is trickled over beds of rock that supports bacteria that can break down organic wastes

triclocarban: *see* trichlorocarbanilide

triclosan: short for trichlorohydroxy diphenyl ether; bacteriostat and deodorant used in soaps

tridemorph: *see* fungicides

tridentate: *see* ligands

triethyl aluminum: Ziegler catalyst; co-catalyst for stereospecific polymerization via Ziegler–Natta catalyst

trifluralin: treflan; preemergence herbicide

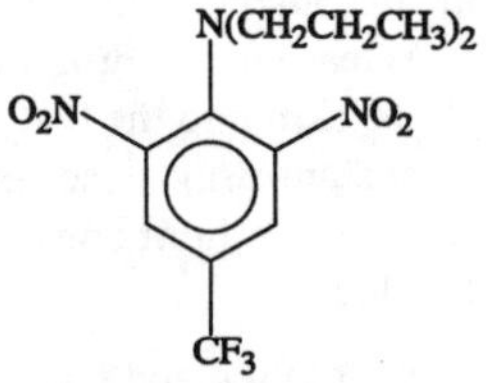

TRIFLURALIN

triglycerides: glycerol esters; fats are triglycerides of long fatty acids

triglyme: *see* ethylene glycol ethers

triisobutyl aluminum: catalyst for stereospecific polymerization

trimellitic anhydride: anhydride of 1,2,4-benzenetricarboxylic acid; produced by gas-phase oxidation of 1,2,4-trimethylbenzene;

use: manufacture of plasticizers and polyimides, plasticizer for polyvinyl chloride

trimethoprim: pyrimidine antibiotic that works synergistically with sulfadrugs

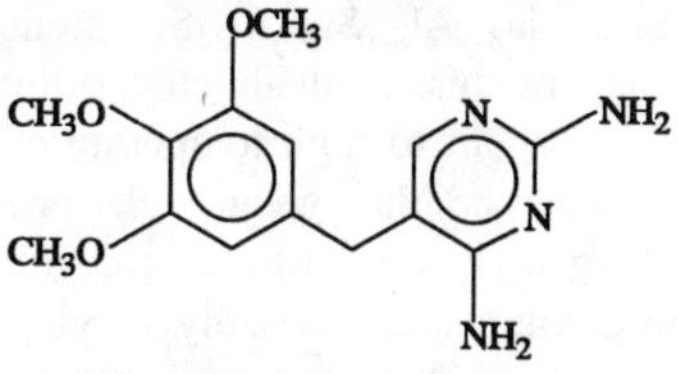

TRIMETHOPRIM

trimethylacetic acid: pivalic acid; $(CH_3)_3$ $CCOOH$; produced by carbonylation from isobutene; use: in production of synthetic oils, paints, and resins; *see* Koch acids

trimethylolpropane: $C_2H_5C(CH_2OH)_3$; hexaglycerol; 1,1,1-*tris*(hydroxymethyl)-propane; produced by condensation of *n*-butyraldehyde with excess formaldehyde; use: glycerol substitute in manufacture such as for varnishes, resins, polyesters, polyurethanes

1,3,5-trinitrophenol: *see* picric acid

2,4,6-trinitrotoluene: TNT; made by nitration from dinitrotoluene; explosive

triol: alcohol having three OH groups

Triolefin process: Phillips Triolefin process; olefin metathesis of propene to yield butene and ethylene using WO_3/SiO_2 catalyst at 370–450°C; other catalysts and process conditions have been used for metathesis with other olefins, e.g., CoO-Mo O_3/Al_2O_3 and 120–210°C at 25 bar; Rhe_2 O_7/Al_2O_3 at room temperature; the commercial production of propylene from ethylene and butene-2 illustrates the reverse metathesis process

trioxane: cyclic trimer of formaldehyde; produced by treating formaldehyde with acid; use: disinfectant, monomer for polyoxymethylene, raw material for organic synthesis

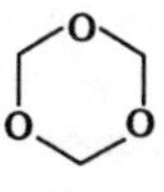

TRIOXANE

tripelennamine: antihistamine; made by sodamide condensation of *N*-benzyl-2-aminopyridine (BAP) with dimethyl aminoethylchloride (DMAEC)

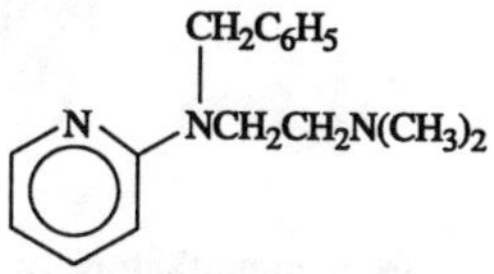

TRIPELENNAMINE

triple superphosphate: TSP; monocalcium phophate; $Ca(H_2PO_4)_2$; contains about 47% P_2O_5; prepared by acidulation of phosphate rock; use: as granulated fertilizer

tripotassium phosphate: detergent builder

trisodium phosphate: detergent builder

tritonal: aluminized trinitrotoluene (TNT); explosive

trona: sodium ore; source of sodium sesquicarbonate

tryptophan: amino acid

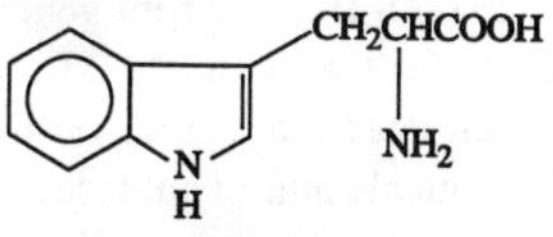

TRYPTOPHAN

tung oil: China wood oil; drying oil derived from seeds of the Chinese tree *Aleurites cordata;* use: for paints and varnishes

Turnbull's blue: *see* ferrocyanate pigments

turpentine: $C_{10}H_{16}$; produced by (1) steam distillation of pine resin; (2) naphtha extraction of pine stumps; or (3) destructive distillation of pine wood; by-product during Kraft pulping of pine wood; use: solvent, insecticide, for synthesis of camphor and menthol, in linaments, in perfumery

283

Twitchell process: old batch process for manufacture of fatty acids; acid hydrolysis of fat in the presence of sulfonated petroleum products; mixture undergoes 2–4 cycles of boiling with live steam; each cycle takes 1–2 days and at the end of each cycle free fatty acids are drawn off

U

Ube process: manufacture of hydroquinone by oxidation of phenol with hydrogen peroxide; *see* hydroquinone

Udex process: recovery of aromatics by liquid–liquid extraction of petroleum reformate

Ugine–Kuhlmann process: manufacture of hydrazine; *see* hydrazine

ultimate elongation: elongation of a material at rupture

ultra-accelerator: in rubber technology, accelerators derived from dithiocarbamic acid (NH_2CSSH)

ultracentrifuge: high-speed centrifuge rotating at 20,000–100,000 rpm generating forces of $\leq 540,000$ times the gravitational force; use: sedimentation studies of high polymers and biologicals and of molecular weight distributions; also for separations of solutes from solutions

ultrafiltration: pressure driven process that is based on liquid flow through a porous membrane; permits solute retention of substances with molecular weights as low as 500 daltons; industrial applications include recovery of cheese whey that was previously discarded, purification of pharmaceuticals, recovery of electrocoat paint, heavy metals, and proteins, and effluent treatment in the paper and pulp industry; used in artificial kidney dialysis

ultrahigh molecular weight HDPE: *see* gel spinning

ultramarine pigments: group of inorganic pigments with the composition $Na_8Al_6Si_6O_{24}S_x$ or $Na_{8-y}Al_{6-y}Si_{6+y}O_{24}S_x$; produced by heating a mixture of alkali, clay, sulfur, and a reducing agent to high temperature; ultramarines can be blue, green violet or red, depending on the chromogenic group S_x; desired color is achieved by varying manufacturing conditions; use: in paints, inks, cosmetics, textiles, paper, plastics; ultramarine blue is used as bluing agent in laundry detergents and shampoos to change the yellowish white to a bluish white

ultrasonication: process using ultrasound

ultrasound: sound vibrations beyond the audible limits; ultrasonic waves cause small bubbles to form, which on collapse induce shock waves that ripple through the liquid, dislodging dirt and particles from regions not easily accessible by ordinary brushing; use: stirring heterogeneous reactions, cleaning laboratory ware and molded plastic parts, atomizing gases; current research involves use of ultrasound as an energy source for chemical reactions, e.g., ultrasound gives Diels–Alder reactions with increased yields under mild conditions

ultraviolet absorber: in polymer technology, additive for plastics that absorbs destructive UV radiation and dissipates it as heat

ultraviolet photoelectron spectroscopy: UPS; a microanalytical technique to determine the elemental and molecular composition of solid surfaces; bombardment with UV light leads to information about a few atomic layers with a millimeter lateral resolution

ultraviolet spectroscopy: absorption spectroscopy using electromagnetic radiation of wavelengths 10–400 nm

umber: naturally occurring brown iron

oxide pigment containing 45–70% Fe_2O_3 and 2–20% MnO_2; burnt umber is more red than raw umber and is made by calcining the ore; largest deposit of umber is found in Cyprus

unbuilt detergent: generally light-duty detergents; *see* builder

undecenoic acid: monounsaturated C11 fatty acid; produced from castor oil by trans-esterification to methylricinolate followed by pyrolysis to methylundecenate, which is hydrolyzed to the free acid; use: intermediate in manufacture of nylon 11, antifungal agent

Unidak process: *see* naphthalene

unigrater: *see* Ducasse shredder

Unipol Process: *see* linear low density polyethylene

unit cell: the smallest crystallographic unit; dimensions of unit cells can be obtained with X-ray diffraction of crystals

unit operation: in chemical engineering, a particular operation used during the industrial production of a chemical, such as filtration, precipitation, heat transfer

unit process: in chemical engineering, a particular process used during the industrial production of a chemical, such as hydrolysis, oxidation, reduction

uracil: one of the four bases present in RNA

URACIL

uranium: U; radioactive metallic element; recovered by mining from uranium oxide deposits, as by-product of phosphoric acid manufacture (wet process) from phosphate rock which contains 0.005–0.03 wt% U_3O_8; used as nuclear fuel

uranium-235: isotope used for fissile fuel and weapons; produced by isotope enrichment of gaseous UF_6 through diffusion or centrifugation; enriched gas is converted to the stable solid UO_2; which is then used as fuel

uranium hexafluoride: UF_6; produced by conversion of UO_2 into UF_4 that is then treated with F_2 to give the hexafluoride; used for U-235 enrichment by gas diffusion or centrifugation; the enriched gas is reconverted to UO_2, which serves as fuel in fission reactors

urban runoff: storm waters from city streets and gutters

urban waste: solid waste, i.e, garbage from cities; *see* recycling solid waste

urea: $CO(NH_2)_2$; manufactured from ammonia and CO_2 in two steps (1) the exothermic formation of ammonium carbamate; (2) by the endothermic conversion to urea; manufacturing processes include Allied Chemical C.P.I., Chemico "Thermo-urea," DuPont, Mitsui Toatsu Chemicals, Pechiney, Snam Progetti, and Stamicarbon process; sold primarily in the form of prills, and also as crystals, flakes or granular material; use: nitrogen fertilizer, raw material for the manufacture of polymers (*see* melamine resins) and organics, slow release fertilizer, additive to soften plastics and to decrease the viscosity of glues, gelatins and starch, in the petroleum industry as agent to separate C15 to C30 n-hydrocarbons in the form of crystalline complexes

urea resins: group of thermosetting amino resins and plastics and urea formaldehyde resins; i.e, based on formaldehyde, urea, melamine (2,4,6 triamino-s-triazine); the resulting polymer networks depend on the precise control of two major reactions: methylolation of the amine with formaldehyde: $RNH_2 + HCHO \rightarrow RNHCH_2OH$ and methylene bridge formation: $RNHCH_2OH + R'NH_2 \rightarrow RNHCH_2NHR' + H_2O$; resins for special applications use various aldehydes, amines and alcohols and additional raw materials including ethylene diamine, 1,3 diaminopropane, alkanolamine, melamine, glyoxal; use: adhesives, paper manufacture, surface coatings and textile finishes (for wash and wear, durable press applications), melamine-formaldehyde, urea-formaldehyde, methylol-carbamate are used in the resins

urethane: carbamic esters, RNHCOOR'; product of the reaction of an isocyanate and an alcohol (RNCO + HOR' $\rightarrow$ RNHCOOR') or a chloroformate with an amine (ROCOCl + H$_2$NR'$\rightarrow$ ROCONHR' + HCl); *see* polyurethanes, isocyanates

urethane block copolymers: polymers prepared from diisocyanates, aliphatic polyols or polyesters and short chain glycol extenders; the urethane link -NHCOO- is formed during the polymerization process; linear urethane polymers are thermoplastic, have good impact strength, good physical properties and processability by either solvent or thermal methods; thermoset polymers are obtained from isocyanates and/or polyols or polyesters with more than two reactive groups; starting materials include isocyanates: toluenediisocyanates (TDI), polymeric methylene diisocyanate (PMDI) (polyfunctional), and 4,4'-methylene-*bis*(phenyl)isocyanate (MDI); polyols or polyethers (e.g., from cyclic oxides such as tetrahydrofuran), polyesters based on aliphatic diacids and glycols; the molecular weight of these ethers or esters is about 2000; applications: linear polyurethanes for elastic fibers (spandex); cross-linked urethanes for foams; flexible foams for furniture and bedding are lightly cross-linked and processed in the presence of a blowing agent; rigid foams for insulation as boards or laminates have a higher concentration of cross-links; these urethanes are also available as liquid precursor systems to be combined or sprayed at the site to be insulated.

urethane polymers: *see* urethane block copolymers

V

vaccine manufacture: vaccines are immunizing agents prepared from killed bacteria or viruses, or from live attenuated viruses; virus is grown in media such as eggs, separated by centrifugation and inactivated with chemicals such as formalin or by heat; the attenuated viruses act by eliciting an immune response (production of antibodies) without causing the disease

vanadium carbide: hard, brittle material; produced by carburization of V$_2$O$_5$ in vacuum at 1700°C; use: particle growth inhibitor in tungsten WC-Co alloys, steel additive

vanadium (V) oxide: vanadium pentoxide; V$_2$O$_5$; produced by extraction from vanadium ores or from ferrophosphorous slag that contains 7–14% vanadium as V$_2$O$_5$; use: oxidation catalyst

van der Waals interactions: weak intermolecular forces due to electronic motions within atoms and molecules

vancomycin: antibacterial drug; restricted to hospital use against staphylococci that are resistant to less toxic antibiotics

vanillin: 3–CH$_3$O-4-HOC$_6$H$_3$CHO; produced (1) by extraction with ethanol from vanilla bean; (2) synthetically from lignin in pulp–mill waste liquors, from eugenol in clove oil or from coniferin isolated from northern pine; use: flavoring agent

vapam: CH$_3$NHCSSNa; carbamate soil fungicide

vapor–liquid chromatography: *see* chromatography

vapor-phase-grown carbon fiber: VGCF; whiskers; high-strength discontinuous fiber; produced by heating a mixture of hydrogen and hydrocarbon in an electric furnace to above 1000°C; fine precursor fibers grow on metal catalyst particles (50 nm or less);

fibers increase in radial size by deposition of thermally decomposed hydrocarbons; *see* carbon fibers

varnish: a colorless transparent coating; dries by evaporation of solvent, followed by oxidation or polymerization; "long-oil" varnish that contains 30 or more gallons of oil/100 lb of resin has the elasticity and weathering resistance needed for outdoor applications; "short-oil" varnish that contains <20 gallons oil/100 lb resin is used to achieve the rubbed look in furniture; "medium-oil" is used for floor protection

vasodilators: pharmaceuticals used to dilate blood vessels, examples: nitroglycerin and isosorbide dinitrate

vat dyes: insoluble, color- and light fast dyes; when the dye is reduced with hydrosulfite and dissolved in strong alkali it has a strong affinity to cellulose fibers; after adsorption on the fibers the reduced dye is air-oxidized to the insoluble form; *see also* anthraquinone dyes

vector: in genetic engineering, agent such as a virus or plasmid that is used to incorporate genes into a cell or an organism

vegetable oils: esters of fatty acids derived from oil-rich trees, fruit and seeds; during refining pigments such as carotenes and odor-producing substances are removed; free fatty acids are saponified with caustic soda and the resulting soap is removed in a continuous process; phosphatides are recovered and sold as lecithin; use: edible oils, raw material for margarine

vehicle: in paint and ink manufacture, the binder plus the solvent used to hold the pigment

verapamil: drug to combat heart irregularities; calcium antagonist that acts by inhibiting the inward flow of Ca^{2+} into heart cells, thus relaxing the muscle; used against pain and irregular heart beat

VERAPAMIL

vermiculite: mica-like mineral; used as filler for plastics

vernolate: thiocarbamate herbicide

versatics: branched C6 to C11 carboxylic acids obtained by Shell from appropriate olefins by reaction with CO and H_2O; *see* Koch acids

vesicles: small sacs or bags; surfactant vesicles typically are spherical bags with a diameter of 50Å consisting of a double layer of surfactant molecules with the hydrophobic tails pointing toward each other; unlike micelles, vesicles do not break upon dilution

Vickers hardness test: *see* hardness

VI improvers: *see* pour-point depressants

vinal: synthetic fiber containing ≥50 wt% vinyl alcohol units; *see* poly(vinyl alcohol)

vinblastine: alkaloid obtained from the periwinkle plant; anticancer agent that acts by interfering with cell division

vinyl acetate: $CH_3COOCH=CH_2$; vinyl acetate monomer; VAM; prepared by vapor-phase reaction of ethylene, acetic acid and oxygen over palladium catalyst; formerly produced by addition of acetic acid to acetylene at 170–250°C using $Zn(OAc)_2$/charcoal catalyst, also by reaction of acetaldehyde and acetic anhydride; use: raw material for poly(vinyl acetate) homo- and copolymers, (mostly with ethylene or vinyl chloride), poly(vinyl alcohol) (PVA), poly(vinyl acetals) with butyraldehyde and formaldehyde; major end uses are adhesives, emulsion paints, paper and textile coatings

vinylacetylene: $CH_2=CHC\equiv CH$; produced

by dimerization of acetylene; use: raw material for manufacture of chloroprene

vinyl chloride: $H_2C=CHCl$; vinyl chloride monomer; VCM; produced from ethylene via thermal dechlorination of ethylenedichloride; use: monomer and comonomer in polymerizations, starting material for vinylidine chloride

vinyl ethers: $ROCH=CH_2$; produced by liquid-phase reaction of alcohol and acetylene under a nitrogen atmosphere or by dehydration of ethylene glycol ether; use: monomer or comonomer, solvent and organic synthesis intermediate

vinyl fluoride: $FHC=CH_2$; vinyl fluoride monomer; VFM; produced by (1) gas phase addition of HF to acetylene in the presence of fluoride catalysts; (2) noncatalyzed addition of HF to vinyl chloride followed by dehydrochlorination at 500–600°C in the presence of Cu powder; use: monomer for production of polyvinyl fluoride; *see* fluorocarbon polymers

vinylidene chloride: $H_2C=CCl_2$; vinylidene dichloride; VDC; produced from vinylchloride by chlorination to 1,1,2-trichloroethane followed by dehydrochlorination with aqueous NaOH or $Ca(OH)_2$ at 100°C; use: comonomer, films

vinylidene fluoride: $H_2C=CF_2$; produced by elimination of HCl from 1,1-difluoro-1-chloroethane; use: polymer, copolymer, e.g., hexafluoropropene/vinylidene fluoride copolymer; *see* fluorocarbon polymers

vinylon: *see* vinyon

vinyl plastisol: dispersion of finely divided poly(vinylchloride) resin in a plasticizer; forms a homogeneous solid when heated

vinylpyridine: 2-vinylpyridine; produced from 2-picoline (2-methylpyridine) via 2-hydroxyethylpyridine; use: comonomer with butadiene and styrene to improve adhesion between synthetic fibers and rubbers

***N*-vinylpyrrolidone**: produced from pyrrolidone (from butyrolactone with methylamine) by addition of acetylene; use: manufacture of polymers for cosmetic and medical applications

N-VINYLPYRROLIDONE

vinyls: *see* vinyl chloride

vinyon: vinylon; synthetic fiber which contains ≥85 wt% vinyl chloride units

virus: infectious agent that requires a host cell to multiply (replicate); composed of DNA or RNA within a protein coat; in genetic engineering when used to transfer genes between cells the virus acts as a vector

visbreaking: mild, once-through thermal cracking process; use: for making specified fuel oils

viscoelasticity: tendency of plastics to respond to stress as if they were elastic and viscous fluids

viscose: dilute NaOH solution of cellulose xanthate; use: production of viscose rayon

viscose rayon: modified cellulose; regenerated cellulose; produced by converting native cellulose to cellulose xanthate by treatment with aqueous NaOH and CS_2; the solid xanthate is dissolved in dilute NaOH to give viscose; this solution is deaerated, aged and filtered before being extruded into a coagulation–regeneration bath of dilute H_2SO_4, Na_2SO_4, $ZnSO_4$, and other additives; further modification occurs while the fiber is being stretched

viscosity: resistance of a fluid to shear motion; a measure of internal friction; in polymers, the solution viscosity depends on polymer molecular weight and solution concentration; *see* isoelectric point viscosity average molecular weight: M_v; *see* molecular weight

viscosity index: VI; empirical number that shows a lubricant's sensitivity to viscosity change with temperature; the higher the number the smaller the change; calculated from kinematic viscosities measured at 40°C and 100°C; in lubricating oils VI improves stabilize the viscosity within optimal limits: poly(methacrylates) prevent oil crystallization at low temperature and thus increase oil fluidity at low temperatures, i.e, they depress the pour-point; to control viscosity at high temperature, one takes advantage of the increasing thickening power of polymer additives with molecular weight

viscosity index improvers: *see* pour-point depressants

Vitamin A: retinol, axerophthol; fat soluble; industrial synthesis starts with the condensation of isobutene, formaldehyde and acetone leading to the first intermediate, 2-methylheptanone; this ketone is a key intermediate in industrial terpene chemistry; in the body vitamin A is synthesized from β-carotene; vitamin A appears to have some antitumor activity; unlike water-soluble vitamins it is toxic in large amounts; use: pharmaceutical, food fortifiers and animal feed; *see* β-carotene

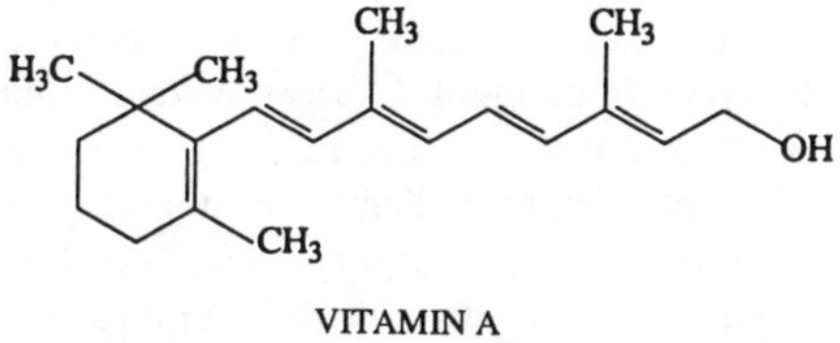

VITAMIN A

Vitamin B complex: mostly water-soluble compounds that are found in rice bran, yeast and wheat germ; most of them can be isolated from natural sources; they generally serve as nonprotein constituents of enzymes (cofactors); use: primarily nutritional additives (1) vitamin B_1; thiamin; thiamine; antineuritic vitamin, present in grains, yeast, milk, legumes and meat; (2) vitamin B_2; Vitamin G, riboflavin; growth promoting factor; produced from distiller's residues; (3) vitamin B_3; niacinamide, nicotinamide, nicotinic acid, vitamin PP; manufactured from niacin (nicotinic acid, pyridine-3-carboxylic acid); in addition to use as nutritional additive, used to stabilize color of cured meat and as brightener in electroplating baths; (4) vitamin B_6; pyridoxine; manufactured from isoquinoline; (5) vitamin B_{12}; cyanocobalamin; cobalt-containing porphyrin; (6) nositol; $C_6H_6(OH)_6$; hexahydroxycyclohexane; widely distributed in animals and plants; a growth factor

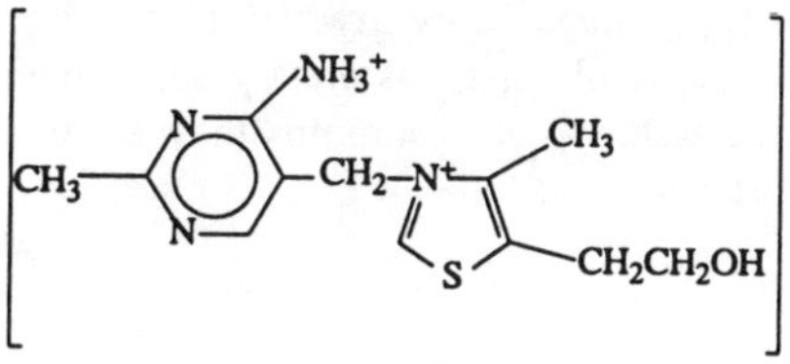

VITAMIN B_1

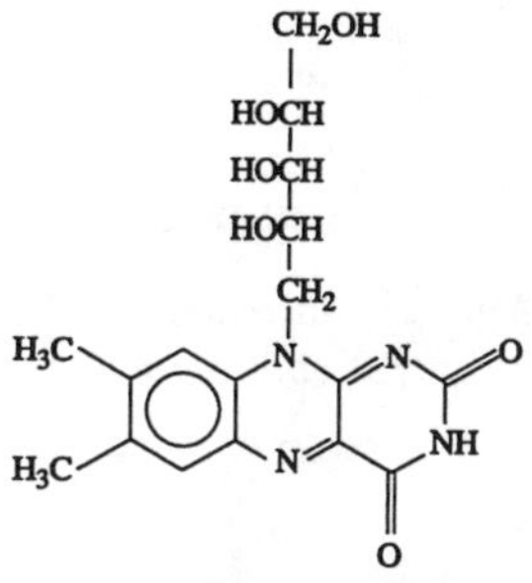

VITAMIN B_2

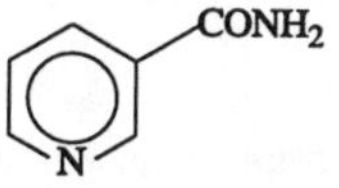

VITAMIN B_3

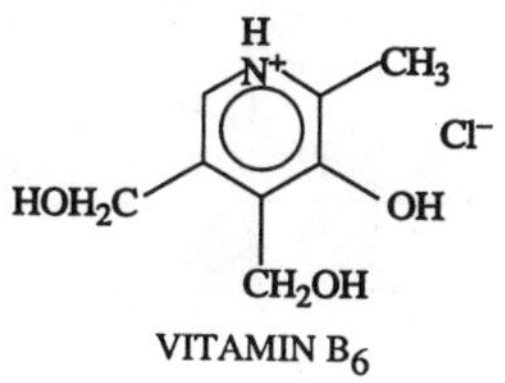

VITAMIN B_6

Vitamin C: ascorbic acid; water soluble, isolated from vegetable and fruit sources, manufactured from glucose; use: nutritional additive,
industrial antioxidant; *see* ascorbic acid

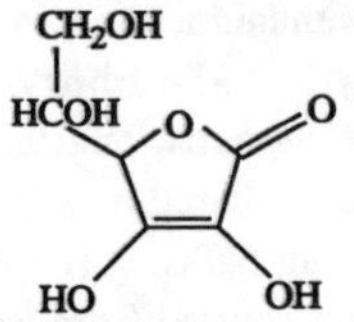

VITAMIN C

Vitamin D group: fat soluble, antirachitic
vitamins; vitamins D_2 and D_3 are the two
economically most important isomers; vitamin D_3 is prepared by irradiation of 7-dehy-
drocholesterol; used as dietary supplement;
unlike water-soluble vitamins they are toxic
in large amounts

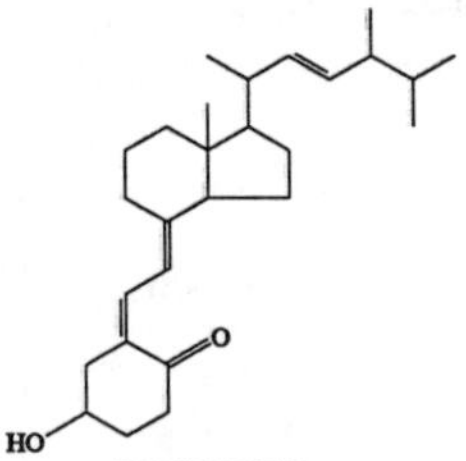

VITAMIN D_2

Vitamin E group: tocopherols; fat soluble;
used as nutritional additive and as antioxidant for fats

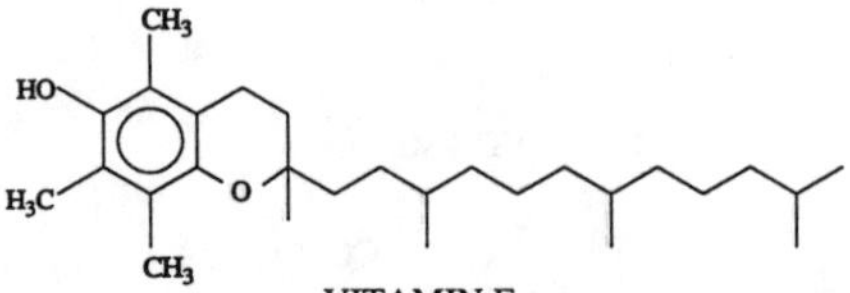

VITAMIN E

Vitamin G: riboflavin; vitamin B_2

Vitamin H: biotin; cofactor for enzymes

VITAMIN H

catalyzing carboxylation and deamination;
found in yeast, egg yolk, liver, milk and molasses; used as nutritional supplement

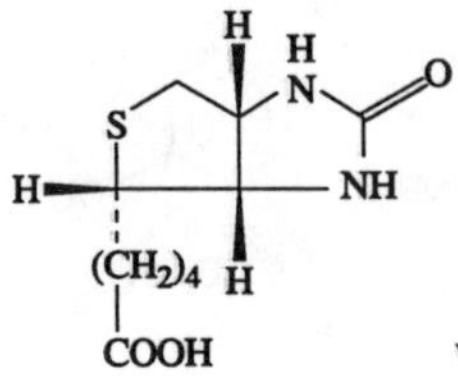

Vitamin K: fat soluble; essential for blood
clotting

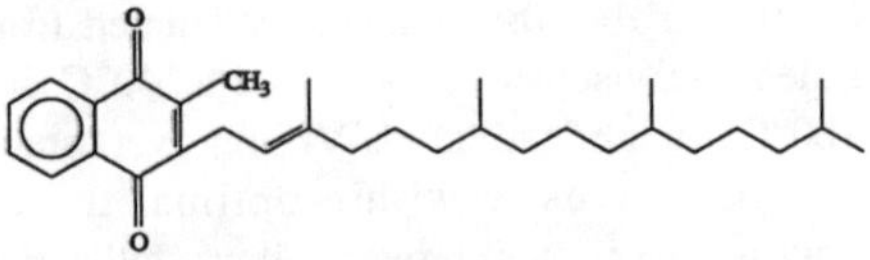

VITAMIN K

vitrify: to render glassy (vitreous)

vitrinite: glossy layers of coal derived
mainly from the woody parts of plants

volt: SI unit used for electromotive force
and for the difference in electrical potential;
1 volt equals 1 joule/coulomb

voluminizing yarns: *see* false twisting
process

vulcanite: *see* hard rubber

vulcanization: curing of latex by cross-
linking; changing from plastic to elastic
properties above the glass transition temperature, T_g; *see* vulcanized rubber

vulcanized rubber: elastomer in which
polymer chains are tied together with sulfur
cross-links; commonly used elastomers include natural rubber (polyisoprene), butadiene rubber, styrene-butadiene rubber, and
nitrile-butadiene rubber; the most common
curative is elemental sulfur whose eight-
membered ring is opened during vulcanization; other curing agents are organic polysulfides and organic peroxides; synthetic
rubbers that contain functional groups (e.g.,
neoprene, butyl rubber) can be cured with
ZnO or MgO; other additives in the formulation of vulcanized rubber are accelerators,
activators, antioxidants and stabilizers; vulcanization changes the viscous or plastic
polymer to a cross-linked elastic rubber

290

W

Wacker process: *see* acetaldehyde

ware: in the ceramic industry, useful shaped or manufactured articles

warfarin: rodenticide; oral anticoagulant, but because of its high toxicity patients must be carefully monitored

WARFARIN

warp: in textile technology, threads in a fabric that run parallel to the selvedge of woven materials; *see* weft

warpage: in solid materials, distortion due to uneven internal stress

wash active alkylsulfonates: *see* alkyl sulfonates

washings: effluents from purifications; can be used for product recovery; potential sources of pollution

water color inks: inks in which the pigments are suspended in a vehicle of dextrin, glycerin, gum arabic, and water

water conditioners: *see* water softeners

water-extended polyesters: WEP; unsaturated polyesters that are extended with water rather than with conventional fillers; curing with a peroxide is an exothermic reaction that molds WEP without heating; WEP behaves like wood but has more shrinkage because of loss of water with time

water gas: mixture of CO, H_2, and CO_2 made from coke by treatment with air and steam; *see* watergas process

water gas process: coal gasification using air and steam

$$C + O_2 \underset{k-1}{\overset{k+1}{\rightleftarrows}} 2CO;$$

$$C + H_2O \underset{k-1}{\overset{k+1}{\rightleftarrows}} CO + H_2;$$

$$CO + H_2O \underset{k-1}{\overset{k+1}{\rightleftarrows}} 2CO_2 + H_2 \text{ (shift reaction)}$$

water glass: $Na_2O(SiO_2)_n$; sodium silicate; produced by fusing sand and soda ash (sodium carbonate) at 1300–1500°C; the melt is quenched with cold water and dissolved with superheated steam; in industrially important silicates n = 2 to 4; use: in manufacture of detergents, starting materials for silica fillers, for synthesis of zeolites, catalysts and silica gels, in adhesives, cements, and as flocculating agents; potassium water glass is used to coat welding electrodes, as binders for plasters and luminescent paints in cathode ray tubes, and for wall impregnation; *see* sodium silicate, alkali silicates

water hardness: water-containing dissolved calcium and magnesium salts; hardness is expressed as grains per gallon (gpg), grains per liter (gpl) or parts per million (ppm) of $CaCO_3$; 1 gpg equals 17.1 ppm; soft water contains <60 ppm $CaCO_3$, very hard water >180 ppm; hard water causes scale deposits in boilers and pipes, metal corrosion and precipitation of soap curd during laundering; Ca and Mg ions in hard water can be removed with water softeners (sequestering agents), ion exchangers, or zeolites

waterless hand cleaners: oil-in-water emulsions that do not require rinsing; use: for removing difficult soils

water mark: symbol imparted during paper manufacture on stock that contains ~70% water

water quality standard: plan for management of water quality for public use (i.e, agriculture, drinking water, industrial, wildlife) encompassing criteria to protect those uses, implementation, and enforcement of rules

water reducible inks: inks based on pigments in a vehicle of modified rosin soap in glycol

water softeners: complex phosphate sequestering agents added to detergents to remove Ca and Mg ions; *see* water hardness

water softening resins: *see* ion exchange

water-soluble silicate: *see* sodium silicate

water vapor transmission: WVT; grams of water vapor passing through a material per day per m^2

Watson characterization factor: in the petroleum industry, an index of the chemical characteristics of crude oil, $K = (T_B)^{1/3}$ / sp. gr. where T_B is the boiling point in degrees Rankine (absolute °F scale, equals °F + 460) and sp. gr. is the specific gravity compared with water at 60°F; factor ranges between 10.5 for highly naphthenic and 13 for highly paraffinic crude oil

watt: SI unit of power; joule/sec

wax cracking: manufacture of higher olefins by cracking C20 to C30 paraffin fractions in the presence of steam at 500–600°C

waxes: high molecular weight hydrocarbons and esters of mono- or dihydric alcohols (C16 to C36) with long chain unbranched fatty acids (C24 to C36); some of the alcohols are sterols; use: lubricants, protective coatings, waterproofing, cosmetics

wax-set inks: inks based on a vehicle of wax-insoluble resin dissolved in a wax-soluble solvent; use: printing on wax paper causes the solvent to migrates onto paper causing the ink to dry

weathering: changes of materials due to exposure to the environment (e.g., light, heat, air, humidity, and biological and chemical agents); to evaluate such changes the effect of weather can be simulated, accelerated, and measured

web: in paper manufacture, the sheet exiting from the fourdrinier machine before it is being cut; in plastic manufacture, a thin sheet in the process of being formed; in extrusion coating one distinguishes between the molten web that issues from the die and the substrate web that refers to the material that is being coated

weber: Wb; SI unit of magnetic flux

weft: in textile manufacture, the threads of the fabric that run perpendicular to the selvedge; *see* warp

Weibull statistics: statistical method that describes the distribution of severity of flaws in a system

weight-average molecular weight: M_w of polymers; it is measured by light scattering or by sedimentation using an ultracentrifuge; these methods depend on the molecular weight and thus are most useful for heavy molecules; M_w correlates with viscosity, toughness of a polymer; *see* molecular weight

welding: fusing two pieces of metal or plastic; in countries outside the United States the term heat sealing is applied for welding plastics

Werner-Pfeider mixer: *see* internal mixers; mixers

wet end: in paper manufacture, the section of the fourdrinier machine where paper is formed, i.e, the front section

wet-felting: production of fiberboard similar to paper manufacture, i.e., use of wet pulp to produce a sheet of desired thickness; *see* fiberboard

wet gas: in petroleum recovery, methane

(natural gas) that comes to the surface together with some heavier hydrocarbons

wet-laid: *see* nonwoven fabrics

wet process phosphoric acid: acidulation of pulverized phosphate rock with sulfuric acid followed by filtration and evaporation; the reaction mixture consists of a slurry of dilute phosphoric acid (equal to ~30% P_2O_5), calcium sulfate crystals (35 wt%) and other impurities; the reaction is exothermic, the temperature is kept at or below 175°F to crystallize calcium sulfate as gypsum, which is filtered off; the hot slurry is cooled in cooling towers; sulfate concentration is maintained within narrow limits by adjusting the sulfuric acid addition; large vacuum filters permit processing of >1000 tons P_2O_5/day; the filter acid containing 30% P_2O_5 is concentrated by vacuum evaporation; the filter cake (phospho-gypsum) is impounded in old mine shafts and sold to farmers as control of salt build-up in irrigated land; by-products of the process include fluorine, which is recovered during the evaporation step as aqueous H_2SiF_6 and uranium, which is extracted from filter strength acid

wet spinning: extrusion of a polymer solution through a spinnerette into a precipitating bath in fiber form; this method of spinning is a slow process, rates of fiber formation, solvent removal, precipitant control and stretching have to be carefully controlled to produce a commercially satisfactory product; use: production of viscose rayon, acrylics

wet strength: strength of paper or fiber that is saturated with water; strength of an adhesive joint after water immersion

wet strength resins: in paper manufacture, additives that are beaten into the pulp (resin-filled paper) or impregnated into the paper sheet (resin-impregnated paper)

wetting agents: surface-active agents, detergents

wet-web process: in paper manufacture, impregnation of the wet (web) paper sheet with water-soluble or water-dispersible resin

whale oil: *see* jojoba oil

wheal rubber: factice containing brown maize oil

whiskers: in fiber technology, thin single crystal fibers; when defect-free, they have extremely high tensile strength; graphite whiskers can be grown by decomposing hydrocarbons (methane) in the gas phase; use: in composites to increase toughness

white cement: Portland cement manufactured from low iron oxide raw materials

white lead: basic lead carbonate; white pigment; used in paint, putty and ceramics; use is restricted because of lead toxicity

white phosphorus: P; elemental phosphorus; produced by electrothermal reduction of apatite ore with coke in the presence of gravel; used for manufacture of pure phosphoric acid, phosphorus halides, oxides and sulfides, red phosphorus

whiteware: fired, partially vitreous ceramic ware

Winkler gasifier: *see* coal gasification

winterization: winterizing, wintering; production of highly unsaturated oils that remain liquid under refrigeration; natural vegetable oils are cooled so that the more saturated oil fractions solidify and can be removed by filtration

Witten process: production of dimethyl-terephthalate from *p*-xylene by stepwise oxidation and esterification

wood flour: pulverized dry wood used as filler for plastics

wood hydrolysis: products of wood hydrolysis include polysaccharides, oligosaccharides, and glucose; process conditions vary

(1) hydrolysis with dilute sulfuric acid and steam at pressures ≤150psi (Madison process, Scholler–Tornesh process); (2) the Bergius-Rheinau process uses dry chips and 40–45 wt% HCl; the acid is recycled and fortified by gaseous HCl from salt-sulfuric acid retorts; the hydrolysate that contains the oligosaccharide is concentrated under vacuum and spray-dried; the intermediary sugars are hydrolyzed for further use: (3) several Japanese processes obtain crystalline glucose as main product; they use H_2SO_4 below 100°C and atmospheric pressure followed by neutralization with $Ca(OH)_2$; the by-product is gypsum that is used in the manufacture of gypsum board

wood naval stores: terpenes produced from shredded wood by solvent extraction; *see* naval stores

wood-plastic composites: produced by impregnation of wood with a water solution of phenol-formaldehyde resin or with liquid vinyl monomers; composites have improved dimensional stability and mechanical properties over wood; phenolic resins swell the wood and penetrate the cell walls; drying and curing produces permanent dimensional stability in the resulting "impreg"; drying without curing gives a highly compressible composite; compressed resin-treated wood is called "compreg"; impregnation with methyl methacrylate or styrene does not darken the wood; these composites are cured by radiation or heat and catalysts; use: for electrical insulators in high tension lines, knife handles, tooling jigs and forming dies, for parquet flooring and sports equipment (wood-vinyl composites)

woodstone: xylolith; material made of saw dust and sorel cement (mixture of MgO and magnesium salts)

wood vinegar: *see* acetic acid

Wulff process: manufacture of acetylene by cracking of hydrocarbons near 1300°C with superheated steam at short reaction times; to prevent coking the process uses two regenerator furnaces that operate at alternate hot and cold temperatures

X

xanthates: $R-O-CS_2^-$; produced by treating alcohols or acids with CS_2; used to scavenge heavy metal ions in waste water originating from electroplating operations, for rayon manufacture; *see* viscose rayon, froth flotation

xanthene dyes: chromophore of xanthene dyes; fluorescein is a representative xanthine dye

A XANTHENE DYE (eosin)

xerography: nonimpact printing method in which the image is transferred electrostatically and fused to the paper by heat or pressure; the process consists of five steps (1) a rotating belt or drum is positively charged; (2) the image to be printed is exposed on the surface of the photoreceptor causing the charge to drain away from the unexposed surface; (3) the surface is sprayed with negatively charged toner particles that adhere to the positively charged latent image; (4) a piece of paper is placed on the surface and is given a positive charge so that the pigmented particles are attracted to the paper; and (5) the powder image is fused to the paper; *see* printing

xetal: latex-containing cationic soaps

XIS process: manufacture *p*-xylene from *p/m*- xylene by isomerization of *m*-xylene in the presence of $Al_2O_3 \cdot SiO_2$ catalyst

X-ray: electromagnetic radiation with wavelengths of 0.01–1 nm; X-rays are produced by high voltage electron beams impinging on a target or by irradiation of the target with X-rays of shorter wavelength; use: medical diagnosis and therapy, for chemical analysis and structure determination

X-ray analytical methods: monochromatic X-rays can either be absorbed or diffracted by samples; absorption analysis is used to identify atoms, while diffraction serves to identify crystalline structures; absorbed X-rays remove electrons from the innermost shell (K shell) of atoms; as electrons from the outer shells (L, M,..) drop inward they emit radiation; these transitions provide a characteristic absorption pattern; to measure X-ray absorption one can use direct X-rays or X-rays of shorter wavelength (fluorescence analysis); the absorption spectra are independent of the chemical state of the atoms; diffraction methods are based on the scattering of X-rays by electrons and yield a specific diffraction pattern that allows the calculation of interatomic and intermolecular distances as well as bond angles in crystals; instrumentation for X-ray diffraction consists of various types of cameras: flat plate and cylindrical cameras are used for unit cell data, small-angle X-ray camera methods provide information concerning the gross structure; goniometers record powder pattern or small-angle intensities; the structure determination via single crystal methods is fully automated; the intensity of all the diffraction spots (usually a few thousand) are recorded and analyzed using computer programs; the methods have broad applications: they can be used to correlate the details of molecular structure with properties of materials; single crystal methods have been used to determine the molecular structure of polymers and such large molecules such as DNA and vitamin B_{12}

X-ray diffraction: *see* X-ray analytical methods

X-ray photoelectron spectroscopy: XPS; *see* electron spectroscopy for chemical analysis

xylenes: $C_6H_4(CH_3)_2$; cracking and catalytic reforming of naphtha (BTX process) yields a mixture of *o, m, p* isomers (35–45%) together with toluene (50%) and benzene (10–15%); use: industrial and laboratory solvents (1) *p*-xylene, *p*-$H_3CC_6H_4CH_3$, is separated from mixed xylene streams by (a) crystallization (m.p. 13.3°C); (b) zeolite molecular sieves; (c) adsorption (*see* Parex process); (d) isomerization of *m*-xylene (*see* XIS process); and (e) disproportionation of toluene; use: production of dimethyl terephthalate and terephthalic acid (intermediates for polyester fibers, plastics and films); (2) *o*-xylene is recovered by fractional distillation; use: raw material for phthalic anhydride, which is an intermediate for plasticizers; (3) most *m*-xylene is returned to the gasoline pool and converted to *p*-xylene; some *m*-xylene is oxidized to isophthalic acid, which with *m*-phenylene diamine gives a specialty polymer

xylenol: dimethylphenol; $(CH_3)_2C_6H_3OH$; 2,6-xylenol is the starting material for the thermoplastic, polyphenylene oxide

Y

Yankee driers: large driers used in manufacture of tissue paper

yarn: assembly of filaments obtained by spinning

yarn spinning: process of converting natural fibers (wool, cotton) or synthetic staple fibers into yarns for weaving or knitting into fabrics, e.g., in the cotton spinning system the harvested cotton fiber is opened, carded, drawn into a roving (oriented lose bundle of staple) and twisted into a yarn; the same process applies to staple fiber blends of natural with synthetic fibers; the synthetics must first be crimped, then cut into staple and uniformly blended

yeast: unicellular species of major industrial importance; use: in fermentation processes (bakery products, alcoholic beverages, ethanol fuel), for cheese production, as source of enzymes in biotechnology, food additive, single cell protein

yield strength: stress at which a permanent deformation of the specimen begins to occur

Young's modulus: tensile stress/tensile strain; relates tension to compression; *see* modulus

Z

Z-calender: calender with four rollers so that material passes through them in the form of the letter Z; *see* calender

zein: polypeptide derived from corn

zeolites: family of natural and synthetic porous minerals; primarily aluminosilicates (also a few aluminophosphates and silicoaluminophosphates, SAPO); differentiated by Si/Al ratios; natural zeolites are products of vulcanic activity; synthetic zeolites are manufactured by crystallization from natural raw materials such as kaolin or from synthetic raw material such as sodium aluminate and silica (water glass); manufacturing variables include stoichiometry, concentration, temperature and shear energy during stirring; zeolites are used as molecular sieves in

separation processes, as ion exchanger in waste water treatment and detergents, as adsorption agent for drying of gases and for removal of CO_2, H_2S, and mercaptans from gases, as catalysts for catalytic reforming and hydrocracking, manufacture of gasoline from methanol (ZSM-5), isomerization of hydrocarbons; used as light weight building material and filler

zero twist: in textile technology, winding up a yarn on a spool without a twist

zidovudine: *see* 3'-azidothymidine

Ziegler alcohols: alfols; linear primary alcohols made by reduction of olefins using H_2 and $(C_2H_5)_3Al$

Ziegler–Natta catalysts: Ziegler catalysts; catalysts made from transition metals and from group I–III metal alkyls, hydrides, and other compounds; these substances catalyze stereoregular polymerization

Ziegler process: *see* Ziegler reaction

Ziegler reaction: manufacture of α-olefins in the presence of $Al(C_2H_5)_3$; *see* Alfen process

zinc chromate: yellow pigments; approximate composition: $(ZnO \cdot K_2O)_4 \cdot (CrO_3)_4 \cdot (H_2O)_3$

zinc oxide: ZnO; Chinese white; zinc white; a white pigment manufactured by (1) vapor phase air-oxidation of vaporized Zn (French process); (2) reduction of roasted zinc ores with coal followed by air-oxidation (American process); (3) precipitation of $ZnCO_3$ or $Zn(OH)_2$ followed by calcination; use: additive in rubber, glass, and ceramics; in paints, glues, copier paper, cosmetics, dental cements and ointments

zinc sulfide: ZnS; a yellowish-white pigment that is a constituent of lithopone (30% ZnS + 70% $BaSO_4$); produced by precipitation from a mixture of $ZnSO_4$ and BaS; used in luminous paints, phosphor for TV and X-ray screens

zirconia: ZrO_2; produced from zircon ($ZrSiO_4$) by fusing with coke and lime; use: refractory material (m.p. 2700°C), solid electrolyte, neutron reflector in nuclear reactors, catalyst, ceramic modifier in partly stabilized zirconia ceramic (PSZ-ceramics) to improve temperature shock resistance

zirkaloy: zirconium alloy used as cladding in nuclear reactors

ZSM-5 zeolite: acid-resistant synthetic zeolite with an Si/Al ratio >10; characterized by two internally crossing channel systems; used as catalyst in the commercial conversion of methanol to gasoline (MTG process)

zwitterion: compounds such as proteins that have an acid, basic and neutral form; *see* isoelectric point

zygote: in genetics, a cell formed from the union of two mature reproductive cells